C From Theory to Practice

C From Theory to Practice

George S. Tselikis | Nikolaos D. Tselikas

CRC Press
Taylor & Francis Group
Boca Raton   London   New York

CRC Press is an imprint of the
Taylor & Francis Group, an **informa** business

CRC Press
Taylor & Francis Group
6000 Broken Sound Parkway NW, Suite 300
Boca Raton, FL 33487-2742

© 2014 by Taylor & Francis Group, LLC
CRC Press is an imprint of Taylor & Francis Group, an Informa business

No claim to original U.S. Government works

Printed on acid-free paper
Version Date: 20140108

International Standard Book Number-13: 978-1-4822-1450-5 (Paperback)

Visit the Taylor & Francis Web site at
http://www.taylorandfrancis.com

and the CRC Press Web site at
http://www.crcpress.com

Contents

Preface

This book is primarily addressed to students who are taking a course on the C language, to those who desire to pursue self-study of the C language, as well as to experienced C programmers who want to test their skills. It could also prove useful to instructors of a C course, who are looking for explanatory programming examples to add in their lectures.

So what, exactly, differentiates this book from the others in the field? This book tests the skills of both beginners and advanced developers by providing an easy-to-read compilation of the C theory enriched with tips and advice as well as difficulty-scaled solved programming exercises.

When we first encountered the C language as students, we needed a book that would introduce us quickly to the secrets of the C language as well as provide in-depth knowledge on the subject—a book with a focus on providing inside information and programming knowledge through practical examples and meaningful advice. That is the spirit that this book aims to capture.

The programming examples are short but concrete, providing programming know-how in a substantial manner. Rest assured that if you are able to understand the examples and solve the exercises, you can safely go on to edit longer programs and start your programming career successfully.

For all of you who intend to deal with computer programming, a little bit of advice coming from our long experience may come in handy:

- Programming is a difficult task that requires a calm mind, a clear development plan, a methodical approach, patience, and luck as well.

- When coding, try to write in a simple and comprehensive way for your own benefit and for those who are going to read your code. Always remember that the debug, support, and upgrade of a code written in a complex way is a painful process.

- Hands-on! To learn the features of a programming language, you must write your own programs and experiment with them.

- Programming is definitely a creative activity, but do not forget that there are plenty of creative, pleasant, and less stressful activities in life. Do not waste your life in front of a computer screen, searching for *losing* pointers, buffer overflows, memory overruns, and buggy conditions. Program for some time and then have some fun. Keep living and do not keep programming.

Enjoy the C flight; it would be safe, with some turbulence, though.

Authors

Dr. George S. Tselikis received his Dipl–Ing and his PhD from the School of Electrical and Computer Engineering of the National Technical University of Athens (NTUA) in 1993 and 1997, respectively. In 1998, he joined the COMET group in the Center for Telecommunications Research at Columbia University, New York and worked as a postdoc research associate. He was a founding member of 4Plus S.A. (1999–2013), where he worked in the development of network protocols and services. He has a long working experience in the telecom area, and his research interests focus on software specification, development, and testing of network protocols and services in wired and wireless networks. During his professional career, he has collaborated with big players in the telecom industry like Siemens, Nokia, and Alcatel. Since 2004, he has been a visiting lecturer in several universities and technical institutes. He teaches courses related to network technologies, protocols, and communications, as well as the C language.

Dr. Nikolaos D. Tselikas received his Dipl–Ing and his PhD from the School of Electrical and Computer Engineering of the National Technical University of Athens (NTUA) in 1999 and 2004, respectively. His research interests focus on the specification and implementation of network services and applications. He has participated in several European and national research projects in this field. During his involvement in the research projects, he has collaborated in software design and development topics with big players in the telecom industry like Ericsson, Siemens, Alcatel, and Vodafone. He has served as a visiting lecturer in the Department of Informatics and Telecommunications, University of Peloponnese, Greece, since 2006 and was appointed lecturer in 2009. He currently serves as an assistant professor at the same university, where he teaches courses related to applications' programming and network services, as well as the C language.

1

Introduction to C

Before getting into the details of C language, this chapter presents, in brief, its history, evolution, strengths, and weaknesses. Then, we'll discuss some basic concepts that we'll need in order to write our first program.

History of C

The C language was developed at Bell Laboratories in the early 1970s by Dennis Richie and others. At that time, the UNIX operating system, also developed at Bell Labs, was written in assembly language. Programs written in assembly are usually hard to debug, maintain, and enhance, and UNIX was no exception. Richie decided to rewrite UNIX's code in another language that would make the execution of these tasks easier. He named the language C because it was the evolution of an earlier language written by Ken Thompson, called B.

The C language continued to evolve during the 1970s, and since then it is widely used by thousands of programmers for the development of various software applications.

ANSI Standard

The rapid expansion of the C language and its increased popularity led many companies to develop their own C compilers. Due to the absence of an official standard, their development relied on the bible of C programmers, the legendary K&R book, written by Brian Kernighan and Dennis Ritchie in 1978.

However, the K&R book was not written in the precise way that a standard requires. Features that were not clearly described could be implemented in different ways. As a result, the same program could be compiled with one C compiler and not with another. In parallel, the C language continued to evolve with the addition of new features and the replacement or obsolescence of existing ones.

The need for the standardization of C language became apparent. In 1983, the American National Standard Institute (ANSI) began the development of C standard that was completed and formally approved, in 1989, as ANSI C or Standard C. This standard describes precisely the features, characteristics, and properties of the C language, and every C compiler must support it. The addition of some new features to the ANSI standard during the late 1990s led to a new standard called C99.

This book describes the C language based on the ANSI/ISO C standard [1].

Advantages of C

Despite the emergence of many programming languages, C still remains competitive and popular in the programming world for several reasons, such as the following:

1. It is a flexible language, which can be used for the development of different kinds of applications, from embedded systems and operating systems to industrial applications. For example, we've used C in the area of communication networks for the development of network protocols and the support of network services.
2. A C program is executed very fast.
3. It is a small language. Its vocabulary consists of a few words with special meanings.
4. It is portable, meaning that a C program may run under different operating systems.
5. It supports structural programming, meaning that a C program may contain functions to perform several tasks.
6. It is a language very close to the hardware.
7. Every C compiler comes with a set of ready-to-use functions, called C standard library. The use of these library functions saves considerable programming effort.
8. Thanks to the popularity of the C language, there are many C compilers available, some of them free of charge.
9. Learning C is the first step toward object-oriented programming. Most of the C features are supported in several object-oriented languages, like C++, Java, and C#.

Disadvantages of C

1. Because the C language does not impose many restrictions on the use of its features, it is an error-prone language. When writing a C program, be cautious because you may insert bugs that won't be detected by the compiler.
2. Although C is a small language, it is not an "easy" to use language. C code can be very hard to understand even if it consists of a small number of lines. After reading this book, check out the International Obfuscated C Code Contest (http://www.ioccc.org) to get a feeling.
3. It is not an object-oriented language.

C Program Life Cycle

The life cycle of a C program involves several steps: writing the source code, its compilation, linking the code produced by the compiler with the code of the used library functions, and executing the program.

Usually, a C compiler provides an integrated development environment that allows us to perform this set of operations without leaving the environment.

Write a C Program

To write a C program, you can use any available text editor. The source code must be saved in a file with extension .c.

When the size of the code is very large, it is a common practice to divide the code into several files in order to facilitate tasks like debugging and maintenance. In such cases, each file is compiled separately.

First C Program

Our first program will be a "rock" version of the program that most programmers begin with. Instead of the classical K&R `"Hello  world"`, our program displays `"Ramones: Hey Ho, Let's Go"`.

```c
#include <stdio.h>
int main()
{
  printf("Ramones: Hey Ho, Let's Go\n");
  return 0;
}
```

The following sections explain the significance of each program line.

#include Directive

The `#include` directive instructs the compiler to include the contents of the specified file in the program before it is compiled. The C standard library contains a number of header files; each contains information about a specific part of the library. For example, the `stdio.h` (standard input output) file contains information about data input and output functions. When you get familiar with C language, you may edit your own header files and include them in your programs.

When the program is compiled, the compiler searches for the included files. The brackets < > instruct the compiler to look into predefined folders. If a file is not found, the compiler will raise an error message and the compilation fails.

Regarding syntax, notice that a directive starts with an # and does not end with a semicolon (;).

main() Function

Every C program must contain a function named `main()`. In C, a function is a series of statements that have been grouped together and given a name. The statements of the program must be enclosed in braces {}. A statement is a command that will be executed when the program runs. Unlike the directives, a statement ends almost always with a semicolon.

The `main()` function is called *automatically* when the program runs. It ends when its last statement is executed, unless an exit function (like the **return**) is called earlier. The word **int** indicates that `main()` must return a status code to the operating system when it terminates. This value is returned with the **return** statement; the value 0 indicates normal termination.

The declaration **int** `main()` is fairly common. However, you may see other declarations like

```
void main()
{
   ...
}
```

The word **void** indicates that the `main()` function doesn't return any value. Although a C compiler may accept this declaration, it is illegal according to the C standard because `main()` must return a value. On the other hand, the declaration

```
main()
{
   ...
}
```

is acceptable because the return type is **int** by default.

This declaration is also acceptable:

```
int main(void)
{
   ...
}
```

The word **void** inside the parentheses indicates that `main()` has no arguments.

In Chapter 11, we'll see another declaration of `main()`, where it accepts arguments.

As discussed, we may use functions of the standard library in our programs. For example, `printf()` is a standard library function that is used for data output. The reason to include the `stdio.h` file is that it contains information about `printf()`. The new line character `'\n'` instructs `printf()` to advance to the next line. We'll discuss more about `printf()` in the next chapter.

Until Chapter 11, where you'll learn how to write other functions, `main()` will be the only function in our programs.

Add Comments

A well-documented program should contain comments to explain its complicated parts and make it easier to understand. A comment begins with the `/*` symbol and ends with `*/`. Comments can extend in more than one line.

The compiler ignores anything included between the /* */ symbols, meaning that the comments do not affect the operation of the program. For example, a comment is added to describe the purpose of the program:

```c
#include <stdio.h>
/* This program calls printf() function to display a message on the
   screen. */
int main()
{
  printf("Ramones: Hey Ho, Let's Go\n");
  return 0;
}
```

Notice that you may see C programs containing one line comments that begin with // instead of the /* symbol. For example,

```c
int main() // This is my first C program
```

Although a compiler may support this syntax, it is not according to the C standard. In fact, beginning a comment with // is a C++ practice that may be supported by some C compilers, but not by others.

Add comments to explain the complicated parts of your program. An explanatory program saves you time and effort when you need to modify it, and the time of other people who may need to understand and evolve your program.

Compilation

After writing the program, the next step is to compile it. The compiler translates the program to a form that the machine can interpret and execute.

Many companies (i.e., *Microsoft* and *Borland*) develop C compilers for *Windows*, while one of the most popular free compilers for *Unix/Linux* systems is *gcc* (GNU Compiler Collection, http://gcc.gnu.org). Another popular compiler is *Bloodshed Dev-C++* (http://www.bloodshed.net), which also provides an integrated environment for the development of C programs.

When the program is compiled, the compiler checks if the syntax of the program is according to the language rules. If the compilation fails, the compiler informs the programmer for the fail reason(s).

If the program is compiled successfully, the compiler produces an object file that contains the source code translated in machine language. By default, it has the same name as the file that contains the source code and its extension is *.obj* (i.e., in *Windows*) or *.o* (i.e., in *UNIX*). For example, if the aforementioned code is saved in the file first.c, the name of the object file would be first.obj (*Windows*) or first.o (*UNIX*).

Common Errors

The most common errors are syntactic. For example, if you don't add the semicolon at the end of `printf()` or you omit a parenthesis or a double quote, the compilation would fail and the compiler would display error messages.

Spelling errors are very common, as well. For example, if you write `studio.h` instead of `stdio.h` or `prinf()` instead of `printf()`, the compilation would fail.

C is a case-sensitive language, meaning that it distinguishes between uppercase and lowercase letters. For example, if you write `Printf()` instead of `printf()`, the compilation would fail.

If the compiler displays many errors, fix the first one and recompile the program. The new compilation may display lesser errors, even none. Also, notice that an error detected by the compiler may not occur in the indicated line, but in some previous line.

It is very important to understand that the compiler detects errors due to the wrong use of the language and not logical errors that may exist within your program. The compiler is not "inside your head" to know what you intend to do. Therefore, a successful compilation *doesn't mean* that your program would operate as you expect. For example, if you want to write a program that displays a message One if the value of the integer variable a is greater than 5, and you write

```
if(a < 5)
printf("One");
```

then, although these lines will be compiled successfully, the program won't display One. This type of error is a logical error (*bug*) not detected by the compiler. The use of the word bug as a synonym of a programming error is credited to the great American mathematician and computer scientist Grace Hopper, when she discovered that a hidden bug inside her computer caused its abnormal operation.

Apart from the error messages, the compiler may display warning messages. If there are only warning messages, the program will be compiled. However, don't ignore them; it may warn you against the potential malfunction of your program.

Linking

In the final step, the object code produced by the compiler is linked with the code of the library functions (like `printf()`) that the program uses.

If the linking is successful, an executable file is created. For example, the default name of the executable file produced by the *gcc* compiler is `a.out`, while a *Microsoft* compiler produces an executable file having the same name with the source file and extension `.exe`.

Run the Program

If the program doesn't operate as you expect, you are in deep trouble. The depth you are in depends on the size of the source code. If it extends to some hundreds of lines, you'll probably find the logical errors in a short time and the debugging procedure won't take long. But if your program consists of thousands of lines, the debugging may become a time-consuming, painful, and very stressful procedure, particularly if your supervisor hangs over your head demanding immediate results.

To avoid such troublesome situations, remember our advice: try to write simple, clear, readable, and maintainable code.

Reference

1. Programming Language C, ISO/IEC, 9899-1990.

2

Data Types, Variables, and Data Output

In order to be able to write programs that actually perform useful tasks that save us time and work, this chapter will teach you how to use data types and variables. We'll also go into more detail on the most important function for outputting data: the `printf()` function.

Variables

The computer's RAM (random access memory) consists of millions of successive storage cells. The size of each cell is one *byte*. For example, an old PC with 16 MB (megabytes) of RAM consists of 16 × 1024 kB (kilobytes), or 16.777.216 memory cells. A newer PC with say 8 GB (gigabytes) of RAM would have 8 × 1024 MB = 8192 × 1024 kB = 8.388.608 × 1.024 = 8.589.934.592 memory cells.

A *variable* in C is a storage location with a given name. The value of a variable is the content of its memory location. A program may use the name of a variable to access its value.

Rules for Naming Variables

There are some basic rules for naming variables. Be sure to follow them or your code won't compile:

1. The name of a variable can contain uppercase letters, lowercase letters, digits, and the *underscore* ' _ ' character.
2. The name must begin with either a letter or the underscore character.
3. The C programming language is *case sensitive*, meaning that it distinguishes between uppercase and lowercase letters. For example, the variable var is different from Var or vAr.
4. The following keywords cannot be used as variable names because they have special significance to the C compiler:

auto	do	goto	signed	unsigned
break	double	if	sizeof	void
case	else	int	static	volatile
char	enum	long	struct	while
const	extern	register	switch	
continue	for	return	typedef	
default	float	short	union	

Some compilers also treat the words **asm**, **far**, and **near** as additional keywords.

Variable Name Conventions

In addition to the rules given earlier, there are a few conventions that are good to follow when naming your variables. While these aren't enforced by the C compiler, these "rules of thumb" will tend to make your programs easier for you to understand, as well as for those who have to read your code after you've written it:

1. Use descriptive names for variables. It's much easier to read a program when the names of the variables indicate their intended use. For example, if you have a variable that you plan to use to hold the sum of some even numbers, name that variable something like sum _ even rather than an arbitrary name like i.

2. When necessary, don't be afraid to use long names to describe the role of a variable. If a variable name is several words long, separate each word with the underscore character (_) for readability. For example, you might call a variable that holds the number of books in a calculation books _ number (instead of booksnumber, or something less readable).

3. By convention, use lowercase letters when naming variables and uppercase letters when defining macros and constants. This is a convention that most C programmers follow; it is not a requirement.

Declaring Variables

Variables must be declared before being used in a program. Declare a variable as follows:

```
data_type name_of_variable;
```

The name _ of _ variable is the variable name. The data _ type should be one of the C-supported data types. For example, the **int** keyword is used to declare integer variables, and the **float** keyword is used to declare floating-point variables; variables that can store values with a fractional part.

Each data type specifies the range of values that may be stored in a variable of that type. The actual size of the types may vary from one machine to another. Table 2.1 shows the usual ranges on a 32-bit system.

For example, to declare variables a and b as integer and floating-point variables, respectively, write

```
int a; /* Declare an integer variable with name a. */
float b; /* Declare a float variable with name b. */
```

Variables of the same type can be declared in the same line, separated with a comma. For example, instead of declaring the variables a, b, and c in three different lines, like so

```
int a;
int b;
int c;
```

TABLE 2.1

C Data Types

Type	Size (Bytes)	Range (Min – Max)
char	1	-128 … 127
short	2	-32.768 … 32.767
int	4	-2.147.483.648…2.147.483.647
long	4	-2.147.483.648…2.147.483.647
float	4	Lowest positive value: $1.17*10^{-38}$
		Highest positive value: $3.4*10^{38}$
double	8	Lowest positive value: $2.2*10^{-308}$
		Highest positive value: $1.8*10^{308}$
long double	8, 10, 12, 16	
unsigned char	1	0 … 255
unsigned short	2	0 … 65535
unsigned int	4	0 … 4.294.967.295
unsigned long	4	0 … 4.294.967.295

you can declare them in a single line, as follows:

```
int a, b, c;
```

Once a variable is declared, the compiler reserves the bytes in memory that it needs in order to store its value. As indicated in the second column of Table 2.1, each type requires specific memory space. For example, the **char** type requires one byte, the **float** and **int** types require four bytes, the **double** type requires eight bytes, and so on.

The memory space that a data type requires may vary from one system to another. For example, the int type may reserve two bytes in one system and four bytes in another. (To determine the number of bytes a data type uses on a particular system, use the sizeof operator, discussed in Chapter 4.)

The **char**, **short**, **int**, and **long** types are used to store integer values, which can be either signed or unsigned. By default these types are signed, which means the leftmost bit of integer variables is reserved for the sign (either positive or negative). If an integer variable is declared as **unsigned**, then it has no sign bit and it may store only positive values. The advantage of this is that unsigned values have a much higher upper limit than their signed counterparts since they don't need to account for any values less than zero.

The **float**, **double**, and **long double** types can be used to declare variables that can store values with a fractional part, that is, floating-point numbers. Unlike the integer types, floating-point types are always signed. The C standard doesn't specify the number of the precision digits since different computers may use different formats to store floating-point numbers. Usually, the precision of the **float** type is 6 digits after the decimal point, while the precision of the **double** type is 15 digits. The range of the **long double** type isn't shown in Table 2.1 because its length varies, with 10 and 16 bytes being the most common sizes. The precision of the **long double** type is at least as much the precision of the **double** type, with 18 digits being a typical precision. Although the **long double** type supports the highest precision, it is rarely used because the precision of the **float** and **double** types is enough, for most applications.

Although C's types may come in different sizes, the C standard requires that **long** not be smaller than **int**, which must not be smaller than **short**. Similarly, **long double** not be smaller than **double**, which must not be smaller than **float**.

Assigning Values to Variables

A variable can be given a value by using the assignment operator =. For example, the following statement assigns the value 100 to the variable a:

```
int a;
a = 100;
```

Alternatively, a variable can be initialized together with its declaration:

```
int a = 100;
```

You can also initialize more than one variable of the same type together with its declaration. For example, the following statement declares a, b, and c variables and assigns them the values 100, 200, and 300, respectively:

```
int a = 100, b = 200, c = 300;
```

You could even write

```
int a = 100, b = a+100, c = b+100;
```

In this example, the assignments take place from left to right, meaning that first the value of a becomes 100, then b becomes 200, and finally c becomes 300.

If an integer value begins with the digit 0, this value will be interpreted as an *octal* (base-8) number. For example, the following statement assigns the decimal value 64 and not the value 100 to the variable a:

```
int a = 0100;
```

Similarly, a value that begins with 0x or 0X is interpreted as a *hexadecimal* (base-16) number. For example, the following statement assigns the decimal value 16 to the variable a:

```
int a = 0x10;
```

Appendix D provides a brief introduction to binary and hex systems.

The value assigned to a variable should be within the range of its type. For example, the statement

```
char ch = 130;
```

does not make the value of ch equal to 130 since the range of the values of the signed **char** type is from –128 to 127. In this case, the value of ch is wrapped around to –126 because 130 is out of the allowed range. This type of bug is very common, so always pay attention to the ranges of your variables!

To assign a floating-point value to a **float** variable, we *use the dot* (.) for the fractional part *and not the comma* (,). For example,

```
float a = 1.24;
```

A floating-point value can be written in scientific notation using the letter E (or e), which represents the power of 10. For example, instead of a = 0.085; we can write a = 85E-3; which is equivalent to $85*10^{-3}$. And a = 1.56e6; is equivalent to a = 1560000; or $1.56*10^6$.

Scientific notation is usually used when a floating-point number is very small or very large to make it easier for the programmer to read and write.

The value assigned to a variable should match the variable type. For example, the statement

```
int a = 10.9;
```

actually sets the value of a to 10 since a has been declared as **int** and not **float**. (The decimal part is completely ignored, and the assigned value is not rounded to 11.) However, the value of a **float** variable can be an integer. For example, you could write **float** a = 50; since that's equivalent to **float** a = 50.0;

Constants

A variable whose value cannot change during the execution of the program is called a *constant*. To declare a constant, precede the type of the variable with the **const** keyword. A constant must be initialized when it is declared and you cannot assign it another value within the program.

For example, the following statement declares the integer variable a as a constant and sets it equal to 10:

```
const int a = 10;
```

If we attempt to change the value of a constant later in a program, for example by writing

```
a = 100;
```

the compiler will raise an error message.

#define Directive

The **#define** directive is used to define a *macro,* a name that, in most cases, represents a numerical value. To define a simple macro, we write

```
#define name_of_macro value
```

For example,

```
#define NUM 100
```

defines a macro named NUM.

When a program is compiled, each macro is replaced by its defined value. For example, in the following program, the values a, b, and c are set to −80, 120, and 300, respectively.

```
#include <stdio.h>
#define NUM 100
int main()
{
  int a, b, c;

  a = 20 - NUM;
  b = 20 + NUM;
  c = 3 * NUM;
  return 0;
}
```

As you can see in this example, macros are typically defined before the main() function and are usually named using all capital letters. Also, note that there is no semicolon at the end of a #define directive.

In general, macros are most helpful when they're used to represent a numeric value that appears many times within your program. Once this value is defined as a macro, if you ever need to change it you only need to change it in one place. For example, to change the value 100 to 300 in the following program without using the NUM macro, we would have to replace it three times instead of one:

```
#include <stdio.h>
int main()
{
  int a, b, c;

  a = 20 - 100;
  b = 20 + 100;
  c = 3 * 100;
  return 0;
}
```

It's simple enough to make these replacements by hand in a small program like this one, but imagine trying to change every value in a program that's thousands of lines long! It's much safer and faster to use a macro to just change a defined value as needed.

We discuss macros in more detail in Chapter 16. For now simply think of them as an alternative to defining constants in your program.

printf() Function

The printf() function is used to print a variable number of data items to the standard output stream (stdout). By default, the stdout stream is associated with the screen.

TABLE 2.2

Escape Sequences

Escape Sequence	Action
\a	Make an audible beep
\b	Delete the last character (equivalent to using the Backspace key)
\n	Advance the cursor to the beginning of the next line (equivalent to using the Enter key)
\r	Move the cursor to the beginning of the current line (equivalent to a carriage return)
\t	Move the cursor to the next tab stop (equivalent to the Tab key)
\\	Display a single backslash (\)
\"	Display double quotes (")

When called, `printf()` can accept several arguments. The first is a format string that determines the output format. If arguments follow the format string, `printf()` displays their values to the `stdout`.

The format string may contain *escape sequences* and *conversion specifications*. It may also contain ordinary characters, which are printed as is to the screen.

Escape Sequences

Escape sequences tell the compiler to perform a specific action, such as move the cursor. An escape sequence consists of a backslash (\) followed by a character. Table 2.2 lists the most common escape sequences.

The following programs show some uses of the escape sequences. Don't forget to enclose the format string in double quotes ("") or the program won't compile.

```c
#include <stdio.h>
int main()
{
  printf("This is\n");
  printf("another C\n");
  printf("program\n");
  return 0;
}
```

Since the escape sequence \n moves the cursor to the next line, the output of this program is

```
This is
another C
program
```

Alternatively, we could use a single `printf()` as follows to produce the same output:

```c
printf("This is\nanother C\nprogram\n");
```

Here's another example:

```c
#include <stdio.h>
int main()
```

```
{
  printf("\a");
  printf("This\b is a text\n");
  printf("This\b\b\b is a text\n");
  printf("This\t is\t a\t text\n");
  printf("This is a\"text\"\n");
  printf("This is a \\text\\\n");
  printf("Sample\rtext\n");
  return 0;
}
```

The output of this program looks like this:

```
Hear a beep.
Thi is a text
T is a text
This is a text
This is a "text"
This is a \text\
textle
```

The last `printf()` writes the string `Sample`, then the `\r` moves the cursor back to the beginning of the line and writes the string `text`. Therefore, the program displays `textle`.

To expand the format string of `printf()` to several lines (typically for purposes of readability), use a backslash (\). For example, the following `printf()` is written over three lines but the output will appear on one line:

```
printf("This printf uses three lines, but the \
message will appear \
on one line ");
```

Conversion Specifications

Conversion specifications always begin with the percent (%) character, and it is followed by one or more characters with special significance. It its simplest form a conversion specification is followed by one special character, called conversion specifier. The conversion specifier must be one of the characters listed in Table 2.3.

The following program illustrates the use of `printf()` to print characters and numbers in various formats:

```
#include <stdio.h>
int main()
{
  printf("%c\n", 'w');
  printf("%s\n", "some text");
  printf("%d\n", -100);
  printf("%f\n", 1.56);
  printf("%e\n", 100.25);
  printf("%X\n", 15);
  printf("%o\n", 14);
  printf("%d%%\n", 100);
  return 0;
}
```

TABLE 2.3

Conversion Specifiers

Conversion Specifier	Meaning
c	Display the character that corresponds to an integer value.
d, i	Display a signed integer in decimal form.
u	Display an unsigned integer in decimal form.
f	Display a floating-point number in decimal form using a decimal point. The default precision is six digits after the decimal point.
s	Display a sequence of characters.
e, E	Display a floating-point number in scientific notation using an exponent. The exponent is preceded by the specifier.
g, G	Display a floating-point number either in decimal form (%f) or scientific notation (%e).
p	Display the value of a pointer variable.
x, X	Display an unsigned integer in hex form: %x displays lowercase letters (a-f), while %X displays uppercase letters (A-F).
o	Display an unsigned integer in octal.
%	Display the character %.

The first printf() displays w and the second prints some text. Notice that a single character must be enclosed in single quotes ' ', while a string (a series of characters) must be enclosed in double quotes "".

The third printf() displays –100, while the next one displays 1.560000 with six decimal places.

The fifth printf() displays 1.002500e+002, which is equivalent to 1.0025×10^2 = 1.0025×100 = 100.25.

The sixth printf() displays F, which represents the number 15 in hex.

The next one displays 16, which represents the number 14 in octal.

Finally, the last printf() displays 100%.

To sum up, the output looks like this:

```
w
some text
-100
1.560000
1.002500e+002
F
16
100%
```

Return Value

printf() returns the number of displayed characters or a negative value if an error occurs. For example, consider the following program:

```
#include <stdio.h>
int main()
{
  printf("%d\n", printf("Test\n"));
  return 0;
}
```

Although functions are discussed in Chapter 11, you should be able to get a sense of what the aforementioned code is doing. The inner `printf()` is executed first, so the program displays `Test`. The `%d` of the outer `printf()` is replaced by the return value of the inner `printf()`. Because `printf()` returns the number of displayed characters, the inner `printf()` returns 5: four characters plus the new line character `'\n'`. Therefore, the program displays

```
Test
5
```

Printing Variables

Variable names follow the last double quote of the format string. When printing more than one variable, separate each with a comma. The compiler will associate the conversion specifications with the names of each of the variables from left to right. Each conversion specification should match the type of the respective variable or the output will be meaningless. Take a look at the following program:

```c
#include <stdio.h>
int main()
{
  int a = 10, b = 20;
  printf("%d + %d = %d\n", a, b, a+b);
  printf("%f\n", a);
  return 0;
}
```

In the first `printf()`, the compiler replaces the first `%d` with the value of a, the second `%d` with the value of b, and the third one with their sum. Therefore, the program displays

```
10 + 20 = 30
```

The second `printf()` displays a "nonsense" value since a wrong conversion specification is used.

The C compiler does not check to see if the number of the conversion specifications equals the number of the variables. If there are more conversion specifications than variables, the program will display nonsense values for any extra specifications. On the other hand, if there are fewer conversion specifications than variables, the program simply will not display the values of the extra variables. For example,

```c
#include <stdio.h>
int main()
{
  int a = 10, b = 20;
  printf("%d and %d and %d\n", a, b);
  printf("%d\n", a, b);
  return 0;
}
```

The first `printf()` uses the `%d` specifier three times though there are only two output variables. The compiler replaces the first `%d` with the value of a, the second with the value of b, and the third with a random value. The program displays

```
Values are 10 and 20 and (a random value)
```

The second `printf()` uses the `%d` specifier once, while there are two output variables. The compiler replaces the first `%d` with the value of a and ignores the second variable. The program displays

```
Val = 10
```

As discussed, once a variable is declared the compiler allocates the required memory to store its value. The variable is initialized with the random value in this memory. For example, the following program displays the arbitrary value of the variable i:

```
#include <stdio.h>
int main()
{
    int i;

    printf("Val = %d\n", i);
    return 0;
}
```

Optional Fields

As shown, the simplest form of the conversion specification begins with the `%` character followed by the conversion specifier. However, a conversion specification may include another four fields, as shown in Figure 2.1. We'll review each of these fields in detail in this section.

Precision

When displaying the value of a floating-point type, we can specify the number of significant digits. The default precision is six digits. To specify another precision, add a period (.) followed by either an integer (to specify a precise number) or an asterisk (*) (in which case the precise number of significant digits is defined by the next argument). To display no significant digits, add a period (.) only.

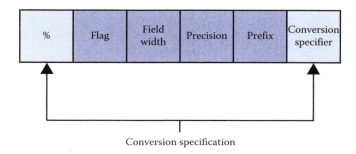

Conversion specification

FIGURE 2.1
The complete `printf()` conversion specification.

If the precision digits are less than the decimal digits, the displayed value is rounded up or down, according to the value of the cutoff digit. If it is less than 5, the displayed value is rounded down, otherwise it is rounded up.

Take a look at the following program:

```
#include <stdio.h>
int main()
{
  float a = 1.2365;

  printf("Val = %f\n", a);
  printf("Val = %.2f\n", a);
  printf("Val = %.*f\n",3, a);
  printf("Val = %.f\n", a);
  return 0;
}
```

The first `printf()` displays the value of a with the default precision of the six digits, that is, `Val = 1.236500`.

The second `printf()` displays the value of a with two precision digits. This value is rounded up since the value of the first nondisplayed digit is 6. The program displays `Val = 1.24`.

The third `printf()` uses the `*` character, so the precision is defined by the next argument, which is 3. The displayed value is rounded up since the value of the first nondisplayed digit is 5. The program displays `Val = 1.237`.

The last `printf()` displays the value of a with no precision digits. Since the value of the first nondisplayed digit is 2, the program rounds it down and displays `Val = 1`.

Therefore, the program displays

```
Val = 1.236500
Val = 1.24
Val = 1.237
Val = 1
```

With this in mind, what is the output of the following program?

```
#include <stdio.h>
int main()
{
  printf("%f", 5.123456789);
  return 0;
}
```

Since the default precision is six digits and the first nondisplayed digit is more than 5, the program displays `5.123457`. If we increase the precision, for example using `%.9f`, the program would display the exact number.

Recall that when using a floating-point number in mathematical expressions (such as in assignments, calculations, or comparisons) and you need precision, use the **double** type. For example, the following program might not store the value `12345.67` into a, but a value close to it, because this value might not be precisely represented by a floating-point number:

```
#include <stdio.h>
int main()
```

```
{
  float a;

  a = 12345.67;
  printf("Val = %f\n", a);
  return 0;
}
```

If you were to write **double** a; instead of **float** a; the program would display the exact value of this calculation.

When displaying a string, we can define how many of its characters will be displayed as with floating-point numbers. If the defined precision exceeds the number of the string's characters, the string is displayed as is. For example,

```
#include <stdio.h>
int main()
{
  char msg[] = "This is sample text";

  printf("%s\n", msg);
  printf("%.6s\n", msg);
  printf("%.30s\n", msg);
  return 0;
}
```

The first printf() uses the %s to display the characters stored into the array msg, while the second one prints the first six characters of the string (including the space character). Since the defined precision (30) exceeds the length of the string, the third printf() displays the entire string as if no precision was specified.

As a result, the program displays

```
This is sample text
This i
This is sample text
```

Field Width

When displaying the value of an integer or floating-point variable, we can define the total number of characters to be displayed by adding an integer (to specify the width of the output field) or an asterisk (*) (in which case the width is defined by the next argument).

> *In the case of floating-point numbers, the defined width should take into account the precision digits and the decimal point.*

If the displayed variable needs fewer characters than the defined width, space characters are added from left to right, and the value is right-justified. If the displayed value needs more characters, the field width will automatically expand as needed. For example, consider this program:

```
#include <stdio.h>
int main()
{
  int a = 100;
  float b = 1.2365;

  printf("%10d\n", a);
```

```
    printf("%10f\n", b);
    printf("%10.3f\n", b);
    printf("%*.3f\n", 6, b);
    printf("%2d\n", a);
    printf("%6f\n", b);
    return 0;
}
```

Since the value of a is 100, the minimum field width should be three in order to hold the three digits. However, since the defined width of 10 (%10d) is seven places larger than the minimum required size, the first printf() displays seven leading spaces and then the number 100.

In order to display the value of b, which is set equal to 1.2365, the minimum field width should be eight characters: six for the precision digits, plus one for the period and one for the integer. The second printf() first displays two spaces and then 1.236500.

At the third printf(), the minimum field width should be five characters since the precision digits are three instead of six. Therefore, the program first displays five spaces and then the rounded-up value 1.237.

Since the fourth printf() uses the * character, the width is defined by the next argument, 6. Therefore, the program first displays one space and then the rounded-up value 1.237.

To display the value of a, the minimum field width should be three. However, since the defined width of 2 (%2d) is less than the required size, it is automatically expanded and the fifth printf() displays 100.

In order to display the value of b, the minimum field width should be eight. Once again, the width is automatically expanded and the last printf() displays 1.236500.

As a result, the program displays

```
       100
  1.236500
     1.237
 1.237
100
1.236500
```

Prefix

To indicate that the displayed value is a **short** integer, we can use the letter h; letter l indicates that the integer is **long**. For example,

```
#include <stdio.h>
int main()
{
   short a = 10;
   long b = 10000;

   printf("%hd %ld\n", a, b);
   return 0;
}
```

Flags

Flags can be used to control the output of numeric values as listed in Table 2.4. To get a better idea of how these flags work, examine the following program:

TABLE 2.4

Flags

Flag	Meaning
-	Left aligns the output value within the defined field width.
+	Prefixes the output positive values with +.
space	Prefixes the output positive values with a space character.
#	Prefixes octal numbers with 0 and hex numbers with 0x or 0X. When used with floating-point numbers, it forces output to contain a decimal point.
0	Pads with zeros until the defined width is reached.

```c
#include <stdio.h>
int main()
{
  int a = 12;

  printf("%-4d\n", a);
  printf("%+4d\n", a);
  printf("% d\n", a);
  printf("%#0x\n", a);
  printf("%#o\n", a);
  printf("%04d\n", a);
  return 0;
}
```

The first `printf()` displays the value 12 left aligned, and the next one adds the sign + since the number is positive.

The third `printf()` prefixes the output value with a space.

The fourth `printf()` displays the number 12 in hex prefixed by 0x, while the next one does the same in octal.

The last `printf()` pads the number with two leading 0s, up to the field width of four.

Therefore, the program displays

```
12
 +12
 12
0xc
014
0012
```

`printf()` *is a powerful function, which provides many ways to format data output. For a more detailed list of its features, see your compiler's documentation.*

Type Casting

C allows the programmer to convert the type of an expression to another type. This conversion is known as *type casting*. Cast expressions have the following form:

```c
(data_type) expression
```

where **data _ type** specifies the type to which the expression should be converted. For example, after declaring the following variables

```
float a, b, c = 2.34;
```

the cast expression a = (**int**)c converts the value of c from **float** to **int** and make a equal to 2. After being used in the cast expression, c will be treated again as **float**. Similarly, with the statement

```
b = (int)(c+4.6);
```

the cast expression (**int**)(c+4.6) converts the result of the addition to type **int** and assigns it to b. Therefore, the value of b becomes 6.

Exercises

2.1 What is the output of the following program?

```
#include <stdio.h>
int main()
{
  int i = 100;

  i = i+i;
  printf("V1:%d V2:%d\n", i+i, i);
  return 0;
}
```

Answer: The statement i = i+i; makes the value of i equal to i = i+i = 100+100 = 200. Inside printf(), the first %d is replaced by the value of the expression i+i = 200+200 = 400, and the second %d by the value of i. Therefore, the program displays V1:400 V2:200.

2.2 What is the output of the following program?

```
#include <stdio.h>
int main()
{
  int i = 30;
  float j = 10.65;

  printf("Values:%f %d\n", i, j);
  return 0;
}
```

Answer: Since the variable i has been declared as an integer, we should use %d to display its value and %f to display the value of j. Because the %d and %f specifiers are used in the wrong order, the program displays "nonsense" values.

2.3 Write a program that declares two integers, assigns both the value 30, and displays their sum, difference, product, and the result of their division and the remainder.

```
#include <stdio.h>
int main()
```

```
{
  int i, j;

  i = j = 30;
  printf("Sum = %d\n", i+j);
  printf("Diff = %d\n", i-j);
  printf("Product = %d\n", i*j);
  printf("Div = %d\n", i/j);
  printf("Rem = %d\n", i%j);
  return 0;
}
```

Comments: The % operator is used to find the remainder in the integer division. The balance of the operators is the same with the ones used in math. Another way to display the math calculations is to use a single printf() and the %d specifier as follows:

```
printf("%d %d %d %d %d\n", i+j, i-j, i*j, i/j, i%j);
```

or even

```
printf("%d+%d=%d,%d-%d=%d,%d*%d=%d,%d/%d=%d,%d%%%d=%d\n",
i, j, i+j, i, j, i-j, i, j, i*j, i, j, i/j, i, j, i%j);
```

2.4 What is the output of the following program?

```
#include <stdio.h>
int main()
{
  int i = 6;
  double j;

  j = i/4;
  printf("Val = %f\n", j);
  return 0;
}
```

Answer: Because the result of the division i/4 is an integer number and not **float** (since both operands are integers), the program displays Val = 1. Had i been declared as a floating-point variable, the program would display 1.500000.
   Also, if we had written i = j/4.0; the program would display 1.500000 since a constant integer written with a decimal point is treated as **float**.

2.5 Write a program that declares two integers, assigns to them the values 20 and 50, and displays the result of their division.

```
#include <stdio.h>
int main()
{
  int i = 20, j = 50;
  double k;

  k = (double)i/j;
  printf("%f %d\n", k, i);
  return 0;
}
```

Comments: Because the cast expression (**double**)i converts the value of i from **int** to **double**, the result of their division is a decimal number. If we had written k = i/j, the value of k would have been 0.

As noted earlier, this conversion is only temporary: i remains an **int** for the rest of the program. This is why we use %d (not %f) to display its value.

2.6 What is the output of the following program?

```c
#include <stdio.h>
int main()
{
    int k;
    float i = 10.9, j = 20.3;

    k = (int)i + j;
    printf("%d %d\n", k, (int)(i + (int)j));
    return 0;
}
```

Answer: The cast expression (**int**)i converts the value of i from **float** to **int**. Since the result of the expression (**int**)i is 10, we have k = 10+20.3 = 30 (not 30.3, since k is declared as integer). If we had written k = i+j, the value of k would have been k = 10.9+20.3 = 31.

Because the value of the cast expression

(**int**)(i+(**int**)j)

is

(**int**)(10.9+(**int**)20.3) = (**int**)(10.9+20) = (**int**)(30.9) = 30

the program displays

30 30

2.7 The following program displays the average of two float numbers. Is there a bug in this code that might cause it to behave differently than you'd expect?

```c
#include <stdio.h>
int main()
{
    double i = 12, j = 5, avg;

    avg = i+j/2;
    printf("Avg = %.2f\n", avg);
    return 0;
}
```

Answer: There is a bug. Because the division operator / has priority over the addition operator +, the division (5/2 = 2.5) is performed first and afterward the addition (12+2.5), which causes the program to display the incorrect value 14.50 instead of 8.50.

This bug is eliminated by enclosing the expression i+j in parentheses. Because parentheses have the highest priority among all C operators, the parenthetical expression is executed first. (See Appendix A for more on the order of operations.) In this case, the right assignment would be

```
avg = (i+j)/2;
```

2.8 What is the output of the following program?

```
#include <stdio.h>
int main()
{
  unsigned char tmp = 255;

  tmp = tmp + 3;
  printf("%d\n", tmp);
  return 0;
}
```

Answer: The new value of `tmp`, which is `tmp = tmp+3 = 255+3 = 258` (100000010 in binary) requires nine bits to be stored. Since the **unsigned char** type allocates eight bits, only the eight low-order bits of the value 258, that is, 00000010, will be stored in `tmp`. Therefore, the program displays 2.

2.9 Write a program that assigns a two-digit positive value to an integer variable and displays the sum of its digits. For example, if the assigned value is 35, the program should display 8.

```
#include <stdio.h>
int main()
{
  int i, j, k;

  i = 35;
  j = i/10;
  k = i - (10*j);
  printf("Sum = %d\n", j+k);
  return 0;
}
```

Comments: The term `i/10` calculates the tens of `i`. Notice that we could skip the declaration of `j` and `k`, and write

```
printf("Sum = %d\n", i/10+(i-(10*(i/10))));
```

Unsolved Exercises

2.1 Write a program that uses a single `printf()` to display the following pattern:

```
*   *
  *
*   *
```

2.2 Write a program that assigns two negative values into two integer variables and uses those variables to display the corresponding positives.

2.3 Fill in the gaps to complete the program in order to display the following output:

```
21
 21
15
25%
A
 a
10
77
077
63
```

```c
#include <stdio.h>
int main()
{
   int x = 21, y = 0xa, z = 077;

   printf("_____\n", x);
   printf("_____\n", x);
   printf("_____\n", x);
   printf("_____\n", x);
   printf("_____\n", y);
   printf("_____\n", y);
   printf("_____\n", y);
   printf("_____\n", z);
   printf("_____\n", z);
   printf("_____\n", z);
   return 0;
}
```

2.4 Fill in the gaps to complete the program in order to display the following output:

```
-12.123
-12.123456789
  -12.123456789
-12.123457
-12.12346
-12
```

```c
#include <stdio.h>
int main()
{
   double x = -12.123456789;

   printf("_____\n", x);
   printf("_____\n", x);
   printf("_____\n", x);
   printf("_____\n", x);
   printf("_____\n", x);
   printf("_____\n", x);
   return 0;
}
```

2.5 Use the flags of `printf()` to fill in the gaps and complete the program in order to display the following output:

```
x + yj = 2-3j
x - yj = 2+3j
y + xj = -3+2j
y - xj = -3-2j

#include <stdio.h>
int main()
{
   int x = 2, y = -3;
   printf("x + yj = _____\n", x, y);
   printf("x - yj = _____\n", x, -y);
   printf("y + xj = _____\n", y, x);
   printf("y - xj = _____\n", y, -x);
   return 0;
}
```

2.6 Write a program that assigns two positive values into two integer variables and displays the remainder of their division. Use only two variables and don't use the % operator.

2.7 Write a program that assigns two positive values into two float variables and displays the integer part of their division and the fractional part. For example, if they are assigned the values 7.2 and 5.4, the program should display 1 and 2.8, since 7.2 = (1×5.4)+1.8.

2.8 Write a program similar to 2.9 (Exercise) for a three-digit positive value.

3

Getting Input with `scanf()`

This chapter focuses on using the `scanf()` function to get input from a user, store that input in variables, and use it in your programs. `scanf()` is a powerful function that offers many ways to read and store data in program variables. We won't cover every aspect of using `scanf()` but we'll give you enough of what you need in order to use it to read data. In this chapter, we'll mainly use `scanf()` to read numeric data and store that data in numeric variables. You'll see other uses of `scanf()` over the next chapters. Just remember as you read that `scanf()` is not an easy function to use; it contains several traps, so use it with great care.

`scanf()` Function

`scanf()` function is used to read data from the *standard input stream* (`stdin`) and store that data in program variables. By default, the `stdin` stream is associated with the keyboard. The input data are read according to a specified format, very similar to the way output is handled with `printf()`.

The `scanf()` function accepts a variety of parameters. The first is a format string enclosed in double quotes `""`, followed by the memory addresses of the variables in which the input data will be stored.

Like `printf()`, the number of conversion specifiers included in the format string should equal the number of variables, and the type of each specifier should match the data type of the respective variable. The conversion characters used in `scanf()` are the same as those used in `printf()`. For example, the `%d` specifier is used to read an integer, while `%f` is used to read a float number.

`scanf()` reads the input data from `stdin`, converts it according to the conversion specifiers, and stores it in the respective variables. The `scanf()` function stops processing the input data once it reads a data item that does not match the type of conversion specifier. Any unread input data as well as any nonmatching items are left in `stdin`.

`scanf()` Examples

In the following example, we use `scanf()` to read an integer and store it in variable `i`:

```
int i;
scanf("%d", &i);
```

The first argument is the `%d` conversion specifier, and the second is the memory address of `i`. Once the user enters an integer and presses *Enter*, this value will be stored in `i`.

The & character that usually precedes the name of a variable is called the *address operator,*
and it indicates the memory location in which the input number will be stored. We'll talk
much more about the & operator when discussing pointers in Chapter 8.

The new line character ('\n') that is generated when the Enter key is pressed is left in
stdin, and it will be the first character read in next call to scanf().

In the following example, we use scanf() to read an integer and a float number
from the keyboard and store them in variables i and j, respectively. The first argument is the format string %d%f, while the next arguments are the memory addresses of
i and j. The %d specifier corresponds to the address of i, while the %f corresponds to
the address of j.

```c
int i;
float j;
scanf("%d%f", &i, &j);
```

When scanf() is used to read more than one number, a *white-space* character is used to
separate the input values. So, in the previous example, if the user enters the values 10 and
4.65 separated by one or more spaces and presses *Enter,* the value of i will become 10 and
the value of j will become 4.65.

If precision is critical in your program, use a **double** variable (instead of **float**) to store
the input value and use the %lf specifier to read it (instead of %f). For example, consider
the following program:

```c
#include <stdio.h>
int main()
{
  float a;

  printf("Enter number: ");
  scanf("%f", &a);
  printf("%f\n", a);
  return 0;
}
```

Because the **float** data type may not always represent float numbers precisely, the program may not display the input value, but a value close to it. It is much safer to declare a
as **double** and use the %lf specifier to read it.

Just like when reading numeric values, we can use scanf() to read characters. The
following example uses scanf() to read a single character and store it in variable ch:

```c
char ch;
scanf("%c", &ch);
```

In the following example, we use `scanf()` to read a string, a series of characters, and store it in the `str` array:

```
char str[100];
scanf("%s", str);
```

If the user enters `sample`, the letters s-a-m-p-l-e will be stored in the respective elements of the `str` array. The value of `str[0]` will be `'s'`, `str[1]` will be `'a'`, and so on.

Notice that the address operator `&` does not precede the `str` since the name of an array may be used as a pointer to the memory address of its first element. We'll discuss the relationship between arrays and pointers in Chapter 8.

For now, just remember that the `&` operator should always precede the name of a numeric variable (such as `int`, `short`, `char`, `float`, and so on). If you forget the `&` operator, the program will behave unpredictably and may even crash. On the other hand, be sure not to add the `&` operator when using pointer variables.

> *If the entered string consists of multiple words (such as* This is text with multi-*ple words), only the first word will be stored into* str *because when* scanf() *is used to read characters it stops reading once it encounters a white-space character. (If you were to write* scanf("%[^\n]",str); *it would read multiple words.) The format string of* scanf() *can take many forms, but a detailed description of all* scanf() *capabilities is beyond the scope of this book.*

Use of Ordinary Characters

In its simplest form, the format string (such as `"%d%d"`) does not contain other characters than the conversion characters. If you add any characters, they must be also entered by the user in the same expected location. If the characters don't match, `scanf()` aborts reading. For instance, note the comma between the `%d` specifiers in the following example:

```
#include <stdio.h>
int main()
{
  int a, b;

  scanf("%d,%d", &a, &b);
  printf("%d %d\n", a, b);
  return 0;
}
```

In this case, the user should also add a comma between the values (such as `12,43`) or `scanf()` will fail and the program won't display the expected values. If the `%d` specifiers were separated by the character `'m'` instead of the comma, such as `%dm%d`, the user should also add an `'m'` between the values (such as `12m43`).

In another example, if a program asks from the user to enter a date using / to separate the values, the format string should be `"%d/%d/%d"` and the user must enter the date in that form (e.g., 3/8/2020).

`scanf()` Return Value

A function may return a value to the calling program, and `scanf()` returns the number of data items that were successfully converted and assigned to program variables. `scanf()` fails if the values entered don't match the types or order of the conversion specifiers.

In the following example, `scanf()` expects an integer first and then a float number:

```
int i;
float j;
scanf("%d%f", &i, &j);
```

In this case, `scanf()` will return 2. Otherwise, if the user enters a different value, like two float numbers or a float number and then an integer, `scanf()` will fail.

The following program checks the returned value of `scanf()` to verify that the input integer is successfully read and stored into num. (You'll learn more about how this program works after reading Chapters 6 and 11.)

```
#include <stdio.h>
int main()
{
  int num;

  printf("Enter number: ");
  while(scanf("%d", &num) != 1)
  {
    printf("Enter number: ");
    while(getchar() != '\n'); /* Consume the unread characters. */
  }
  printf("Inserted value:%d\n", num);
  return 0;
}
```

If the returned value is not 1, the data conversion failed. For example, if the user enters a character and not an integer, `scanf()` will fail and the **while** loop will prompt the user to enter another value. Notice that if the user enters a floating point number only the integer part will be stored into num. The rest will remain in stdin.

For the sake of brevity and simplicity, we won't check `scanf()`'s return value throughout this book. Instead, we'll assume that the user's input is valid. Nevertheless, remember that a robust program should always check its return value to verify that no read failure occurred.

Exercises

3.1 Write a program that reads an integer and a float number and displays the triple of their sum.

```
#include <stdio.h>
int main()
{
  int i;
  float j, sum;

  printf("Enter numbers: ");
  scanf("%d%f", &i, &j);
```

```
    sum = i+j;
    printf("%f\n", 3*sum);
    return 0;
}
```

Comments: There is no need to declare the variable sum. Instead, we could write printf("%f\n", 3*(i+j));

3.2 Write a program that reads the name, the code, and the price of a product and displays them.

```
#include <stdio.h>
int main()
{
    char name[100]; /* Declare an array of 100 characters. */
    int code;
    float prc;

    printf("Enter name: ");
    scanf("%s", name);

    printf("Enter code: ");
    scanf("%d", &code);

    printf("Enter price: ");
    scanf("%f", &prc);

    printf("\nN:%s\tC:%d\tP:%f\n", name, code, prc);
    return 0;
}
```

3.3 Write a program that reads two integers and displays their sum, difference, product, the result of their division, and the remainder.

```
#include <stdio.h>
int main()
{
    int i, j;

    printf("Enter 2 integers (the second should not be 0):");
    scanf("%d%d", &i, &j);

    printf("Sum = %d\n", i+j);
    printf("Diff = %d\n", i-j);
    printf("Product = %d\n", i*j);
    printf("Div = %f\n", (float)i/j); /* Typecast of i from int to float
      to display the decimal part of the division. */
    printf("Rem = %d\n", i%j);
    return 0;
}
```

Comments: The % operator is used to find the remainder of the division of two integer operands. If a user enters the value 0 as the second integer, the program may crash since division by 0 is impossible. (This can be avoided with the use of an **if** statement, as we'll see in Chapter 5.)

3.4 Write a program that reads the prices of three products and displays their average.

```c
#include <stdio.h>
int main()
{
  float i, j, k, avg;

  printf("Enter 3 prices: ");
  scanf("%f%f%f", &i, &j, &k);

  avg = (i+j+k)/3; /* The parentheses are necessary to perform the
    addition first and then the division. */
  printf("Average = %f\n", avg);
  return 0;
}
```

Comments: Without using the variable avg, we could write

```c
printf("Average = %f\n", (i+j+k)/3);
```

3.5 Write a program that reads the radius of a circle and displays its area and perimeter.

```c
#include <stdio.h>

#define PI 3.14159

int main()
{
  double radius;

  printf("Enter radius: ");
  scanf("%lf", &radius);
  printf("%f %f\n", PI*radius*radius, 2*PI*radius);
  return 0;
}
```

3.6 Write a program that reads a float number and displays the previous and next integers.

```c
#include <stdio.h>
int main()
{
  float i;

  printf("Enter number: ");
  scanf("%f", &i);
  printf("%f is between %d and %d\n", i, (int)i, (int)(i+1));
  return 0;
}
```

3.7 Write a program that reads two integers and swaps their values.

```c
#include <stdio.h>
int main()
{
  int i, j, temp;

  printf("Enter numbers: ");
  scanf("%d%d", &i, &j);
```

```
    temp = i;
    i = j;
    j = temp;
    printf("%d %d\n", i, j);
    return 0;
}
```

3.8 Continuing the previous exercise, write a program that reads three floats, stores them in three variables, and rotates them one place right. For example, if the user enters the numbers 1.2, 3.4, and 5.6 and they are stored in variables d1, d2, and d3, the program should rotate their values one place right, so that d1, d2, and d3 become 5.6, 1.2, and 3.4, respectively.

```
#include <stdio.h>
int main()
{
    double d1, d2, d3, temp;

    printf("Enter numbers: ");
    scanf("%lf%lf%lf", &d1, &d2, &d3);

    temp = d1;
    d1 = d2;
    d2 = temp;

    temp = d3;
    d3 = d1;
    d1 = temp;
    printf("%f %f %f\n", d1, d2, d3);
    return 0;
}
```

3.9 Write a program that reads a float and an integer positive number and displays the remainder of their division. For example, if the user enters the numbers 7.24 and 4, the program should display 3.24.

```
#include <stdio.h>
int main()
{
    int num2, div;
    double num1;

    printf("Enter float and int: ");
    scanf("%lf%d", &num1, &num2);
    div = num1/num2; /* Assume that the user enters the numbers 7.24
        and 4. Since div is declared as integer it becomes 1. */
    printf("%f\n", num1 - (div*num2));
    return 0;
}
```

Comments: Without using the variable div, we could write

```
printf("%f\n", num1 - (int(num1)/num2*num2));
```

3.10 What is the output of the following program when the user enters an integer and presses Enter?

```c
#include <stdio.h>
int main()
{
  char ch;
  int i;

  printf("Enter number: ");
  scanf("%d", &i);

  printf("Enter character: ");
  scanf("%c", &ch);

  printf("Int = %d and Char = %c\n", i, ch);
  return 0;
}
```

Answer: After the user enters an integer and presses the Enter key, the generated new line character ('\n') is stored in stdin. Since the second scanf() is used to read a character, it gets the new line character from stdin and stores it automatically into the ch variable without letting the user enter any other character. The program displays the entered integer and terminates. However, if we reverse the read order, the program would execute correctly (remember, white space before a numeric value is ignored).

3.11 Write a program that reads two positive float numbers and displays the sum of their integer and decimal parts. Use only two float variables. For example, if the values 1.23 and 9.56 are entered, the program should display 10 and 0.79.

```c
#include <stdio.h>
int main()
{
  double i, j;

  printf("Enter numbers: ");
  scanf("%lf%lf", &i, &j);
  printf("%d %f\n", (int)i+(int)j, (i-(int)i) + (j-(int)j));
  return 0;
}
```

Comments: We are using the (int) typecast to get the integer part of each float variable. For example, the term (i-(int)i) is equal to the decimal part of the variable i.

3.12 Write a program that reads an integer and converts it to multiples of 50, 20, 10, and 1. For example, if the user enters the number 285, the program should display 5*50,1*20,1*10,5*1.

```c
#include <stdio.h>
int main()
{
  int i, n_50, n_20, n_10, n_1, rem;

  printf("Enter number: ");
```

```
scanf("%d", &i); /* Assume that the user enters the number 285, so
  i = 285. */
n_50 = i/50; /* n_50 = 285/50 = 5. */
rem = i%50; /* The rem = 285 % 50 = 35 should be analysed in 20s,
  10s and singles. */

n_20 = rem/20; /* n_20 = 35/20 = 1. */
rem = rem%20; /* The rem = 35 % 20 = 15 should be analysed in 10s
  and singles. */

n_10 = rem/10; /* n_10 = 15/10 = 1. */
n_1 = rem%10; /* n_1 = 15%10 = 5. */

printf("%d*50,%d*20,%d*10,%d*1\n", n_50, n_20, n_10, n_1);
return 0;
}
```

Comments: Without using the variables n_50, n_20, n_10, n_1, and rem, we could write printf("%d*50,%d*20,%d*10,%d*1\n", i/50, i%50/20, i%50%20 /10, i%50%20%10);

3.13 Write a program that reads a two-digit positive integer and displays its reverse. For example, if the user enters the number 72, the program should display 27. Use a single integer variable.

```
#include <stdio.h>
int main()
{
  int i;

  printf("Enter number between 10 and 99: ");
  scanf("%d", &i);
  printf("%d\n", 10*(i%10) + i/10);
  return 0;
}
```

Comments: The term (i/10) calculates the tens of i, while the term (i%10) calculates its units. The reverse is produced by multiplying the units by 10 and adding the tens.

3.14 Write a program similar to the previous one for a three-digit positive integer.

```
#include <stdio.h>
int main()
{
  int i;

  printf("Enter number between 100 and 999: ");
  scanf("%d", &i);
  printf("%d\n", 100*(i%10) + 10*(i%100/10) + i/100);
  return 0;
}
```

Comments: The term (i%10) calculates the units of i, (i%100/10) calculates its tens, and the (i/100) its hundreds.

Unsolved Exercises

3.1 Write a program that reads a bank deposit, the annual interest rate as percentage, and displays what would be the total amount one year after.

3.2 Write a program that reads the ages of a father and his son and displays in how many years the father will have the double age of his son's and their ages at that time, as well.

3.3 Write a program that reads the number of students who passed and failed in the exams and displays the percentages. For example, if the user enters 12 and 8, the program should display

```
Success Ratio: 60%
Fail Ratio: 40%
```

3.4 Suppose that a customer in a store bought plates and cups. Write a program that reads the number of the plates and the price of one plate, the number of the cups and the price of one cup, and the amount the customer paid. The program should display the change the customer got back.

3.5 Write a program that reads an octal number, a hexadecimal number, and a decimal integer and display their sum in decimal. For example, if the user enters 20, 3f, and 9, the program should display 88.

3.6 Write a program that reads an integer indicating some number of seconds and converts it to hours, minutes, and seconds. For example, if the user enters 8140, the program should display 2h 15m 40s. Use only one variable.

3.7 Write a program that reads a two-digit positive integer and duplicates its digits. For example, if the user enters 12, the program should display 1122.

4

Operators

Now that you've been introduced to the concepts of C data types, variables, and constants, it is time to learn how to manipulate their values. C provides a rich set of operators that are used to build expressions and perform operations on them. In this chapter, we'll present some of C's most important operators. We'll introduce the rest gradually over the next several chapters, and you'll see how to apply them when building more complex programs.

Assignment Operator

The = operator is used to assign a value to a variable. For example, the statement a = 10; assigns the value 10 to the variable a, while the statement a = k; assigns the value of the variable k to a.

When used in a chained assignment, the assigned value is stored in all variables in the chain, from right to left. For example, in this statement

```
int a, b, c;
a = b = c = 10;
```

the variables c, b, and a are assigned the value 10, in that order. If the variable and the assigned value are not of the same type, the value is first converted to the variable's type. For example, what would the value of b be in the following?

```
int a;
float b;
b = a = 10.22;
```

Since a is an **int** variable, it is assigned the value 10, and this value is stored in b. Therefore, the value of b is 10, not 10.22 as you might expect.

Arithmetic Operators

The arithmetic operators +, −, *, and / are used to perform addition, subtraction, multiplication, and division, respectively.

When both operands are integers, the operator / cuts off the decimal part. For example, suppose that a and b are **int** variables with values 3 and 2, respectively. Then, the result of a/b is 1. If either of the operands is a floating point variable or constant, the decimal part isn't truncated. For example, the result of a/5.0 is 0.6 since the constant 5.0 is considered **float**.

The % operator is used to find the remainder of the division of two integer operands. Both operands should be integers or the program won't compile.

Increment and Decrement Operators

The increment operator ++ is used to increment the value of a variable by 1. It can be added before *(prefix)* or after *(postfix)* the variable's name. Both forms increment the variable by one, but they do so at different times. For example,

```c
#include <stdio.h>
int main()
{
  int a = 4;

  a++; /* Equivalent to a = a+1; */
  printf("%d\n", a);
  return 0;
}
```

This code outputs 5 since the statement a++ is equivalent to a = a+1.

When the ++ operator is used in postfix form, the increment is performed after the current value of the variable is used in the expression. For example,

```c
#include <stdio.h>
int main()
{
  int a = 4, b;

  b = a++;
  printf("a = %d b = %d\n", a, b);
  return 0;
}
```

The statement b = a++; stores the current value of a into b first and then increments the value of a by 1. As a result, the program displays a = 5 b = 4.

If used in the prefix form, the variable's value is immediately incremented, and the new value is used in the evaluation of the expression. For example,

```c
#include <stdio.h>
int main()
{
  int a = 4, b;

  b = ++a;
  printf("a = %d b = %d\n", a, b);
  return 0;
}
```

The statement b = ++a; immediately increments the value of a and then stores that value into b. The program displays a = 5 b = 5.

Like the increment operator, the decrement operator -- can be added either before or after the variable's name. Both forms decrement the variable's value by 1, but, like the increment operator, they do so at different times. For example,

```c
#include <stdio.h>
int main()
{
  int a = 4;

  a--; /* Equivalent to a = a-1; */
  printf("Num = %d\n",a);
  return 0;
}
```

The output of this program is Num = 3 since the statement a-- is equivalent to a = a-1.

The rules that apply for the postfix and prefix usage of the increment operator ++ also apply for the decrement operator --. For example,

```c
#include <stdio.h>
int main()
{
  int a = 4, b;

  b = a--;
  printf("a = %d b = %d\n", a, b);

  b = --a;
  printf("a = %d b = %d\n", a, b);
  return 0;
}
```

With the statement b = a--; the current value of a is stored in b and then its value is decremented by 1. The first printf() displays a = 3 b = 4.

Next, with the statement b = --a; the value of a is immediately decremented by 1 and then its value is stored in b. The second printf() displays a = 2 b = 2.

The operators ++ and -- can be combined in the same expression. For example,

```c
#include <stdio.h>
int main()
{
  int a = 1, b = 2, c = 3;

  printf("%d\n", (++a)-(b--)+(--c));
  return 0;
}
```

With the operation ++a the current value of a is first increased and then used in the expression. With the operation b-- the current value of b is first used and then its value is decremented by 1. Finally, with the operation --c the value of c is first decreased and then used in the expression.

Therefore, the result of (++a)-(b--)+(--c) is 2-2+2 = 2 and the values of a, b, and c become 2, 1, and 2, respectively.

Relational Operators

The relational operators >, >=, <, <=, !=, ==, are used to compare two operands and determine their relationship. As we'll see in next chapters, the relational operators are mostly used in **if** statements and iteration loops. For example, in the following **if** statements,

- **if**(a > 10) the > operator is used to check whether the value of a is greater than 10
- **if**(a >= 10) the >= operator is used to check whether the value of a is greater than or equal to 10
- **if**(a < 10) the < operator is used to check whether the value of a is less than 10
- **if**(a <= 10) the <= operator is used to check whether the value of a is less than or equal to 10
- **if**(a != 10) the != operator is used to check whether the value of a is not equal to 10
- **if**(a == 10) the == operator is used to check whether the value of a is equal to 10

The outcome of an expression using relational operators is either 1 or 0. An expression is considered true when it has any nonzero value; otherwise, it is considered false. For example, the outcome of the expression (a > 10) is 1 only if the value of a is greater than 10, otherwise it is 0.

*Do not confuse the == operator with the = operator. The == operator is used to check whether two expressions have the same value, while the = operator is used to assign a value to a variable. For instance, the statement **if**(a==10) checks whether a is equal to 10 or not, while the statement **if**(a=10) assigns the value 10 to a. We'll see more examples like this in the next chapter.*

Exercises

4.1   What is the output of the following program?

```c
#include <stdio.h>
int main()
{
  int a = 3, b = 5, c;

  a = (a > 3) + (b <= 5);
  b = (a == 3) + ((b-2) >= 3);
  c = (b != 1);
  printf("%d %d %d\n", a, b, c);
  return 0;
}
```

Answer: The value of the expression (a > 3) is 0 since a is 3. The value of the expression (b <= 5) is 1 since b is 5. Therefore, a = 0+1 = 1.

The value of the expression (a == 3) is 0 since a is 1. The value of the expression ((b–2) >= 3) is 1 since b–2 = 5–2 = 3. Therefore, b = 0+1 = 1.

The value of the expression (b != 1) is 0 since b is 1. Therefore, c becomes 0. As a result, the program displays 1 1 0.

4.2 What is the output of the following program?

```
#include <stdio.h>
int main()
{
  int a = 4, b = 5, c = 3;

  printf("%d\n", a < b > c);
  return 0;
}
```

Answer: The relational operators are left associative and, as shown in Appendix A, the operators < and > have the same precedence, so the program first checks to see whether a is less than b. The value of this expression is 1 since a is 4 and b is 5. Therefore, the expression a < b > c is equivalent to 1 > c, which is not true since the value of c is 3. As a result, the program displays 0.

4.3 What is the output of the following program?

```
#include <stdio.h>
int main()
{
  int a = 2, b = 2, c = 2;

  printf("%d\n", a == b == c);
  return 0;
}
```

Answer: First, the program checks to see whether a is equal to b. The value of this expression is 1 since both a and b equal 2. Therefore, the expression a == b == c is equivalent to 1 == c, which is not true since the value of c is 2 and the program displays 0. If the value of c were 1 instead of 2, the program would display 1.

Not Operator !

The *not operator* (!) is *unary*, which means it only acts on a single operand and produces either 0 or 1. Specifically, if the value of an expression exp is nonzero, then the value of !exp is 0. Conversely, if its value is zero, then the value of !exp is 1. For example,

```
#include <stdio.h>
int main()
{
  int a = 4;

  printf("Num = %d\n", !a);
  return 0;
}
```

This will display Num = 0 since a is 4. If the value of a were 0, the program would display Num = 1.

As we'll see in the next chapter, the ! operator is mostly used in **if** statements to check whether the value of an expression is true or false. For example,

- The statement **if**(!a) is equivalent to **if**(a == 0)
- The statement **if**(a) is equivalent to **if**(a != 0)

Exercises

4.4   What would be the output of the previous program if we print !!a instead of !a?
Answer: Since a is 4, the value of !a is 0. Therefore, the expression !!a is equivalent to !0. As a result, the program displays 1.

4.5   What is the output of the following program?

```
#include <stdio.h>
int main()
{
  int a = 4, b = 3, c = 5;

  printf("%d\n", (a < b) == !(c > b));
  return 0;
}
```

Answer: The value of the expression (a < b) is 0 since a is 4 and b is 3. Similarly, the value of the expression (c > b) is 1. Therefore, the value of the expression !(c > b) is 0. Since the two expressions have the same value, the program displays 1.

Compound Operators

Compound operators are used to shorten certain statements—usually arithmetic ones—using the following form: exp1 op= exp2; This statement is equivalent to exp1 = exp1 op (exp2). Notice that, for reasons of priority, the compiler encloses the expression exp2 in parentheses.

Usually, the op operator is one of the arithmetic operators: +, −, *, %, /, though it can also be any of the bit operators: &, ^, |, <<, >>, as we'll describe later in this chapter. For example,

```
#include <stdio.h>
int main()
{
  int a = 4, b = 2;

  a += 6;
  a *= b+3;
  a -= b+8;
  a /= b;
  a %= b+1;
  printf("Num = %d\n", a);
  return 0;
}
```

The statement a += 6; is equivalent to a = a+6, which is equal to 10.
   The next statement a *= b+3; is equivalent to a = a*(b+3) = 10*(2+3) = 50. Notice that the expression b+3 is automatically enclosed in parentheses.
   The statement a -= b+8; is equivalent to a = a-(b+8) = 50-(2+8) = 40.
   The statement a /= b; is equivalent to a = a/b = 40/2 = 20.
   Finally, the statement a %= b+1; is equivalent to a = a%(b+1) = 20%(2+1) = 2.
   The final output of the program is Num = 2.

Exercise

4.6   What is the output of the following program?

```
#include <stdio.h>
int main()
{
   int a = 4, b = 3, c = 1;

   a += b -= ++c;
   printf("%d %d %d\n", a, b, c);
   return 0;
}
```

Answer: The compound operators are executed from right to left, as shown in Appendix A. Therefore, the expression a += b -= ++c; is equivalent to a = a+(b -= ++c); By itself, the expression (b -= ++c) is equivalent to b = b-(++c). The value of c is first incremented and becomes 2, therefore b = b-2 = 3-2 = 1. Now, the expression a = a+(b -= ++c) becomes equivalent to a = a+1 = 4+1 = 5. As a result, the program displays 5 1 2.

Logical Operators

Like the relational operators, the logical operators can also be used to form logical expressions, which produce either 1 or 0 as their result.

Operator &&

The && operator is the logical *AND* operator. It works like the AND gate in electronic circuit designs. Specifically, the value of an expression that contains && operators only is 1 if all operands of the expression are true. Otherwise, the result is 0.
   For example, if we write a = (10 == 10) && (5 > 3); the value of a becomes 1 since both operands are true.
   If we write a = (13 < 8) && (10 == 10) && (5 > 3); the value of a becomes 0 since the first operand is false.

The logical operators are left associative, so the compiler evaluates the operands from left to right. In expressions with multiple && operators, if the compiler finds an operand with a false value, it makes the value of the entire expression false and it does not evaluate the rest of the operands. For example,

```c
#include <stdio.h>
int main()
{
  int a = 10, b = 20, c;

  c = (a > 15) && (++b > 15);
  printf("%d %d\n", c, b);
  return 0;
}
```

The second operand (++b > 15) is not evaluated since the first one (a > 15) is false. The program displays 0 20.

Operator ||

The || operator is the logical *OR* operator. It works like the OR gate in electronic circuit designs. Specifically, the value of an expression that contains || operators only is 1 if one or more operands of the expression are true. If all operands of the expression are false, then the result is 0.

For example, if we write a = (10 == 10) || (3 > 5); the value of a becomes 1 since the first operand is true.

If we write a = (10 != 10) || (3 > 5); the value of a becomes 0 since all operands are false.

In expressions with multiple || operators, if the compiler finds an operand with a true value, it makes the value of the entire expression true and it does not evaluate the rest of the operands. For example,

```c
#include <stdio.h>
int main()
{
  int a = 10, b = 20, c;

  c = (a > 5) || (++b > 15);
  printf("%d %d\n", c, b);
  return 0;
}
```

The second operand (++b > 15) is not evaluated since the first one (a > 5) is true. The program displays 1 20.

Exercises

4.7   What is the output of the following program?

```c
#include <stdio.h>
int main()
```

```
{
  int a = 10, b = -10;

  if(a > 0 && b < -10)
    printf("One ");
  else
    printf("Two ");

  if(a > 10 || b == -10)
    printf("One ");
  else
    printf("Two ");
  return 0;
}
```

Answer: Although we have yet to discuss **if-else** statements, you should be able to get a sense of what the aforementioned code is doing. These types of conditional statements are the most common way to use logical operators.

Since b is –10, the second term in the first **if** statement is false and the program outputs Two. Since the second term in the second **if** statement is true, the program outputs One.

4.8 What is the output of the following program?

```
#include <stdio.h>
int main()
{
  int a = 1;

  printf("%d\n", (a > 0) && (--a > 0));
  return 0;
}
```

Answer: With the statement --a the value of a becomes 0. Since the operand (--a > 0) is false, the program displays 0.

4.9 What is the output of the following program?

```
#include <stdio.h>
int main()
{
  int a = 4, b = 5, c = 6;

  printf("%d\n", (a-4) || (b++ == c));
  return 0;
}
```

Answer: The value of the expression (a-4) is 0 since a is 4. The value of the expression (b++ == c) is 0 since the value of b (5) is first compared to the value of c (6) and then increased. Since both operands are false, the program displays 0.

If we had ++b instead of b++, the program would display 1.

4.10 What is the output of the following program?

```
#include <stdio.h>
int main()
```

TABLE 4.1

Tax Income

Income ($)	Tax (%)
0 – 25.000	0
25.001 – 35.000	15
>35.000	30

```
{
  int a = 1, b = 3, c = -4;

  printf("%d", !(a > b) && (a-1 == b-3) && (c%2 || (a+b+c)));
  return 0;
}
```

Answer: The value of the expression (a > b) is 0 since a is 1 and b is 3. Therefore, the value of the first operand !(a > b) is 1.

The value of the second operand (a-1 == b-3) is 1 since 0 equals 0. The value of the expression c%2 is 0 since c is -4. The value of the expression (a+b+c) is 1+3-4 = 0. Therefore, the value of the third operand is 0. Since the third operand is false, the result of the entire expression is false and the program displays 0.

4.11 Write a program that reads an annual income and calculates the tax according to Table 4.1.

```
#include <stdio.h>
int main()
{
  float tax, a;

  printf("Enter income: ");
  scanf("%f", &a);

  tax = (a > 25000 && a <= 35000)*(a-25000)*0.15 + (a > 35000) *
    ((a-35000)*0.3 + 10000*0.15);
  printf("Tax = %.2f\n", tax);
  return 0;
}
```

Comments: If the user enters an amount greater than 25000 and less than or equal to 35000, the value of the expression (a > 25000 && a <= 35000) is 1, while the value of (a > 35000) is 0. The tax is calculated for the part of the amount over than 25000. If the input value is greater than 35000, the reverse is true. If the income is greater than 35000, the first term calculates the tax for the part of the amount over 35000, while the second term calculates the tax for the part of the amount from 25000 to 35000.

If you had difficulty in understanding this solution, don't worry. Such kind of exercises would be much easier solved after learning the **if** statement.

Comma Operator

The comma (,) operator can be used to merge several expressions to form a single expression. An expression using the comma operator has the form

```
exp1, exp2, exp3,…
```

The comma operator means "evaluate the expression `exp1` first, then evaluate `exp2`, then `exp3`, through the last expression." Since the comma operator is left associative, the expressions are sequentially evaluated from left to right. The result of the entire expression is the result of the last expression evaluated. For example,

```
#include <stdio.h>
int main()
{
  int b;

  b = 20, b = b+30, printf("Num = %d\n", b);
  return 0;
}
```

Since the expressions are executed from left to right, the value of `b` becomes 50 and the program displays `Num = 50`.

In fact, there is little reason to use a comma operator to merge expressions because the resulting code can be more complex and harder to evaluate. For example, as we'll see in Chapter 6, the most common use of the comma operator is in the controlling expressions of a **for** statement:

```
int a, b;
for(a = 1, b = 2; b < 10; a++, b++)
```

The first controlling expression first performs `a = 1` and then `b = 2`. The third expression first performs `a++` and then `b++`.

Exercise

4.12 What is the output of the following program?

```
#include <stdio.h>
int main()
{
  int a, b;

  a = (b = 20, b = b+30, b++);
  printf("a = %d, b = %d\n", a, b);
  return 0;
}
```

Answer: Since the expressions `b = 20, b = b+30` are executed from left to right, the value of `b` becomes 50. With the expression `b++` the current value of `b` is first stored in `a` and then its value is incremented. The final output of the program is `a = 50 b = 51`.

The parentheses are necessary since the comma operator has the lowest precedence of any C operator, including the assignment (=) operator.

sizeof Operator

The `sizeof` operator is used to determine the number of bytes required to store a value of a specific type. For example, the following program displays how many bytes allocate in memory the C data types.

```
#include <stdio.h>
int main()
{
  printf("Char = %d bytes\n", sizeof(char));
  printf("Short int = %d bytes\n", sizeof(short));
  printf("Int = %d bytes\n", sizeof(int));
  printf("Long = %d bytes\n", sizeof(long));
  printf("Float = %d bytes\n", sizeof(float));
  printf("Double = %d bytes\n", sizeof(double));
  return 0;
}
```

In the following program, the `sizeof` operator calculates how many bytes allocate the program variables.

```
#include <stdio.h>
int main()
{
  char c;
  int i;
  float f;

  printf("%d %d %d\n", sizeof(c), sizeof(i), sizeof(f));
  return 0;
}
```

The program outputs 1 4 4.

enum Type

The `enum` type is used to define an enumeration type, which is a set of named integer constant values. In the simplest case, it is declared as follows:

```
enum tag {enumeration_list};
```

The `tag` is an optional label that identifies the enumeration list. For example, the statement

```
enum seasons {AUTUMN, WINTER, SPRING, SUMMER};
```

declares the `seasons` enumeration type and the enumeration integer constants AUTUMN, WINTER, SPRING, and SUMMER.

By default, the value of the first constant is set to 0, though the values of the constants may be explicitly set, like so

```
enum seasons {AUTUMN=10, WINTER=20, SPRING=30, SUMMER=40};
```

If a constant is not explicitly assigned a value, it will be initialized to the value of the previous constant plus one. As a result, the values of AUTUMN, WINTER, SPRING, and SUMMER constants in the first example become 0, 1, 2, and 3, respectively (it is possible to initialize two or more constants with the same value).

We may also declare enumeration variables using the enumeration tag like so

```
enum tag variable_list;
```

For example, the following statement

```
enum seasons s1, s2;
```

declares variables s1 and s2 as enumeration variables of the type seasons. Alternatively, we may declare them together with the declaration of the enumeration type, as follows:

```
enum seasons {AUTUMN, WINTER, SPRING, SUMMER} s1, s2;
```

The important thing with the enumeration variables is to remember that C treats them as integer variables and the names of the enumeration list as constant integers.

For example, the following program shows how ordinary variables, enumeration variables, and constants may be mixed together.

```
#include <stdio.h>
int main()
{
  int next_seas;
  enum seasons {AUTUMN = 1, WINTER, SPRING, SUMMER} s;

  printf("Enter season [1-4]: ");
  scanf("%d", &s);

  if(s == SUMMER)
    next_seas = AUTUMN;
  else
    next_seas = s+1;

  printf("Next season = %d\n", next_seas);
  return 0;
}
```

The program reads an integer, stores it into the enumeration variable s, and displays the number that corresponds to the next season.

Bitwise Operators

Bitwise operators are used to access the individual *bits* of an integer variable or constant. The value of a bit is either 0 or 1. Bitwise operators are particularly useful in several applications, such as data coding in communication protocols, data encryption, and in low-level applications that communicate with the hardware.

> *When performing bitwise operations, it's usually safer to declare your variables as* **unsigned**. *If you choose not to, make sure you account for the sign bit when performing your calculations.*

& Operator

The & operator performs the *bitwise AND* operation on all corresponding bits of its two operands. The result bit is 1 only if both corresponding bits are 1. Otherwise, the bit is set to 0. For example, the result of 19 & 2 is 2

```
  00010011  (19)
& 00000010  (2)
  -------------
  00000010  (2)
```

| Operator

The | operator performs the *bitwise inclusive OR* operation on all corresponding bits of its two operands. The result bit is 0 only if both corresponding bits are 0. Otherwise, the bit is set to 1. For example, the result of 19 | 6 is 23:

```
  00010011  (19)
| 00000110  (6)
  -------------
  00010111  (23)
```

^ Operator

The ^ operator performs the *bitwise exclusive OR (XOR)* operation on all corresponding bits of its two operands. The result bit is 1 only if both corresponding bits are different. Otherwise, the bit is set to 0. For example, the result of 19 ^ 6 is 21:

```
  00010011  (19)
^ 00000110  (6)
  -------------
  00010101  (21)
```

~ Operator

The complement operator ~ performs the *bitwise NOT* operation on all bits of its single operand. It means that it reverses the 1s to 0s and vice versa. For example, the result of ~19 in a 32-bit system is

```
~ 00000000 00000000 00000000 00010011  (19)  =
  11111111 11111111 11111111 11101100
```

Exercise

4.13 A simple way to encrypt data is to use the ^ operator with a secret key. Write a program that reads an integer that corresponds to the secret key and another one that will be encrypted. The encryption is performed by applying the ^ operator on them. Then, the program should use once more the ^ operator to decrypt the encrypted result.

```c
#include <stdio.h>
int main()
{
    int num, key;

    printf("Enter key: ");
    scanf("%d", &key);

    printf("Enter number: ");
    scanf("%d", &num);

    num = num ^ key;
    printf("Encrypted : %d\n", num);

    num = num ^ key;
    printf("Original: %d\n", num);
    return 0;
}
```

Comments: The decrypted number is the input number.

Shift Operators

The shift operators shift the bits of an integer variable or constant left or right. The >> operator shifts the bits to the right, while the << operator shifts the bits to the left. As explained later, both operators take a right operand, which indicates how many places the bits will be shifted.

>> Operator

The expression i >> n shifts the bits in i by n places to the right. The n high-order bits that are inserted on the left are set to 0. For example, what value would a have in the following code?

```c
unsigned int a, b = 35;
a = b >> 2;
```

Since the number 35 is 00100011 in binary, the result of 00100011 >> 2 is 00001000 (decimal 8) and that will be the value of a. As you can see, the two rightmost bits 1 and 1 of the original number are "shifted off" and two zero bits are entered in positions 7 and 8. Notice that the value of b remains 35.

Since a bit position corresponds to a power of 2, shifting a positive integer n places to the right divides its value by 2^n.

<< Operator

The expression `i << n` shifts the bits in `i` by `n` places to the left. The `n` low-order bits inserted on the right are set to 0. For example, what value would `a` have in the following code?

```
unsigned int a, b = 35;
a = b << 2;
```

Since the number 35 is 00100011 in binary, the result of 00100011 << 2 is 0010001100 (decimal 140) and that will be the value of `a`. As you can see, the bits of the original number are shifted two places to the left and two zero bits are added in positions 1 and 2. As with the previous example, the value of `b` does not change.

Since a bit position corresponds to a power of 2, shifting a positive integer n places to the left multiplies its value by 2^n.

When storing shifting results in a variable, make sure the type of the variable is large enough to hold the value.

For example, we see some unexpected results when the variable can't store the value.

```
#include <stdio.h>
int main()
{
  unsigned char a = 1;

  a <<= 8; /* Equivalent to a = a << 8; */
  printf("Val = %d\n", a);
  return 0;
}
```

The expression `a <<= 8` shifts the value of `a` by eight places to the left and inserts eight zero bits at the right. The result of the shifting is 100000000.

However, the variable `a` is declared as **unsigned char** (one byte long), which means that it can hold only the value of the eight low-order bits. Therefore, the program displays `Val = 0` and not `Val = 256` as expected. If the variable `a` were declared as **int** (four bytes long), the program would display `Val = 256`.

Exercises

4.14 What is the output of the following program?

```
#include <stdio.h>
int main()
{
  printf("%d\n", ~(~0 << 4));
```

```
    return 0;
}
```

Answer: In a 32-bit system, the result of ~0 is the number with all 32 bits set to 1s. Therefore, the value of (~0 << 4) is the number 11111111 11111111 11111111 11110000

Because the ~ operator reverses its bits, the number becomes

00000000 00000000 00000000 00001111 and the program displays 15.

4.15 What is the output of the following program?

```
#include <stdio.h>
int main()
{
  unsigned int i = 10;

  if((i >> 4) != 0)
    printf("One\n");
  else
    printf("Two\n");

  printf("Val = %d\n", i);
  i = 1;
  if((i << 3) == 8)
    printf("One\n");
  else
    printf("Two\n");

  printf("Val = %d\n", i);
  return 0;
}
```

Answer: The first **if** statement checks to see whether the value of i shifted four places to the right isn't zero. Since the number 10 (binary 00001010) shifted four places to the right evaluates to 00000000, the program displays Two. The **if** statement does not change the value of i to 0; it only checks to see if the value of i shifted four places to the right is 0 or not. Because i remains 10, the program displays Val = 10.

The second **if** statement checks to see if the value of i shifted three places to the left is 8. Because the number 1 (binary 00000001) shifted three places to the left evaluates to 00001000 (decimal 8), the program displays One.

As with the previous example, because the value of i remains the same, the program displays Val = 1.

4.16 What is the output of the following program?

```
#include <stdio.h>
int main()
{
  char a = 8;

  a <<= 4;
  printf("Val = %d\n", a);
  return 0;
}
```

Answer: The expression a <<= 4 shifts the value of a four places to the left and makes it 10000000. Since the variable a is declared as **char**, the leftmost bit is reserved for the sign of the number. If its value is 1, the number is negative, otherwise it is positive, so the program displays Val = –128. If the variable a were declared as **unsigned char**, the program would display Val = 128.

*Remember, it's safer to perform bitwise operations only on **unsigned** types.*

4.17 Write a program that reads an integer and displays a message to indicate whether it is even or odd.

```
#include <stdio.h>
int main()
{
  int num;

  printf("Enter number: ");
  scanf("%d", &num);

  if(num & 1)
    printf("The number %d is odd\n", num);
  else
    printf("The number %d is even\n", num);
  return 0;
}
```

Comments: An integer number is even or odd depending on whether its last bit is 0 or 1. If it is 0, the number is even; otherwise, it is odd.

The **if** statement tests whether the last bit is 0 or 1. For example, suppose that the variable num is coded in the binary system as

```
xxxxxxxx xxxxxxxx xxxxxxxx xxxxxxxx
```

where each "x" represents a bit with a value of either 0 or 1. Then, the result of num & 1:

```
    xxxxxxxx xxxxxxxx xxxxxxxx xxxxxxxx
&   00000000 00000000 00000000 00000001
    -----------------------------------
    00000000 00000000 00000000 0000000x
```

is equal to the value of the last bit. If it is 1, the expression in the **if** statement is true and the program displays a message that the number is odd. Otherwise, it displays a message that the number is even.

4.18 The bitwise operators and the shift operators are often used in data coding in network communications. For example, some specific bits in the header of a Transport Control Protocol (TCP) packet are coded as depicted in Figure 4.1.

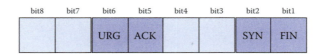

FIGURE 4.1
Part of the header of a TCP packet.

FIN (bit1): If it is 1, it indicates the release of the TCP connection.
SYN (bit2): If it is 1, it indicates the establishment of the TCP connection.
ACK (bit5): If it is 1, it indicates the acknowledgment of data reception.
URG (bit6): If it is 1, it indicates the transport of urgent data.

Write a program that reads the values of URG, ACK, SYN, and FIN bits and encodes this information in a program variable. Then, the program should decode the value of this variable and display the values of the respective bits.

```
#include <stdio.h>
int main()
{
  unsigned int temp, urg, ack, syn, fin;

  printf("Enter FIN bit: ");
  scanf("%d", &fin);
  printf("Enter SYN bit: ");
  scanf("%d", &syn);

  printf("Enter ACK bit: ");
  scanf("%d", &ack);

  printf("Enter URG bit: ");
  scanf("%d", &urg);

  temp = fin + (syn << 1) + (ack << 4) + (urg << 5);
  printf("\nEncoding: %d\n", temp);

  fin = temp & 1;
  syn = (temp >> 1) & 1;
  ack = (temp >> 4) & 1;
  urg = (temp >> 5) & 1;
  printf("FIN = %d, SYN = %d, ACK = %d, URG = %d\n", fin, syn,
    ack, urg);
  return 0;
}
```

4.19 Write a program that reads an integer in [0,255], then it swaps the two quads of its binary digits and displays the new number. For example, if the user inserts the number 10 (binary 00001010), then the program should display the number 160 (binary 10100000).

```
#include <stdio.h>
int main()
{
  unsigned int num, temp;

  printf("Enter number [0-255]: ");
  scanf("%d", &num);

  if(num >= 0 && num <= 255)
  {
    temp = num & 0xF; /* The value of the four low bits is stored in
      temp. 0xF is coded as 00001111 in binary. */
    temp <<= 4; /* Shift temp by four places to the left. */
    temp += num >> 4; /* Add to temp the value of num shifted by four
      places to the right. */
```

```
    printf("Num_1 = %d Num_2 = %d\n", num, temp);
  }
  else
    printf("Number should be in [0-255]\n");
  return 0;
}
```

4.20 Usually, applications that read data from the hardware are needed to figure out the values of specific bits. For example, write a program that reads an integer and a bit position and displays the value of the respective bit.

```
#include <stdio.h>
int main()
{
  unsigned int num, pos;

  printf("Enter number: ");
  scanf("%d", &num);

  printf("Enter bit position [1-32]: ");
  scanf("%d", &pos);
  if(pos >= 1 && pos <= 32)
    printf("The value of bit%d is %d\n", pos, (num >> (pos-1)) & 1);
  else
    printf("Bit position should be in [1-32]\n");
  return 0;
}
```

4.21 What is the output of the following program?

```
#include <stdio.h>
int main()
{
  unsigned char ch = 3;

  ch = ((ch&1) << 7) | ((ch&2) << 5) | ((ch&4) << 3) | ((ch&8) << 1) |
    ((ch&16) >> 1) | ((ch&32) >> 3) | ((ch&64) >> 5) | ((ch&128) >> 7);
  printf("%d\n", ch);
  return 0;
}
```

Answer: Suppose that ch is coded in the binary system as $x_8x_7x_6x_5x_4x_3x_2x_1$, where each x_i can be either 1 or 0. Consequently, (ch&1) << 7 is evaluated as

$$x_8x_7x_6x_5x_4x_3x_2x_1$$

$$\&\ \ 0\ 0\ 0\ 0\ 0\ 0\ 0\ 1$$

$$0\ 0\ 0\ 0\ 0\ 0\ 0\ x_1 << 7 = x_1\ 0\ 0\ 0\ 0\ 0\ 0\ 0$$

Similarly, (ch&2) << 5 is evaluated as

$$x_8x_7x_6x_5x_4x_3x_2x_1$$

$$\&\ \ 0\ 0\ 0\ 0\ 0\ 0\ 1\ 0$$

$$0\ 0\ 0\ 0\ 0\ 0\ x_2\ 0 << 5 = 0\ x_2\ 0\ 0\ 0\ 0\ 0\ 0$$

Following the same logic:

```
(ch&4)   <<  3 is evaluated as 0  0  x₃ 0  0  0  0  0
(ch&8)   <<  1 is evaluated as 0  0  0  x₄ 0  0  0  0
(ch&16)  >>  1 is evaluated as 0  0  0  0  x₅ 0  0  0
(ch&32)  >>  3 is evaluated as 0  0  0  0  0  x₆ 0  0
(ch&64)  >>  5 is evaluated as 0  0  0  0  0  0  x₇ 0
(ch&128) >>  7 is evaluated as 0  0  0  0  0  0  0  x₈
```

Therefore, the expression is evaluated as

$$(x_1\ 0\ 0\ 0\ 0\ 0\ 0\ 0)\ |\ (0\ x_2\ 0\ 0\ 0\ 0\ 0\ 0)\ |\ (0\ 0\ x_3\ 0\ 0\ 0\ 0\ 0)\ |\ (0\ 0\ 0$$
$$x_4\ 0\ 0\ 0\ 0)\ |\ (0\ 0\ 0\ 0\ x_5\ 0\ 0\ 0)\ |\ (0\ 0\ 0\ 0\ 0\ x_6\ 0\ 0)\ |\ (0\ 0\ 0\ 0\ 0\ 0$$
$$x_7\ 0)\ |\ (0\ 0\ 0\ 0\ 0\ 0\ 0\ x_8)\ =\ x_1 x_2 x_3 x_4 x_5 x_6 x_7 x_8$$

As you can see, the program reverses the bits of the variable ch and displays the corresponding value. Notice also that the bits' reversion takes place without the use of any other auxiliary variable.

The value of ch (binary 00000011) is reversed to 192 (binary 11000000) and the program displays 192.

Operator Precedence

Each operator has a *precedence* associated with it that determines the order in which operators are evaluated in an expression. In an expression that contains several operators, the operators with higher precedence are evaluated first. For example, since the * operator has higher precedence than the + and – operators, the result of 7+5*3–1 is 21.

When operators of equal precedence appear in the same expression, the operators are evaluated according to their *associativity*. For example, since the * and / operators have the same precedence and are *left associative*, the result of 7*4/2*5 is 70. Since the evaluation takes place from left to right, the multiplication (7*4 = 28) is executed first, then the division (28/2 = 14), and last the multiplication (14*5 = 70).

Appendix A presents the precedence table. Parentheses have the highest precedence, so an expression in parentheses is evaluated before any others. Parentheses may be used in an expression to override operator precedence and force some parts of an expression to be evaluated before other parts.

Since it is not easy to memorize the entire precedence table, use parentheses when you are unsure of the evaluation order. The use of parentheses, even when it is not necessary to do so, increases the readability of code and clarifies the way that expressions are evaluated. For example, it is clearer to write
a = 7+(5*3)–1; rather than a = 7+5*3–1;

Unsolved Exercises

4.1 Write a program that reads the prices of three products and displays a message to indicate if any of them costs more than $100 or not.

4.2   Suppose that a customer in a store buys some things. If the total cost is less than $100, no discount is due, otherwise a 5% is given. Write a program that reads the total cost and displays the amount to be paid.

4.3   Write a program that reads the minimum score required to pass the exams, the grades of three students, and displays how many of them succeeded. (*Note*: a grade to be valid should be <= 10.)

4.4   Continue the previous exercise and find the average grade of those who passed the exams. (*Note*: suppose that one student succeeded, at least.)

4.5   Write a program that reads two integers, it stores them into two variables, and uses the ^ operator and the formula $(x\char94 y)\char94 x$ = y and $(x\char94 y)\char94 y$ = x to swap their values. Use only two variables.

4.6   Write a program that reads an integer and displays the sum of the digits in the positions 2, 4, 6 and 8. For example, if the user enters 170 (bin: 10101010), the program should display 4. Use only one variable.

4.7   Write a program that reads an integer and rotates its bytes one place right. For example, if the user enters 553799974 (bin: 00100001|00000010|01010001|00100110) the program should display: 639697489 (bin: 00100110|00100001|00000010|010 10001).

4.8   Write a program that reads an integer (i.e., a) in [0, 255] and the number of shifting bits (i.e., n). The program should display the sum of

(a) Shifting the input number n places to the left and add the "shifted-off" bits to the right of the number

(b) Shifting the input number n places to the right and add the "shifted-off" bits to the left of the number

For example, if the user enters 42 (bin: 00101010) and 3, the program should display the sum of

00101010 << 3 = 01010**001** = 81 (the shifted-off bits are 001)

and

00101**010** >> 3 = **010**00101 = 69 (the shifted-off bits are 010)

4.9   Write a program that reads a positive integer and rounds it up to the next highest power of 2. For example, if the user enters 35, the program should display 64, which is the next highest power of 2 since 2^5 = 32 < 35 < 64 = 2^6.

5

Program Control

Up to this point we have seen that program's statements are executed from top to bottom, in the order that they appear inside the source code. However, in real programming, certain statements should be executed only if specific conditions are met. This chapter will teach you how to use **if** and **switch** selection statements to control the flow of your program and make it select a particular execution path from a set of alternatives. It also introduces the conditional operator (?:), which can be used to form conditional expressions.

if Statement

The **if** statement controls program flow based on whether a condition evaluates to true or false. The simplest form of **if** statement is

```
if(condition)
{
  ... /* block of statements. */
}
```

If the value of the condition is true, the block of statements between the braces will be executed. For example,

```
int x = 3;
if(x != 0)
{
  printf("x isn't zero\n");
}
```

Since x is not 0, the value of the **if** condition is true and the program displays x isn't zero.

> **Tip**
>
> if(x != 0) *is equivalent to* if(x)

If the value of the condition is false, the block of statements won't be executed. For example, the following code displays nothing:

```
int x = -3;
if(x == 0)
{
  printf("x is zero\n");
}
```

Tip

```
if(x == 0) is equivalent to if(!x)
```

If the block of statements consists of a single statement, the braces can be omitted. In other words, the previous code could be written as

```
if(x == 0)
  printf("x is zero\n");
```

Common Errors

A common error is to add a semicolon (;) at the end of the **if** statement, as you usually do with the most statements. The semicolon is handled as a statement that does nothing (null statement) and the compiler terminates the **if**. For example, in the following code,

```
int x = -3;
if(x > 0);
  printf("x is positive\n");
```

the ; terminates the **if** statement and the program continues with the printf() call. Therefore, the output is always x is positive regardless of the value of x.

Another common error is to confuse the == operator with the = operator inside an **if** condition. Remember that the == operator checks whether two expressions have the same value, while the = operator assigns a value to a variable. For example, this code

```
int x = -3;
if(x = -2)
  printf("x equals -2\n");
```

outputs x equals -2, although x is –3. If we had written **if**(x == -2) nothing would have been displayed.

The following code

```
int x = 0;
if(x = 0)
  printf("x equals zero\n");
```

displays nothing since the assignment of 0 in x makes the condition false.

if-else Statement

An **if** statement may have an **else** clause, as shown here:

```
if(condition)
{
  … /* block of statements A. */
}
```

```
else
{
  ... /* block of statements B. */
}
```

If the condition is true, the first block of statements will be executed; if not, the second block will run. For example, the following program displays x is negative or zero because x is less than or equal to 0.

```
int x = -3;
if(x > 0)
{
  printf("x is positive\n");
}
else
{
  printf("x is negative or zero\n");
}
```

And, since both blocks consist of a single statement, we could omit the braces:

```
int x = -3;
if(x > 0)
  printf("x is positive\n");
else
  printf("x is negative or zero\n");
```

Nested if Statements

An if statement can contain any kind of statement, including other if and else statements. For example, the following program contains two nested if statements:

```
#include <stdio.h>
int main()
{
  int a = 10, b = 20, c = 30;

  if(a > 5)
  {
    if(b == 20) /* nested if statement. */
      printf("1 ");

    if(c == 40) /* nested if statement. */
      printf("2 ");
    else
      printf("3 ");
  }
  else
    printf("4\n");

  return 0;
}
```

Since a is greater than 5, the program executes the block of statements beginning with **if**(b == 20) and displays 1 and 3.

In a program with nested **if** statements, each **else** statement is associated with the nearest **if** statement that does not contain an **else**. For example, consider the following program:

```c
#include <stdio.h>
int main()
{
  int a = 5;

  if(a != 5)
    if(a-2 > 5)
      printf("One\n");
  else
    printf("Two\n");
  return 0;
}
```

According to this rule, the **else** statement should be associated with the nearest **if** statement that does not contain an **else**. This is the second **if** statement. As a result, the program displays nothing because the first **if** statement is false and it has no **else** associated with.

Notice how the indenting can trick you into thinking that the **else** statement is associated with the first **if** statement. By aligning each **else** statement with the corresponding **if** (indenting the statements and adding braces to **if-else** statements even if you don't have to) can make the program easier to read. For example, here's how we could make the previous program more readable by aligning the correct **if** with the corresponding **else**:

```c
#include <stdio.h>
int main()
{
  int a = 5;

  if(a != 5)
  {
    if(a-2 > 5)
      printf("One\n");
    else
      printf("Two\n");
  }
  return 0;
}
```

To test a series of **if** statements and stop when a condition is true, use nested **if** statements, like this (the comments label the blocks of statements):

```c
if(condition_A)
{
  … /* block of statements A. */
}
else if(condition_B)
{
  … /* block of statements B. */
}
```

```
else if(condition_C)
{
  … /* block of statements C. */
}
.
.
.
else
{
  … /* block of statements N. */
}
… /* the program's next statements. */
```

In this listing, once a true condition is found, the corresponding block of statements is executed, while the remaining **else if** conditions are not checked and the program continues with the execution of the first statement following the last **else**.

For example, if condition_A is true, the block of statements A is executed; if not, condition_B is checked. If condition_B is true, the block of statements B is executed; if not, condition_C is checked. If condition_C is true, the block of statements C is executed; if not, the next condition is checked, and so on. If no condition is true, the block of statements N of the last **else** will be executed.

The following program reads an integer and displays a message according to the value of that integer:

```
#include <stdio.h>
int main()
{
  int a;

  printf("Enter number: ");
  scanf("%d",&a);

  if(a == 1)
    printf("One\n");
  else if(a == 2)
    printf("Two\n");
  else
    printf("Something else\n");

  printf("End\n");
  return 0;
}
```

If the user enters 1, the program will display One; if the number 2 is entered, the program will display Two. If an integer other than 1 or 2 is entered, the program will display Something else. If a true condition is found, the remaining conditions are not checked. In any case, the program continues with the last printf() call and displays End.

The block of statements in the last **else** *is executed only if all previous conditions are false. Nevertheless, the last* **else** *statement is not mandatory. If it is missing and all conditions are false, the program continues with the execution of the first statement after the last* **else if**.

For example, in the previous program, if the number 4 is entered and the last **else** statement was missing, the program would display End.

Exercises

5.1   Write a program that reads an integer and displays its absolute value.

```c
#include <stdio.h>
int main()
{
  int num;

  printf("Enter number: ");
  scanf("%d", &num);

  if(num >= 0)
    printf("The absolute value is %d\n", num);
  else
    printf("The absolute value is %d\n", -num);
  return 0;
}
```

5.2   Write a program that reads two integers and displays them in ascending order.

```c
#include <stdio.h>
int main()
{
  float i, j;

  printf("Enter grades: ");
  scanf("%f%f", &i, &j);

  if(i < j)
    printf("%f %f\n", i, j);
  else
    printf("%f %f\n", j, i);
  return 0;
}
```

5.3   Write a program that reads two integers and displays their relationship without using an **else** statement.

```c
#include <stdio.h>
int main()
{
  int i, j;

  printf("Enter numbers: ");
  scanf("%d%d", &i, &j);

  if(i < j)
    printf("%d < %d\n", i, j);
```

```
    if(i > j)
      printf("%d > %d\n", i, j);
    if(i == j)
      printf("%d = %d\n", i, j);
    return 0;
}
```

5.4   What is the output of the following program?

```
#include <stdio.h>
int main()
{
  int a = 10, b = 20;

  if(a && a/b)
    printf("One\n");
  else
    printf("Two\n");
  return 0;
}
```

Answer: The result of a/b is 0 (false), not 0.5, because both a and b are integer variables. Therefore, the value of the expression (a && a/b) is false and the program displays Two.

5.5   What is the output of the following program?

```
#include <stdio.h>
int main()
{
  int i = 10, j = 20, k = 0;

  if(i = 40)
    printf("One ");
  if(j = 50)
    printf("Two ");
  if(k = 60)
    printf("Three ");
  if(k = 0)
    printf("This is the end... ");
  printf("%d %d %d\n", i, j, k);
  return 0;
}
```

Answer: This is an example of how using the = operator instead of the == operator can produce unexpected results.

The first **if** condition does not check to see if the value of i is 40, but instead makes i equal to 40. Since a nonzero (true) value is assigned to i, the **if** condition becomes true and the program displays One.

Similarly, the second **if** condition makes j equal to 50 and displays Two.

The third **if** condition makes k equal to 60 and displays Three. Then, the last **if** condition makes k equal to 0. Since a zero (false) value is assigned to k, the **if** condition becomes false and the printf() call is not executed.

The program displays One Two Three 40 50 0

5.6 Write a program that reads two floats (such as a and b) and displays the solution of
the equation *a*x + b = 0.*

```c
#include <stdio.h>
int main()
{
  double a, b;

  printf("Enter numbers: ");
  scanf("%lf%lf", &a, &b);

  if(a == 0)
  {
    if(b == 0)
      printf("Infinite solutions\n");
    else
      printf("There is no solution !!!\n");
  }
  else
    printf("The solution is %f\n", -b/a);
  return 0;
}
```

5.7 What is the output of the following program?

```c
#include <stdio.h>
int main()
{
  float a = 3.1;

  if(a == 3.1)
    printf("Yes ");
  else
    printf("No ");

  printf("%.9f\n", a-3.1);
  return 0;
}
```

Answer: Although the obvious answer is Yes, did the program output No and a nonzero value? Why is that the case?

Recall from Chapter 2 that when using a floating-point number in mathematical
expressions it is safer to use the **double** type. Therefore, if a had been declared as
double, the program would display Yes 0.000000000.

In fact, when you test a floating-point value for equality, you can never be sure
(even if you are using a **double** type) because there is always the chance of a potential rounding error.

5.8 Write a program that reads the temperature (in Fahrenheit degrees) and displays the
corresponding description, as follows:

(a) <=32: Intense cold
(b) (32–59]: Cold
(c) (59–77]: Warm
(d) (77–95]: Heat
(e) >95: Intense heat

```
#include <stdio.h>
int main()
{
  double i;

  printf("Enter Fahrenheit degrees: ");
  scanf("%lf", &i);

  if(i > 95)
    printf("Intense heat\n");
  else if(i > 77) /* Since the previous if checks values greater than
    95, this else-if statement is equivalent to: else if(i > 77 && i
    <= 95). */
    printf("Heat\n");
  else if(i > 59)
    printf("Warm\n");
  else if(i > 32)
    printf("Cold\n");
  else /* corresponds to degrees <= 32 */
    printf("Intense cold\n");
  return 0;
}
```

5.9   What is the output of the following program?

```
#include <stdio.h>
int main()
{
  int a, b;

  a = b = 9;
  if(++a == b++)
    printf("One\n");
  else
    printf("Two\n");
  printf("%d %d\n", a, b);
  return 0;
}
```

Answer: The prefix form of the ++ operator immediately increases the value of a by 1 and makes it 10. The postfix form of the ++ operator increases the value of b, after this value is used in the expression.

   Therefore, since a is 10 and b is still 9, the program displays Two. After the comparison is performed, b becomes 10 and the program displays 10 10.

5.10   Write a program that reads a man's height and weight and calculates his body mass index (bmi) using the formula bmi = weight/height^2. The program should display the bmi and a corresponding message, according to Table 5.1.

```
#include <stdio.h>
int main()
{
  float bmi, height, weight;

  printf("Enter height (in meters): ");
  scanf("%f", &height);
```

TABLE 5.1

Weight Categories

Mass Index	Result
BMI < 20	Lower than normal weight
20 <= BMI <= 25	Normal weight
25 < BMI <= 30	Overweight
30 < BMI <= 40	Obese
40 < BMI	Extremely obese

```c
printf("Enter weight (in kgrs): ");
scanf("%f", &weight);

bmi = weight/(height*height);
printf("BMI: %.2f\n", bmi);

if(bmi < 20)
  printf("Under normal weight\n");
else if(bmi <= 25) /* Since the previous if checks values up to 20,
  this else-if statement is equivalent to: else if(bmi >= 20 && bmi
  <= 25). */
  printf("Normal weight\n");
else if(bmi <= 30)
  printf("Overweight\n");
else if(bmi <= 40)
  printf("Obese\n");
else
  printf("Serious obesity\n");

printf("According to your height the bounds of normal weight are
  [%.2f-%.2f]\n", 20*height*height, 25*height*height); /* The lower
  and the upper limit of BMI for a normal weight man are 20 and 25,
  respectively. */
return 0;
}
```

5.11 What is the output of the following program?

```c
#include <stdio.h>
int main()
{
  int a = 5, b = 7, c = 3;

  if(a < b < c)
    printf("One\n");
  else
    printf("Two\n");
  return 0;
}
```

Answer: The < and > operators have the same precedence and are left associative. Therefore, in the expression (a < b < c), the program first checks to see whether a is less than b. The value of this expression is true (1) because a is 5 and b is 7.

Therefore, the expression (a < b < c) becomes equivalent to (1 < c), which is true since c is 3.

As a result, the program displays One.
To compare b against a and c, we would use the && operator like this:

```
if(a < b && b < c)
```

5.12 Write a program that reads three integers. If the sum of any two of them is equal to the third one, it should display the integers within [0, 10]. Otherwise, the program should read another three integers and display how many of them are multiples of 6.

```
#include <stdio.h>
int main()
{
  int i, j, k, cnt;

  printf("Enter numbers: ");
  scanf("%d%d%d", &i, &j, &k);

  if((i+j == k) || (j+k == i) || (i+k == j))
  {
    if(i >= 0 && i <= 10)
      printf("%d\n", i);
    if(j >= 0 && j <= 10)
      printf("%d\n", j);
    if(k >= 0 && k <= 10)
      printf("%d\n", k);
  }
  else
  {
    printf("Enter numbers: ");
    scanf("%d%d%d", &i, &j, &k);

    cnt = 0;
    if(i%6 == 0)
      cnt++;
    if(j%6 == 0)
      cnt++;
    if(k%6 == 0)
      cnt++;

    printf("%d\n", cnt);
  }
  return 0;
}
```

5.13 Write a program that reads the prices of four products and displays the largest one.

```
#include <stdio.h>
int main()
{
  double i, j, k, m, max;

  printf("Enter prices: ");
  scanf("%lf%lf%lf%lf", &i, &j, &k, &m);

  if(i > j)
    max = i;
  else
    max = j;
```

```
  if(k > max)
    max = k;
  if(m > max)
    max = m;
  printf("Max = %f\n", max);
  return 0;
}
```

Comments: To display the smallest of the four prices, replace the > operator with <.

5.14 Suppose that two PCs reside in the same IP network. Write a program that reads their IP addresses (version 4) and displays if they are configured correctly, meaning if they can communicate. The form of the IP address is x.x.x.x, where each x is an integer within [0, 255].

The value of the first byte of an IP address defines its class, as shown here:

(a) Class A: [0, 127]
(b) Class B: [128, 191]
(c) Class C: [192, 223]

If the two IP addresses indicate different classes, the PCs can't communicate. If they belong to the same class, we compare their network byte(s). The byte(s) to be compared are defined according to their class, as follows:

(a) Class A: first byte
(b) Class B: first two bytes
(c) Class C: first three bytes

If they are the same, the PCs can communicate.

```
#include <stdio.h>
int main()
{
  int a1, a2, a3, a4, b1, b2, b3, b4;

  printf("Enter first IP address: ");
  scanf("%d.%d.%d.%d", &a1, &a2, &a3, &a4);

  printf("Enter second IP address: ");
  scanf("%d.%d.%d.%d", &b1, &b2, &b3, &b4);

  if(a1 < 128)
  {
    if(a1 == b1)
      printf("Class A: Correct Configuration\n");
    else
      printf("Class A: Wrong Configuration\n");
  }
  else if(a1 < 192)
  {
    if(a1 == b1 && a2 == b2)
      printf("Class B: Correct Configuration\n");
    else
      printf("Class B: Wrong Configuration\n");
  }
```

```
    else if(a1 < 224)
    {
      if(a1 == b1 && a2 == b2 && a3 == b3)
        printf("Class C: Correct Configuration\n");
      else
        printf("Class C: Wrong Configuration\n");
    }
    else
      printf("Error: Wrong class\n");

    return 0;
}
```

Comments: Notice that we put on purpose the dot in `scanf()` so that the user inputs the IP address in its familiar form.

5.15 Write a program that reads the grades of three students and displays them in ascending order.

```
#include <stdio.h>
int main()
{
  float i, j, k;

  printf("Enter grades: ");
  scanf("%f%f%f", &i, &j, &k);

  if(i <= j && i <= k)
  {
    printf("%f ", i);
    if(j < k)
      printf("%f %f\n", j, k);
    else
      printf("%f %f\n", k, j);
  }
  else if(j <= i && j <= k)
  {
    printf("%f ", j);
    if(i < k)
      printf("%f %f\n", i, k);
    else
      printf("%f %f\n", k, i);
  }
  else
  {
    printf("%f ", k);
    if(j < i)
      printf("%f %f\n", j, i);
    else
      printf("%f %f\n", i, j);
  }
  return 0;
}
```

Comments: We first compare each grade with the other two in order to find and display the smallest one. Next, we compare the two remaining grades and display them in ascending order.

Conditional Operator ?:

The conditional operator ?: allows a program to perform one of two actions depending on the value of an expression. Expressions that contain the ?: operator have the form

```
exp1 ? exp2 : exp3;
```

Since the conditional operator takes three operands, it is often referred to as the *ternary* operator. A conditional expression is evaluated in steps. The expression exp1 is evaluated first. If it is true, exp2 is evaluated and its value becomes the value of the conditional expression. If the value of exp1 is false, exp3 is evaluated and its value becomes the value of the conditional expression.

Simply put, a conditional expression is a sort of an **if-else** statement:

```
if(exp1)
  exp2;
else
  exp3;
```

For example, the following program reads an integer and if it is greater than 10 the program displays One, otherwise Two:

```
#include <stdio.h>
int main()
{
  int a;

  printf("Enter number: ");
  scanf("%d", &a);
  (a > 10) ? printf("One\n") : printf("Two\n");
  return 0;
}
```

The value of a conditional expression can be stored to a variable, as shown here:

```
int a = 5, k;
k = (a > 0) ? 100 : -1;
```

First, the expression (a > 0) is evaluated. If it is true, k becomes 100, otherwise it becomes –1. Since a is 5, k becomes 100.

The conditional operator is mostly used to replace simple **if-else** statements. For example, the following **if-else** statement

```
if(a > b)
  max = a;
else
  max = b;
```

can be replaced with

```
max = (a > b) ? a : b;
```

Several conditional expressions may be merged by replacing the expression that follows the colon (:) with another conditional expression. Next, exp3 is replaced with the add1 ? add2 : add3; expression as shown here:

```
k = exp1 ? exp2 : add1 ? add2 : add3;
```

If exp1 is true, k will be equal to the value of exp2 expression. If exp1 is false, the value of add1 is checked. If the value of add1 is true, k will be equal to the value of add2. Otherwise, it will be equal to the value of add3. The equivalent chain of **if-else** statements is

```
if(exp1)
  k = exp2;
else if(add1)
  k = add2;
else
  k = add3;
```

Exercises

5.16 What is the output of the following program?

```
#include <stdio.h>
int main()
{
  int a, b = 0;

  a = b ? 50 : 200;
  printf("Val = %d\n", a);
  return 0;
}
```

Answer: The statement a = b ? 50 : 200; is equivalent to a = (b != 0) ? 50 : 200; Since b is 0, a becomes 200 and the program displays Val = 200.

5.17 Write a program that reads a float number and stores its value in a program's variable only if the input value belongs in the set [5,10]. Otherwise, the variable should be set to –1. Use the conditional operator.

```
#include <stdio.h>
int main()
{
  float i, num;

  printf("Enter grade: ");
  scanf("%f", &num);

  i = (num >= 5 && num <= 10) ? num : -1;
  printf("%f\n", i);
  return 0;
}
```

5.18 Write a program that reads an integer and displays a message to indicate whether it is positive or negative. If the integer is 0, the program should display "Zero". Use the conditional operator.

```
#include <stdio.h>
int main()
{
  int num;

  printf("Enter number: ");
  scanf("%d", &num);

  num > 0 ? printf("Pos\n") : (num < 0) ? printf("Neg\n") :
  printf("Zero\n");
  return 0;
}
```

Comments: Alternatively, we could write printf("%s\n", num > 0 ? "Positive" : num < 0 ? "Negative" : "Zero");

5.19 The following program reads the prices of three products and displays the largest price. Use the conditional operator ?: to replace the **if-else** statements with a single printf() to display the largest price.

```
#include <stdio.h>
int main()
{
  float i, j, k;

  printf("Enter prices: ");
  scanf("%f%f%f", &i, &j, &k);

  if(i >= j && i >= k)
    printf("Max = %f\n", i);
  else if(j >= i && j >= k)
    printf("Max = %f\n", j);
  else
    printf("Max = %f\n", k);
  return 0;
}
```

Answer: printf("Max = %f\n", (i >= j && i >= k) ? i : (j >= i && j >= k) ? j : k);

To display the smallest of the three integers, just replace the > operator with <.

5.20 Write a program that reads a student's grade as a number from 0 through 10, and displays a corresponding message, according to the following:

(a) Grade belongs to [7.5–10], the program prints "A"
(b) Grade belongs to [5–7.5), the program prints "B"
(c) Grade belongs to [0–5), the program prints "Better luck next time"

If the grade is not in [0–10], the program should display an informative message. Use the conditional operator.

Note: The right parenthesis ")" means that the right number is not included in the indicated set.

```
#include <stdio.h>
int main()
{
  float grd;

  printf("Enter grade: ");
  scanf("%f", &grd);

  printf("%s\n", (grd >= 7.5 && grd <= 10) ? "A" : (grd >= 5 &&
    grd < 7.5) ? "B" : (grd >= 0 && grd < 5) ? "Better luck next
    time" : "Grade should be within [0-10]");
  return 0;
}
```

switch Statement

The **switch** statement can be used instead of **if-else-if** statements when we want to test the value of an expression against a series of values and handle each case differently. The general form of the **switch** statement is

```
switch(expression)
{
  case constant_1:
    /* block of statements, which will be executed if the value of the
      expression is equal to constant_1. */
  break;

  case constant_2:
    /* block of statements, which will be executed if the value of the
      expression is equal to constant_2. */
  break;
  ...
  case constant_n:
    /* block of statements, which will be executed if the value of the
      expression is equal to constant_n. */
  break;

  default:
    /* block of statements, which will be executed if the value of the
      expression is not equal to anyone of the pre-mentioned constants. */
  break;
}
```

The expression must be an integer variable or expression, and the values of all constant_1, constant_2,..., constant_n must be integer constant expressions with different values.

When the **switch** statement is executed, the value of expression is compared with constant_1, constant_2,..., constant_n in sequence. If a match is found, the group of statements of the matching **case** clause will be executed and the rest of the **case** clauses will not be checked. If the value of expression does not match any of the choices, the group of statements of the **default** clause will be executed. In either case, the **break** statement terminates the execution of the **switch** and the program continues with the first statement following the **switch**.

The **default** case is not mandatory. If it is missing and the value of expression does not match any of the **case** values, the **switch** statement terminates and the program continues with the first statement following the **switch**.

The program in the next listing reads an integer and displays a message according to its value. For example, if the user enters 1, the program displays One; if 2 is entered, the program displays Two; and if a value other than 1 or 2 is entered, the program displays the message Something else. The **break** statement terminates the **switch** and the End message is displayed.

```c
#include <stdio.h>
int main()
{
  int a;

  printf("Enter number: ");
  scanf("%d", &a);

  switch(a)
  {
    case 1:
      printf("One\n");
    break;

    case 2:
      printf("Two\n");
    break;

    default:
      printf("Something else\n");
    break;
  }
  printf("End\n");
  return 0;
}
```

If the **break** statement is missing from the matching **case**, the program will continue with the execution of the next **case** statements, until a **break** statement is met. For example, consider the following program:

```c
#include <stdio.h>
int main()
{
  int a = 1;

  switch(a)
  {
    case 1:
      printf("One\n");

    case 2:
      printf("Two\n");
```

```
  case 3:
    printf("Three\n");
  break;

  default:
    printf("Something else\n");
  break;
}
printf("End\n");
return 0;
}
```

Since a is 1, the statement in the first **case** is executed and the program displays One. However, since there is no **break**, the **switch** statement is not terminated and the program continues with the execution of the statements of the next **case** clauses, until a **break** is met. Therefore, the program will display Two and Three. The **break** statement terminates the **switch** and the program displays End.

Therefore, the output of this program is One Two Three End

If there are two or more **case** clauses with the same block of statements, they can be merged together like so:

```
case constant_1:
case constant_2:
case constant_3:
  /* block of statements that will be executed if the value of the
    expression matches any of constant_1, constant_2 or constant_3. */
break;
```

For example, the following program reads an integer that represents a month (1 for January, 12 for December) and displays the season in which the month belongs:

```
#include <stdio.h>
int main()
{
  int month;

  printf("Enter month [1-12]: ");
  scanf("%d", &month);

  switch(month)
  {
    case 1:
    case 2:
    case 12:
      printf("Winter\n"); /* The merging of the three cases means that if
        the input value is 1, 2, or 12, the message Winter will be
        displayed.*/
    break;

    case 3:
    case 4:
    case 5:
      printf("Spring\n");
    break;
```

```
    case 6:
    case 7:
    case 8:
      printf("Summer\n");
    break;

    case 9:
    case 10:
    case 11:
      printf("Autumn\n");
    break;

    default:
      printf("Error: Wrong month\n");
    break;
  }
  return 0;
}
```

switch versus if

The main disadvantage of using the **switch** instead of **if** is that **switch** can check only whether the value of an expression and the **case** constants are equal, whereas **if** can check any kind of condition and not only for equality. Moreover, when using **switch**, the values of expression and **case** constants must be integers. Characters are treated as small integers so they may be used, but floating-point numbers and strings are not permitted.

On the other hand, the use of the **switch** statement in place of cascaded **if-else-if** statements may make the program easier to read.

Exercises

5.21 What is the output of the following program?

```
#include <stdio.h>
int main()
{
  int a = 1;

  switch(a)
  {
    case 1:
      printf("One\n");
    return 0;

    case 2:
      printf("Two\n");
    break;
  }
  printf("End\n");
```

```
    return 0;
}
```

Answer: Since a is 1, the program will display One. The **return** statement terminates the main() function, that is, the program itself. As a result, the last printf()
is not called and the program displays One.

5.22 Write a program that simulates a physical calculator. The program should take as
input the symbol of an arithmetic operator and two integers and display the result of
the arithmetic operation.

```
#include <stdio.h>
int main()
{
  char sign;
  int i, j;

  printf("Enter math sign and two integers: ");
  scanf("%c%d%d", &sign, &i, &j);

  switch(sign)
  {
    case '+':
      printf("Sum = %d\n", i+j);
    break;

    case '-':
      printf("Diff = %d\n", i-j);
    break;

    case '*':
      printf("Product = %d\n", i*j);
    break;

    case '/':
      if(j != 0)
        printf("Div = %.3f\n", (float)i/j);
      else
        printf("Second num should not be 0\n");
    break;

    default:
      printf("Unacceptable operation\n");
    break;
  }
  return 0;
}
```

Comments: As we'll see in Chapter 9, the constants of type **char** are treated as small
integers and declared within single quotes ' '.

5.23 Write a program that displays the area of a square or a circle based on the user's
choice. If the user enters 0, the program should read the side of the square and
display its area. If the user enters 1, the program should read the radius of the circle
and display its area.

```
#include <stdio.h>
int main()
```

```c
{
  int sel;
  double len;

  printf("Enter choice (0:square 1:circle): ");
  scanf("%d", &sel);

  switch(sel)
  {
    case 0:
      printf("Enter side length: ");
      scanf("%lf", &len);
      if(len <= 0)
      {
        printf("Error: Wrong length\n");
        return 0;
      }
      printf("Square area is %f\n", len*len);
    break;

    case 1:
      printf("Enter radius: ");
      scanf("%lf", &len);
      if(len <= 0)
      {
        printf("Error: Wrong length\n");
        return 0;
      }
      printf("Circle area is %f\n", 3.14*len*len);
    break;

    default:
      printf("Error: Wrong choice\n");
    break;
  }
  return 0;
}
```

5.24 Write a program that takes as input a person's sex and height and displays the corresponding description for his or her height, according to Table 5.2.

```c
#include <stdio.h>
int main()
```

TABLE 5.2

Height Categories

Sex	Height (ft)	Result
Male	< 5.5	Short
Male	>= 5.5 and < 6.00	Normal
Male	>= 6.00	Tall
Female	< 5.25	Short
Female	>= 5.25 and < 5.75	Normal
Female	>= 5.75	Tall

```
{
  int sex;
  float height;

  printf("Enter sex (0:male - 1:female): ");
  scanf("%d", &sex);

  printf("Enter height (in feet): ");
  scanf("%f", &height);

  switch(sex)
  {
    case 0:
      if(height < 5.50)
        printf("Result: Short\n");
      else if(height < 6.00)
        printf("Result: Normal\n");
      else
        printf("Result: Tall\n");
    break;

    case 1:
      if(height < 5.25)
        printf("Result: Short\n");
      else if(height < 5.75)
        printf("Result: Normal\n");
      else
        printf("Result: Tall\n");
    break;

    default:
      printf("Error: Wrong input\n");
    break;
  }
  return 0;
}
```

5.25 Write a program that calculates the cost of transporting a passenger's luggage, according to Table 5.3. The program reads the type of the passenger's class and the weight of his or her luggage and displays the cost.

TABLE 5.3

Transport Cost

Class	Weight (lb)	Cost ($)
Economy	<25	0
	25–40	1.50 for each pound over 25
	>40	2.00 for each pound over 40
Business	<35	0
	35–50	1.25 for each pound over 35
	>50	1.50 for each pound over 50
VIP	<60	0
	>60	30 (fixed cost)

```c
#include <stdio.h>
int main()
{
  int clas;
  float cost, weight;

  printf("Enter class (1-Eco, 2-Business, 3-VIP): ");
  scanf("%d", &clas);

  printf("Enter weight: ");
  scanf("%f", &weight);

  cost = 0; /* All cases where the passenger pays nothing. */
  switch(clas)
  {
    case 1:
      if(weight > 40)
        cost = 22.5 + 2*(weight-40); /* 22.5 = 1.5 * 15. */
      else if(weight > 25) /* Since the previous "if" checks if the
        weight is more than 40lb, this else-if statement checks
        whether 25 < weight <= 40. */
        cost = 1.5*(weight-25);
    break;

    case 2:
      if(weight > 50)
        cost = 18.75 + 1.5*(weight-50); /* 18.75 = 1.25 * 15. */
      else if(weight > 35)
        cost = 1.25*(weight-35);
    break;

    case 3:
      if(weight > 60)
        cost = 30;
    break;

    default:
      printf("Error: Wrong traffic class\n");
    break;
  }
  printf("Total cost = %.2f\n", cost);
  return 0;
}
```

Unsolved Exercises

5.1 Write a program that reads two **double** numbers and displays the absolute value of their sum.

5.2 Write a program that reads three integers and checks if they are in successive increase order (i.e., 5, 6, and 7 or -3, -2, and -1).

5.3 A water supply company charges the water consumption, as follows:

(a) Fixed amount of $10
(b) For the first 30 cubic meters, $0.6/meter
(c) For the next 20 cubic meters, $0.8/meter
(d) For the next 10 cubic meters, $1/meter
(e) For every additional meter, $1.2/meter

Write a program that reads the water consumption in cubic meters and displays the bill.

5.4 Write a program that reads the grades of three students in the lab part and their grades in the theory part. The final grade is calculated as `lab_grd*0.3 + theory_grd*0.7`. The program should display how many students got a grade between 8 and 10. Don't use more than three variables.

5.5 Write a program that reads four integers and displays the pair with the largest sum. For example, if the user enters `10, -8, 17, 5`, the program should display `10+17 = 27`.

5.6 Write a program that reads the time in the normal form h:m:s and displays how much time is left until midnight (i.e., 24:00:00).

5.7 Use the `?:` operator to replace the **if-else** statements in 5.8 (Exercise) and use a single `printf()`, as well.

5.8 The 13-digit International Standard Book Number (ISBN) is a unique code that identifies a book commercially. The last digit is a check digit used for error detection. To calculate its value, each digit of the first twelve digits is alternately multiplied, from left to right, by 1 or 3. The products are summed up and divided by 10. The check digit is the remainder of the division subtracted from 10. If it is 10, it becomes 0. For example, assume that the first twelve digits are `978960931961`.

(a) `(9*1 + 7*3 + 8*1 + 9*3 + 6*1 + 0*3 + 9*1 + 3*3 + 1*1 + 9*3 + 6*1 + 1*3) = 126`
(b) The check digit `= 10 - (126 % 10) = 10 - 6 = 4`

Write a program that reads a 13-digit ISBN and checks the last digit to see if it is valid or not.

5.9 Write a program that reads three integers and uses the `switch` statement to support three cases. If the user's choice is `1`, it should check if the integers are different and display a message. If it is `2`, the program should check if any two of them are equal, and if it is `3`, the program should display how many of them fall in [-5, 5].

5.10 Write a program that reads the numerators and the denominators of two fractions and a math sign as an integer (i.e., `1`: addition, `2`: subtraction, `3`: multiplication, `4`: division) and uses the `switch` statement to display the result of the math operation.

5.11 Write a program that reads the current year, the year of birth, and uses the `switch` statement to display the age in words (assume that the age hasn't more than two digits). For example, if the user enters `2013` and `1988`, the program should display `twenty-five`.

6

Loops

Often, programs contain blocks of code to be executed more than once. A statement whose job is to repeatedly execute the same block of code as long as the value of a controlling expression is true is called an *iteration loop*. In this chapter, you'll learn about C's iteration statements: **for**, **while**, and **do-while**, which allow us to set up loops, as well as **break**, **continue**, and **goto** statements, which can be used to transfer control from one point of a program to another.

for Statement

for statement is one of C's three loop statements. It is typically used when the number of iterations is predetermined. The general form of the **for** statement is:

```
for(init_exp; cond_exp; update_exp)
{
  /* a block of statements (the loop body) that is repeatedly executed as
    long as cond_exp evaluates to true. */
}
```

The expressions init_exp, cond_exp, and update_exp can be any valid C expressions.

In practice, a **for** statement works like this:

1. The **init_exp** is executed only once, just before the first iteration. Typically, **init_exp** initializes a variable used in the other **for** expressions.

2. The cond_exp is evaluated. If its value is false, the **for** loop terminates. If it is true, the block of statements, called the loop body, is executed.

3. The update_exp is executed to initiate the next loop iteration. Typically, update_exp changes the value of a variable used in cond_exp.

4. Steps 2 and 3 are repeated until the value of cond_exp becomes false.

For example, here is how we could use **for** to print the numbers from 0 to 4.

```
#include <stdio.h>
int main()
{
  int a;

  for(a = 0; a < 5; a++)
  {
```

```
  printf("%d ", a);
  }
  return 0;
}
```

Let's see what happens when the **for** statement is executed:

1. The statement a = 0; is executed.
2. The condition a < 5 is checked. Since it is true, printf() is executed and the program displays 0.
3. The a++ statement is executed and a becomes 1. The condition a < 5 is checked again. Because it is still true, printf() is executed again and the program displays 1.

So far, the program has displayed 0 and 1. This process is repeated until a becomes 5 at which point the condition a < 5 becomes false and the **for** loop terminates. Each loop iteration prints the current value of a.

As a result, the output of the aforementioned listing would be 0 1 2 3 4.

As with the **if** statement, if the loop body consists of a single statement, the braces can be omitted. In other words, the previous **for** loop could be written as

```
for(a = 0; a < 5; a++)
  printf("%d ", a);
```

Now, what would the following code output?

```
for(a = 0; a > 1; a++)
{
  printf("%d\n", a);
  printf("End\n");
}
```

Here, since a is 0, the value of the condition a > 1 is false and nothing would be displayed.

Expressions in the **for** statement can contain more than one statement when we use the comma (,) operator. For example,

```
int a, b;
for(a = 1, b = 2; b < 10; a++, b++)
```

Here, the first expression assigns values to both a and b, and the last one increases both of their values.

*As with **if** statement, a common error is to add accidentally a semicolon (;) at the end of the **for** statement.*

For example, let's see the following program:

```
#include <stdio.h>
int main()
{
  int i;
```

```
for(i = 2; i < 7; i+=3);
  printf("%d\n", i);
  return 0;
}
```

As with the **if** statement, the semicolon at the end of the **for** statement or in the following line indicates a *null* statement, a statement that does nothing. Therefore, the compiler assumes that the loop body is empty and the printf() will be executed only once, after the condition i < 7 becomes false. Let's trace the iterations:

First iteration. Since i is 2, the condition i < 7 (2 < 7) is true. Since the loop body is empty, the next statement to be executed is i+=3 (i = i+3 = 2+3 = 5).

Second iteration. Since i is 5, the condition i < 7 (5 < 7) is still true and the statement i+=3 makes the value of i equal to 8.

Third iteration. Since i is 8, the condition i < 7 (8 < 7) becomes false and the **for** statement ends.

Then, the printf() is executed and the program outputs 8.

Omitting Expressions

The three expressions of a **for** statement are optional. In fact, C allows us to omit any or all of them. For example, in the following listing, because a is initialized before the **for** statement, the first expression can be omitted:

```
int a = 0;
for(; a < 5; a++)
```

Notice that the semicolon before the second expression must be present. In fact, the two semicolons must always be present even if both the first and second expressions are omitted.

In the same way, we can put the third expression in the loop body and remove it from the **for** statement. For example, in the following listing, the third expression a++ is moved inside the body:

```
for(a = 0; a < 5;)
{
  printf("%d ", a);
  a++;
}
```

If the conditional expression is missing, the compiler treats it as always true, and the **for** loop never ends. Such a loop is called *infinite*. For example, this **for** statement creates an infinite loop:

```
for(a = 0; ; a++)
```

Most programmers omit all three expressions in order to create an infinite loop, like:

```
for(;;)
```

If both the first and third expressions are missing, the **for** statement is equivalent to a **while** statement. For example, this **for** statement

```
for (; a < 5;)
```

is equivalent to

```
while (a < 5)
```

Exercises

6.1 Write a program that displays the integers from 10 down to 1.

```c
#include <stdio.h>
int main()
{
    int i;

    for (i = 10; i >= 1; i--)
        printf("%d\n", i);
    return 0;
}
```

Comments: The **for** loop is executed 10 times and terminates when i becomes 0.

6.2 What is the output of the following program?

```c
#include <stdio.h>
int main()
{
    int i, j = 10;

    for (i = 0; j < 10; i++)
        printf("One\n");
    return 0;
}
```

Answer: Since the condition j < 10 (10 < 10) is false, the **for** loop is not executed and the program displays nothing.

And what would be the output of the program if j was initialized to 3? In this case, since the condition j < 10 (3 < 10) is always true, the **for** loop runs forever. Therefore, the program displays One continuously and will never end, reducing the computer's performance significantly.

6.3 Write a program that reads the grades of 10 students and displays those within [5, 10].

```c
#include <stdio.h>
int main()
{
    int i;
    float grd;

    for (i = 0; i < 10; i++)
    {
        printf("Enter grade: ");
```

```
    scanf("%f", &grd);

    if(grd >= 5 && grd <= 10)
      printf("Grade = %f\n", grd);
  }
  return 0;
}
```

6.4 What is the output of the following program?

```
#include <stdio.h>
int main()
{
  int i;

  for(i = 12; i > 2; i-=5)
    printf("%d ", i);
    printf("\nEnd = %d\n", i);
  return 0;
}
```

Answer: Since the **for** statement does not contain braces, the compiler assumes that the loop body consists of one statement, which is the first printf(). Notice how the indenting can trick you into thinking that the second printf() is also included in the loop body. In fact, this is executed only once, after the **for** loop ends. Specifically, these are the loop iterations:

First iteration. Since i is 12, the condition i > 2 (12 > 2) is true and the program displays 12.

Second iteration. With the statement i-=5 (i = i-5), i becomes 7. Therefore, the condition i > 2 (7 > 2) is true and the program displays 7.

Third iteration. With the statement i-=5 (i = i-5), i becomes 2. Therefore, the condition i > 2 (2 > 2) is false and the **for** loop terminates.

Then, the second printf() prints the current value of i. The program outputs

```
12 7
End = 2
```

6.5 Write a program that reads two numbers that represent miles (i.e., a and b) and a third number (i.e., step). The program should display one column with the miles and a second with the corresponding kilometers starting from a up to b (assume that a < b) with a step of step. Note that 1 mile = 1.6 km.

```
#include <stdio.h>
int main()
{
  double i, a, b, step;

  printf("Enter miles interval: ");
  scanf("%lf%lf", &a, &b);

  printf("Enter step: ");
  scanf("%lf", &step);
```

```
    printf("MILE\t\t KLM\n");
    printf("------------------------\n");
    for(i = a; i < b; i += step)
      printf("%.2f\t\t %.2f\n", i, 1.6*i);
    return 0;
}
```

break Statement

We have already discussed how **break** terminates the execution of a **switch** statement, but it can also be used to terminate a **for**, **while**, or **do-while** loop and transfer control to the first statement after the loop. For example,

```
#include <stdio.h>
int main()
{
  int i;

  for(i = 1; i < 10; i++)
  {
    if(i == 5)
      break;

    printf("%d ", i);
  }
  printf("End = %d\n", i);
  return 0;
}
```

As long as i is not 5, the **if** condition is false and the loop displays the values of i from 1 to 4. When i becomes 5, the **break** statement terminates the **for** loop and the program continues with the outer printf(). Therefore, the program displays 1 2 3 4 End = 5.

continue Statement

The **continue** statement can be used within a **for**, **while**, or **do-while** loop. While the **break** statement completely terminates a loop, the **continue** statement terminates the current loop iteration and continues with the next iteration, skipping the rest of the statements in the loop body. For example, consider the following program:

```
#include <stdio.h>
int main()
{
  int i;

  for(i = 1; i < 10; i++)
  {
    if(i < 5)
      continue;
```

```
    printf("%d ", i);
  }
  return 0;
}
```

As long as i is less than 5, the **if** condition is true and the **continue** statement terminates the current iteration. The rest of the body loop, that is, printf(), is skipped and the program continues with i++, which initiates the next loop iteration.

When i becomes greater or equal to 5, the **continue** statement is not executed and the program displays the value of i.

Therefore, the program displays 5 6 7 8 9.

Exercises

6.6 What is the output of the following program?

```
#include <stdio.h>
int main()
{
  int i, j = 5;
  for(i = 0; i+j == 5; j++)
  {
    printf("One\n");
    i = 4;
    j = 1;
  }
  printf("Val1 = %d Val2 = %d\n", i, j);
  return 0;
}
```

Answer: Let's trace the loop iterations:

First iteration. Since i becomes 0, the condition i+j == 5 (0+5 = 5 == 5) is true and the program prints One. Then, i and j become 4 and 1, respectively.

Second iteration. Since the j++ statement makes j equal to 2, the condition i+j == 5 (4+2 = 6 == 5) becomes false and the loop terminates.

The program displays the current values of i and j, which are 4 and 2, respectively.

6.7 Write a program that displays the sum of all numbers from 1 to 200.

```
#include <stdio.h>
int main()
{
  int i, sum;

  sum = 0; /* To calculate the sum, we initialize the sum variable to
    0, which is the neutral element of the addition. */
  for(i = 1; i <= 200; i++)
    sum += i;
```

```
  printf("Sum = %d\n", sum);
  return 0;
}
```

Comments: We'll explain the first three loop iterations:

> *First iteration.* Since i is 1, the condition i <= 200 is true (1 < 200) and sum becomes sum = sum+i = 0+1 = 1.
>
> *Second iteration.* Since the i++ statement makes i equal to 2, the condition i < 200 is true (2 < 200) and sum becomes sum = sum+i = 1+2 = 3.
>
> *Third iteration.* Similarly, sum becomes sum = sum+i = 3+3 = 6.

Therefore, the first three iterations calculate the sum 1+2+3. Similarly, the next iterations calculate the sum of the next numbers up to 200.

6.8 Write a program that displays the product of all odd numbers from 1 to 20.

```
#include <stdio.h>
int main()
{
  int i, prod;

  prod = 1; /* To calculate the product, we initialize the prod
    variable to 1, which is the neutral element of the
    multiplication. */
  for(i = 1; i < 20; i+=2) /* To calculate the product of odd numbers,
    we increase the step by two. */
      prod *= i;

  printf("Product = %d\n", prod);
  return 0;
}
```

Comments: We'll explain the first three loop iterations:

First iteration. Since i is 1, the condition i < 20 is true (1 < 20) and prod becomes prod = prod*i = 1*1 = 1.

Second iteration. Since the i+=2 statement makes i equal to 3, the condition i < 20 is true (3 < 20) and prod becomes prod = prod*i = 1*3 = 3.

Third iteration. Similarly, prod becomes prod = prod*i = 3*5 = 15.

Therefore, the first three iterations calculate the product 1*3*5. Similarly, the next iterations calculate the product of the next odd numbers up to 20.

6.9 Write a program that reads an integer and displays its multiplication table. For example, if the user enters 5, the output should be 1*5 = 5, 2*5 = 10,..., 10*5 = 50. The program should force the user to enter an integer within [1, 10].

```
#include <stdio.h>
int main()
{
  int i, num;

  for(;;) /* Exit from the infinite loop when the user enters a number
    in [1,10]. */
```

```
    {
      printf("Enter number [1-10]: ");
      scanf("%d", &num);
      if((num >= 1) && (num <= 10))
        break;
    }
    for(i = 1; i <= 10; i++)
      printf("%d * %d = %d\n", i, num, i*num);
    return 0;
}
```

6.10 A test consists of 10 multiple choice questions each of which has three possible answers. The first answer is worth three points, the second one point, and the third two points. Write a program that uses the **switch** statement to read the test taker's 10 answers and display their final score.

```
#include <stdio.h>
int main()
{
    int i, ans, points;

    points = 0;
    for(i = 0; i < 10; i++)
    {
      printf("Enter answer [1-3]: ");
      scanf("%d", &ans);

      switch(ans)
      {
        case 1:
          points += 3;
        break;

        case 2:
          points += 1;
        break;

        case 3:
          points += 2;
        break;

        default:
          printf("Error: Wrong answer ...\n");
        break;
      }
    }
    printf("Candidate gets %d points in total\n", points);
    return 0;
}
```

6.11 Write a program that reads the grades of 100 students on a test and before it terminates it displays the average of the passed and failed students. A student passes the exams if his/her grade is equal to or greater than 5. If the user enters a grade out of [0, 10], the program should ignore it and display an error message. If the user enters −1, the insertion of grades should end.

```c
#include <stdio.h>
int main()
{
  int i, suc, fail;
  float grd, sum_suc, sum_fail;

  suc = fail = 0;
  sum_suc = sum_fail = 0;
  for(i = 0; i < 100; i++)
  {
   printf("Enter grade: ");
   scanf("%f", &grd);

   if(grd == -1)
     break;
   if(grd > 10 || grd < 0)
   {
     printf("Wrong grade, try again …\n");
     i--; /* Since the input grade is out of [0,10], the grade is
        ignored and we decrease the value of i to repeat the input
        process. */
     continue;
   }
   if(grd >= 5)
   {
     suc++;
     sum_suc += grd;
   }
   else
   {
     fail++;
     sum_fail += grd;
   }
  }
  if(suc)
   printf("Avg(+) = %.2f\n", sum_suc/suc);
  else
   printf("Everybody failed\n");

  if(fail)
   printf("Avg(-) = %.2f\n", sum_fail/fail);
  else
   printf("None failed\n");
  return 0;
}
```

6.12 Write a program that reads two integers and displays the sum of the integers between them. For example, if the user enters 3 and 8, the program should display 22 because 4+5+6+7 = 22. The program should check which one of the two input numbers is the greater and act accordingly.

```c
#include <stdio.h>
int main()
{
  int i, j, sum;

  printf("Enter numbers: ");
```

```
    scanf("%d%d", &i, &j);
    sum = 0;
    if(i < j)
    {
      for(i = i+1; i < j; i++)
        sum += i;
    }
    else if(j < i)
    {
      for(j = j+1; j < i; j++)
        sum += j;
    }
    printf("Sum = %d\n", sum);
    return 0;
}
```

6.13 What is the output of the following program?

```
#include <stdio.h>
int main()
{
  int i;
  for(i = 3; i && i-1; i--)
    printf("%d\n", i);
  return 0;
}
```

Answer: The **for** loop terminates when the condition i && i-1 becomes false; when i becomes 1. Therefore, the **for** loop is executed twice and the program displays 3 and 2.

6.14 Write a program that reads an integer in [0,170] and displays its factorial. *Note:* The factorial of a positive integer n, where n ≥ 1, is defined as n! = 1*2*3* ... *n, whereas the factorial of 0 equals 1 (0! = 1).

```
#include <stdio.h>
int main()
{
  int i, num;
  double fact;

  printf("Enter number within [0, 170]: ");
  scanf("%d", &num);

  if(num >= 0 && num <= 170)
  {
    fact = 1; /* To calculate the factorial of num, we initialize the
      fact variable to 1. */
    for(i = 1; i <= num; i++)
      fact = fact * i;
    /* If the user enters 0, the for loop won't be executed because
      the condition i <= num (1 <= 0) would be false. In such case,
      the program would display the initial value of the fact
      variable, which is 1 (and this is correct, since 0! = 1). */
    printf("Factorial of %d is %e\n", num, fact);
  }
  else
```

```
      printf("Error: Number should be within [0, 170]\n");
   return 0;
}
```

Comments: The variable `fact` is declared as **double** in order to calculate the factorial of as many numbers as possible. Because the maximum value that can be stored
in a double variable is $\sim 10^{308}$, the input is constrained to 170 since the factorial of 171
exceeds 10^{308}.

  Had we used the **int** type instead of **double**, the program would have displayed
only the factorials of numbers from 0 to 12 correctly.

6.15   Write a program that reads the initial population of a country and its annual population growth (as a percentage). Then, the program should read the number of years
and display the new population for each year.

```c
#include <stdio.h>
int main()
{
  int i, years, pop, pop_incr;
  double rate;

  printf("Enter population: ");
  scanf("%d", &pop);

  printf("Enter increase rate (%%): ");
  scanf("%lf", &rate);

  printf("Enter years: ");
  scanf("%d", &years);

  printf("\nYear\tIncrease\tPopulation\n");
  printf("---------------------------------------\n");

  for(i = 1; i <= years; i++)
  {
   pop_incr = pop * rate / 100; /* Calculation of population's
     increase. */
   pop += pop_incr; /* Calculation of new population. */
   printf("%d\t%d\t\t%d\n", i, pop_incr, pop);
  }
  return 0;
}
```

6.16   What is the output of the following program?

```c
#include <stdio.h>
int main()
{
  int i, j;

  for(i = 10, j = 2; i != j; i-=2, j+=2)
   printf("%d %d\n", i, j);
  return 0;
}
```

Answer: The **for** loop terminates when i equals j after the second iteration, when
both become 6. Therefore, the **for** loop is executed twice and the program displays
10 2 and 8 4.

6.17 What is the output of the following program?

```
#include <stdio.h>
int main()
{
   int i;

   for(i = 0; i ? 0 : i+1; i++)
     printf("%d\n", i);
   return 0;
}
```

Answer: In the first loop iteration, the value of i ? 0 : i+1 is i+1 = 1, so the **for** loop is executed and the program displays 0. In the second iteration, i becomes 1, so the value of the expression becomes 0 and the loop terminates.

6.18 Write a program that reads the number of students in a class and their grades on a test. The program should display the average test grade of the class, the minimum and maximum test grade, and how many students got the same maximum grade, as well. (Assume that the minimum grade is 0 and the maximum is 10.)

```
#include <stdio.h>
int main()
{
  int i, studs_num, times;
  float grd, min_grd, max_grd, sum_grd;

  printf("Enter number of students: ");
  scanf("%d", &studs_num);

  if(studs_num <= 0)
  {
    printf("Wrong number of students\n");
    return 0; /* Program termination. */
  }
  printf("Enter grade [0-10]: ");
  scanf("%f", &grd); /* We assume that the user enters a number
    within [0,10]. */

  min_grd = max_grd = grd;
  sum_grd = grd;
  times = 1; /* First appearance of the maximum grade. */

  for(i = 1; i < studs_num; i++)
  {
    printf("Enter grade [0-10]: ");
    scanf("%f", &grd);

    if(grd < min_grd)
      min_grd = grd;
    if(grd > max_grd)
    {
      max_grd = grd;
      times = 1; /* First appearance of the new maximum grade. */
    }
    else if(max_grd == grd)
      times++;

    sum_grd += grd;
```

```
    }
    printf("Min = %.2f, Max = %.2f (appeared %d times) Avg = %.2f\n",
      min_grd, max_grd, times, sum_grd/studs_num);
    return 0;
}
```

6.19 What is the output of the following program?

```
#include <stdio.h>
int main()
{
    int i;

    for(i = 0; i < 3; printf("%d ", ++i))
      ; /* Empty block of statements. */
    return 0;
}
```

Answer: As discussed previously, the expressions in a **for** statement can be any valid C expression. As such, this code uses printf() in place of the third expression.

Before printf() is executed, i is increased by one. The **for** loop is executed three times and the program displays 1 2 3.

6.20 Write a program that reads 8 bits (each bit is 0 or 1) and displays the corresponding decimal number, assuming that the bits are entered from left to right. For example, if the user enters 10000000 the program should display 128.

```
#include <stdio.h>
int main()
{
    int i, num, dig;

    num = 0;
    for(i = 7; i >= 0; i--)
    {
      printf("Enter digit (0 or 1): ");
      scanf("%d", &dig);

      num += (dig << i);
    }
    printf("The decimal value = %d\n", num);
    return 0;
}
```

6.21 What is the output of the following program? Remember that printf() returns the number of displayed characters.

```
#include <stdio.h>
int main()
{
    int i;

    for(i = 0; printf("%d", i++) < 2;)
      ;
    printf("\nEnd = %d\n", i);
    return 0;
}
```

Answer: In the first loop iteration, printf() prints the current value of i (which is 0), and then i is increased by 1. Since printf() returns the number of displayed characters, the returned value is 1. Therefore, the **for** loop continues because the condition is true (1 < 2).

The same happens with all numbers up to 10. When i becomes 10, printf() prints 10 and then i becomes 11. Since two characters are printed, printf() returns 2, which makes the condition false (2 < 2) and the **for** loop terminates.

Therefore, the program displays

```
012345678910
End = 11
```

6.22 Write a program that reads an integer and displays a message to indicate whether it is prime or not. Remember, a prime number is any integer greater than 1 with no divisor other than one and itself.

```c
#include <stdio.h>
int main()
{
  int i, num;

  printf("Enter positive number: ");
  scanf("%d", &num);

  if(num >= 0)
  {
    for(i = 2; i <= num/2; i++)
    {
      if(num % i == 0)
      {
        printf("%d is not prime\n", num);
        return 0; /* When a divisor is found, there is no need to
          search for other divisors, and the return statement
          terminates the program immediately. */
      }
    }
    printf("The number %d is prime\n", num);
  }
  else
    printf("Error: enter positive number\n");
  return 0;
}
```

Comments: Because any number N has no divisor greater than N/2, the program checks to see if any integer from 2 up to half of the entered number divides the given integer. For example, if the user enters the number 10, the program checks to see if there is a number from 2 to 5 that divides 10. If a divisor is found, the program terminates.

Nested Loops

When an iteration loop is included in the body of another loop, each iteration of the outer loop triggers the full completion of the nested loop. We'll use **for** statements to explain the nested loops. Nested **while** and **do-while** loops are executed in a similar way. For example, let's trace the loop iterations in the following program:

```c
#include <stdio.h>
int main()
{
  int i, j;

  for(i = 0; i < 2; i++)
  {
    printf("One ");
    for(j = 0; j < i; j++)
      printf("Two ");
  }
  return 0;
}
```

First iteration of the outer loop (i = 0): Since the condition (i < 2) is true, the program displays One and the inner loop is executed.

　First iteration of the inner loop (j = 0): Since the condition (j < i) is false, the loop is not executed.

Second iteration of the outer loop (i = 1): Another One is displayed and the inner loop is executed.

　First iteration of the inner loop (j = 0): Since the condition (j < i) is true, the program displays Two.

　Second iteration of the inner loop (j = 1): Now, the condition is false and the loop terminates.

Third iteration of the outer loop (i = 2): Since the condition (i < 2) is false, the loop terminates.

The program displays One One Two

Tip *The **break** statement always terminates the loop in which it belongs to.*

Consider the following program:

```c
#include <stdio.h>
int main()
{
  int i, j;
  for(i = 0; i < 2; i++)
  {
    for(j = 0; j < 2; j++)
    {
      if(i+j == 1)
        break;

      printf("Two ");
    }
    printf("One ");
  }
```

```
  printf("\nVal1 = %d   Val2 = %d\n", i, j);
  return 0;
}
```

1st iteration of the outer loop (i = 0): Since the condition (i < 2) is true, the inner loop is executed.

1st iteration of the inner loop (j = 0): Since the **if** condition is false, the program displays Two.

2nd iteration of the inner loop (j = 1): Now, the **if** condition is true; therefore, the **break** statement terminates the inner loop and the program displays One.

2nd iteration of the outer loop (i = 1): The inner loop is executed again.

1st iteration of the inner loop (j = 0): Since the **if** condition is true, the **break** statement terminates the inner loop and the program displays One.

3rd iteration of the outer loop (i = 2): Since the condition (i < 2) is false, the loop terminates.

The program also displays the current values of i and j, which are 2 and 0, respectively. Therefore, the program outputs:

```
Two One One
Val1 = 2   Val2 = 0
```

Exercises

6.23 What is the output of the following program?

```
#include <stdio.h>
int main()
{
  int i;
  for(i = 0; i < 2; i++)
  {
    printf("One ");
    for(i = 0; i < 2; i++)
      printf("Two ");
  }
  printf("Val = %d\n", i);
  return 0;
}
```

Answer: Let's trace the iterations:

First iteration of the outer loop (i = 0): Since the condition (i < 2) is true, the program displays One and the inner loop is executed.

First iteration of the inner loop (i = 0): Since the condition (i < 2) is true, the program displays Two.

Second iteration of the inner loop (i = 1): Another Two is displayed.

Third iteration of the inner loop (i = 2): Now, the condition becomes false and the loop terminates.

Second iteration of the outer loop (i++): The value of i became 2 in the inner **for** loop; therefore, the i++ statement makes it 3. Since the condition (i < 2) is false, the loop terminates.

The program outputs

```
One Two Two
Val = 3
```

6.24 Write a program that displays the multiplication table from 1 to 10.

```
#include <stdio.h>
int main()
{
  int i, j;
  for(i = 1; i <= 10; i++)
  {
    for(j = 1; j <= 10; j++)
      printf("%d * %d = %d\n", i, j, i*j);

    printf("\n");
  }
  return 0;
}
```

6.25 Write a program that produces the following output.

```
*
**
***
****
*****
```

```
#include <stdio.h>
int main()
{
  int i, j;
  for(i = 0; i < 5; i++)
  {
    for(j = 0; j <= i; j++)
      printf("* ");
    printf("\n"); /* Add a new line to display the next star ('*')
      characters. */
  }
  return 0;
}
```

Comments: Let's trace the first two iterations of the outer loop.

First iteration of the outer loop (i = 0): Since the condition (i < 5) is true, the inner loop is executed.

First iteration of the inner loop (j = 0): Since the condition (j <= i) is true, the program displays '*'.

Second iteration of the inner loop (j = 1): Now, the condition is false; therefore, the loop terminates and the following printf() adds a new line character.

Second iteration of the outer loop (i = 1): Since the condition (i < 5) is true, the inner loop is executed.

First iteration of the inner loop (j = 0): Since the condition (j <= i) is true, the program displays '*'.

Second iteration of the inner loop (j = 1): Another '*' is displayed.

Third iteration of the inner loop (j = 2): Now, the condition is false; therefore, the loop terminates and a new line character is added.

So far, the program has displayed:

```
*
**
```

The program will produce the desired output.

6.26 Write a program that reads the grades of five students in three different courses and displays the average grade of each in the three courses, as well as the average grade of all students in all courses.

```c
#include <stdio.h>

#define LESSONS  3
#define STUDENTS 5

int main()
{
  int i, j;
  float grd, stud_grd, class_grd;

  class_grd = 0;
  for(i = 0; i < STUDENTS; i++)
  {
    printf("***** Student %d\n", i+1);
    stud_grd = 0; /* This variable holds the sum of a student's
      grades in all courses. It is initialized to 0 for each one. */
    for(j = 0; j < LESSONS; j++)
    {
      printf("Enter grade for lesson %d: ", j+1);
      scanf("%f", &grd);
      stud_grd += grd;
      class_grd += grd; /* This variable holds the sum of all
        grades. */
    }
    printf("Average grade for student_%d is %.2f\n",
      i+1, stud_grd / LESSONS);
  }
```

```
    printf("\nAverage class grade is %.2f\n",
      class_grd / (STUDENTS * LESSONS));
    return 0;
}
```

6.27 Write a program that reads an integer and displays the prime numbers that are less than or equal to it.

```c
#include <stdio.h>
int main()
{
  int i, j, num;

  printf("Enter number: ");
  scanf("%d", &num);

  for(i = 2; i <= num; i++)
  {
    for(j = 2; j <= i; j++)
    {
      if(i%j == 0)
      {
        if(i == j)
          printf("Prime number = %d\n", i);
        else
          break;
      }
    }
  }
  return 0;
}
```

Comments: Here's an example of how this program works. Assume that the user enters the number 4, and therefore num = 4. Next, we check to see if the numbers 2, 3, 4 are prime or not and we display only the primes. The inner loop checks to see which integers up to the tested number are primes.

First iteration of the outer loop (i = 2).
> *Execution of the inner loop (for j = 2). We see if i, which is 2, is prime. Since 2%2 is 0, the **if** condition is true and because the divisor j is equal to i (i=j=2), the printf() shows that the number 2 is prime.*

Second iteration of the outer loop (i = 3).
> *Execution of the inner loop (for j = 2 and j = 3). We see if i, which is 3, is prime. As before, printf() shows that the number 3 is prime because 3's only divisors are 1 and the number 3 itself.*

Third iteration of the outer loop (i = 4).
> *Execution of the inner loop (for j = 2, j = 3 and j = 4). We see if i, which is 4, is prime. The **if** condition is true when j is 2 because 4%2 is zero. Therefore, the tested number is not prime because the divisor we've found (the number 2) is not equal to i. Since the number is not prime, **break** terminates the loop.*

while Statement

The **while** statement is the simplest way to create iteration loops in C. It is mostly used when the number of the iterations is unknown. Its form is

```
while(condition)
{
  /* block of statements (loop body) that is repeatedly executed as long
     as the condition remains true. */
}
```

When a **while** statement is executed, the value of condition is evaluated first. If its value is false, the **while** loop is not executed. If it is true, the block of statements is executed and the value of condition is tested again. If this value became false, the **while** loop terminates. If not, the block of statements is executed again. This process repeats until the value of the condition becomes false.

As with the **for** and **if** statements, if the loop body consists of a single statement, the braces can be omitted.

The following program uses a **while** statement to print the integers from 1 to 10.

```
#include <stdio.h>
int main()
{
  int i = 1;

  while(i <= 10)
  {
    printf("%d\n", i);
    i++;
  }
  return 0;
}
```

The following program uses a **while** statement to print the integers from 10 down to 1.

```
#include <stdio.h>
int main()
{
  int i = 10;

  while(i != 0)
  {
    printf("%d\n", i);
    i--;
  }
  return 0;
}
```

Tip

while(i != 0) *is equivalent to* **while**(i) *and*

while(!i) *is equivalent to* **while**(i == 0)

The **while** statement is a special case of the **for** statement, in which the first and last expressions are missing and only the condition is present. For example, the **while** statement of the first program can be replaced with **for**(;i <= 10;).

If the condition is always true the loop is executed forever unless its body contains a statement (such as **break**) to terminate it. For example, the following loop will never end.

```
while(1)
  printf("One\n");
```

Any constant value other than 0 is considered to be always true.

By convention, most programmers use the value 1 to create an infinite **while** loop.

As with the **for** statement, a common error is to add a semicolon at the end of the **while** statement, in which case the compiler assumes that the loop body is empty. For example, the following creates an infinite loop because the statement a--; will never be executed.

```
int a = 10;
while(a != 0);
  a--;
```

Exercises

6.28 Write a program that reads an integer continuously and displays them. If the user enters 0, the insertion of numbers should terminate. Note that the number 0 must not be displayed.

```
#include <stdio.h>
int main()
{
  int i = 1; /* Initialize with a nonzero value, just to be sure
    that the loop will be executed. */
  while(i != 0)
  {
    printf("Enter number: ");
    scanf("%d", &i);
    if(i != 0)
      printf("Num = %d\n", i);
  }
  return 0;
}
```

6.29 How many times is the next **while** loop executed?

```
#include <stdio.h>
int main()
{
  int a = 256, b = 4;

  while(a != b)
```

```
    b = b*b;
  return 0;
}
```

Answer: Let's trace the iterations:

First iteration. The condition (a != b) is true because a = 256 and b = 4, and therefore the statement b = b*b = 4*4 = 16 is executed.

Second iteration. The condition is still true because a = 256 and b = 16, and therefore the statement b = b*b = 16*16 = 256 is executed.

Third iteration. Since a = 256 and b = 256, the condition becomes false and the loop ends.

As a result, the **while** loop is executed twice.

6.30 What is the output of the following program?

```
#include <stdio.h>
int main()
{
  int i = -2;

  while(i-6)
  {
    printf("One ");
    i++;
    while(!(i+1))
    {
      printf("Two ");
      i--;
    }
    i += 2;
  }
  return 0;
}
```

Answer: As with the nested **for** loops, each iteration of the outer **while** loop triggers the full completion of the nested loop.

Notice that the statement **while**(i-6) is equivalent to **while**(i-6 != 0) and the statement **while**(!(i+1)) is equivalent to **while**(i+1 == 0). Let's analyze the first iteration of the outer loop:

First iteration of the outer loop (i = -2). Since i is -2, the condition (i-6 != 0) is true, therefore, the program displays One and i becomes -1.

  First execution of the inner loop. Since i is -1, the condition (i+1 == 0) is true; therefore, the program displays Two and i becomes -2.

  Second execution of the inner loop. Since i is -2, the condition (i+1 == 0) is false and the loop terminates.

Then, the statement i+=2 makes i equal to 0.

Eventually, the program outputs One Two One One.

6.31 Write a program that reads an integer continuously and displays "Hello" as many times as the value of the integer. If the user enters a negative number, the insertion of integers should end and the program should display the total number of the displayed "Hello".

```c
#include <stdio.h>
int main()
{
  int i, num, times;

  times = 0;
  while(1)
  {
    printf("Enter number: ");
    scanf("%d", &num);

    if(num < 0)
      break;

    for(i = 0; i < num; i++)
      printf("Hello\n");

    times += num;
  }
  printf("Total number is = %d\n", times);
  return 0;
}
```

6.32 Write a program that reads an integer and displays the number of its digits and their sum. For example, if the number is 1234, the program should display 4 and 10 (1+2+3+4 = 10).

```c
#include <stdio.h>
int main()
{
  int num, sum, dig;

  sum = dig = 0;

  printf("Enter number: ");
  scanf("%d", &num);
  if(num < 0)
    num = -num; /* Make it positive. */
  else if(num == 0)
    dig = 1; /* Check the case of 0. */
  while(num > 0)
  {
    sum += num % 10;
    num = num / 10;
    dig++;
  }
  printf("%d digits and their sum is %d\n", dig, sum);
  return 0;
}
```

6.33 Write a program that reads the prices of a shop's products continuously until the user enters –1. The program should display the minimum price, the maximum, and the average of those within [5,30], before it terminates. Assume that none of the products costs more than $100.

```c
#include <stdio.h>
int main()
{
  int set_prc;
  float min, max, prc, sum_prc;

  set_prc = 0; /* This variable counts the products whose price is
    within the specified set. */
  sum_prc = 0; /* This variable holds the sum of the prices within
    the specified set. */
  min = 100;
  max = 0;
  while(1)
  {
    printf("Enter price: ");
    scanf("%f", &prc);

    if(prc == -1)
      break;

    if(prc >= 5 && prc <= 30)
    {
      sum_prc += prc;
      set_prc++;
    }
    if(max < prc)
      max = prc;
    if(min > prc)
      min = prc;
  }
  printf("\nMin = %f, Max = %f\n", min, max);
  if(set_prc != 0)
    printf("Avg = %.2f\n", sum_prc/set_prc);
  else
    printf("None product is included\n");
  return 0;
}
```

6.34 Write a program that reads an integer in [0,255] continuously and displays it in binary. For example, if the user enters 32, the program should display 00100000. For any value out of [0,255] the program should terminate.

```c
#include <stdio.h>
int main()
{
  int i, num;

  while(1)
  {
    printf("\nEnter number: ");
    scanf("%d", &num);

    if(num < 0 || num > 255)
      break;
```

```
      for(i = 0; i < 8; i++) /* Check the value of each bit. */
    {
      (num & 128) ? printf("1") : printf("0");
      num <<= 1; /* Shift all bits one place left. */
    }
  }
  return 0;
}
```

6.35 Write a program that reads a positive integer and displays the maximum positive integer n for which the sum $1^2 + 2^2 + 3^2 + \cdots + n^2$ is less than the given number.

```
#include <stdio.h>
int main()
{
  int i, num, sum;

  printf("Enter number: ");
  scanf("%d", &num);
  if(num <= 0)
  {
    printf("Error: The number should be positive\n");
    return 0;
  }
  sum = 0;
  i = 1;
  while(1)
  {
    sum += i*i;
    if(sum >= num)
      break;
    i++;
  }
  printf("The last number is = %d\n", i-1); /* The number i-1 is the
    last number where the value of sum is still less than the given
    number. */
  return 0;
}
```

6.36 Write a program that reads a float number (i.e., a) and an integer (i.e., b) and displays the result of a^b.

```
#include <stdio.h>
int main()
{
  int b, exp, tmp;
  double a, val;

  printf("Enter number and power: ");
  scanf("%lf%d", &a, &b);

  val = 1; /* Must be initialized to 1. */
  tmp = 0;
  exp = b;
  if(b < 0) /* If the exponent is negative, make it positive. */
```

```
    exp = -exp;
  while(tmp < exp)
  {
    val = val * a;
    tmp++;
  }
  if(b < 0)
    printf("%f power %d = %f\n", a, b, 1/val);
  else
    printf("%f power %d = %f\n", a, b, val);
  return 0;
}
```

6.37 The final grade of a student in a course is calculated as 30% of the exercise's grade and as 70% of the exam's grade, only if both grades are greater than or equal to 5; otherwise, the final grade will be their minimum. Write a program that reads continuously pairs of grades (exercises and exam grades) and displays the final grade for each student, until the user enters a pair of grades containing the value −1. Before it ends, the program should display the average grade of all students in the course. The program should check that all given grades belong in [0,10].

```
#include <stdio.h>
int main()
{
  int studs;
  float sum_grd, stud_grd, grd_exc, grd_exam;

  studs = 0;
  sum_grd = 0;

  while(1)
  {
    printf("-------------------------------\n");

    printf("Enter exercise grade [0-10]: ");
    scanf("%f", &grd_exc);

    printf("Enter exam grade [0-10]: ");
    scanf("%f", &grd_exam);

    if(grd_exc == -1 || grd_exam == -1)
      break;
    if((grd_exc < 0) || (grd_exc > 10) ||
      (grd_exam < 0) || (grd_exam > 10))
    {
      printf("Error: Grade should be in [0-10]\n");
      continue;
    }
    studs++;
    if(grd_exc >= 5 && grd_exam >= 5)
      stud_grd = 0.3*grd_exc + 0.7*grd_exam;
    else
    {
      if(grd_exc <= grd_exam)
        stud_grd = grd_exc;
```

```
      else
         stud_grd = grd_exam;
      }
      printf("Student grade = %.2f\n", stud_grd);
      sum_grd += stud_grd;
   }
   if(studs)
      printf("\nAverage grade = %.2f\n", sum_grd/studs);
   return 0;
}
```

6.38 Write a program that displays the following choices:

1. *Add numbers.*
2. *Subtract numbers.*
3. *Multiply numbers.*
4. *Divide numbers.*
5. *Exit.*

The program should read the user's choice, prompt him to enter two integers, and then display the result of the selected operation. This process should be repeated continuously until the user chooses "Exit".

```
#include <stdio.h>
int main()
{
   int i, j, sel;

   while(1)
   {
    printf("\n\nMenu selections\n");
    printf("---------------\n");
    printf("1. Add numbers\n");
    printf("2. Subtract numbers\n");
    printf("3. Multiply numbers\n");
    printf("4. Divide numbers\n");
    printf("5. Exit\n");

    printf("\nSelect an option: ");
    scanf("%d", &sel);
    if(sel == 5) /* Program termination. */
      return 0;

    printf("\nEnter numbers: ");
    scanf("%d%d", &i, &j);

    switch(sel)
    {
      case 1:
        printf("Sum = %d\n", i+j);
      break;

      case 2:
        printf("Diff = %d\n", i-j);
      break;
```

```
    case 3:
      printf("Product = %d\n", i*j);
    break;
    case 4:
      if(j != 0)
        printf("Div= %.2f\n", (float)i/j);
      else
        printf("Error: Second number should not be zero\n");
    break;

    default:
      printf("Wrong input\n");
    break;
    }
    getchar();
  }
  return 0;
}
```

6.39 Write a program that reads continuously a month number (1 = Jan, 12 = Dec), the day that the month begins (1 = Mon, 7 = Sun), and displays the calendar for that month. If the selected month is February, the program should prompt the user to enter the month's number of days, that is, 28 or 29. If the given month is out of [1,12], the program should terminate.

```
#include <stdio.h>
int main()
{
  int i, mon, mon_days, day, rows;

  while(1)
  {
   printf("\n\nEnter month: ");
   scanf("%d", &mon);

   if(mon < 1 || mon > 12)
     break;

   if(mon == 2)
   {
     printf("Enter Feb days: ");
     scanf("%d", &mon_days);
   }
   else if(mon==4 || mon == 6 || mon == 9 || mon == 11)
     mon_days = 30;
   else
     mon_days = 31;

   printf("Enter start day (1=Mon,7=Sun): ");
   scanf("%d", &day);

   printf("Mon\tTue\tWed\tThu\tFri\tSat\tSun\n");
   for(i = 1; i < day; i++) /* Add some spaces up to the first day
     of the month to format the output. */
     printf("\t");
```

```
    rows = 0;
    for(i = 1; i <= mon_days; i++)
    {
      printf("%d\t", i);
      if(i == 8-day+(rows*7))
      {
        printf("\n");
        rows++;
      }
    }
  }
  return 0;
}
```

Comments: The **if** statement within the last **for** statement checks to see if the last day of the week is reached or not. If it is reached, a new line character is added. The rows variable counts how many day rows have been displayed so far.

For example, if the user selects Wed as the first month day (day=3), the first new line should be added when i equals 8-day+(rows*7)=8-3+(0*7)=5. Then, the value of rows becomes 1. The next new line characters will be added when i becomes 12, 19, and 26, respectively.

do-while Statement

The **do-while** statement is similar to the **while** statement except that the value of condition is checked *after* each execution of the loop body, not before. Therefore, a **do-while** loop is executed at least once. It has the form

```
do
{
  /* block of statements (loop body) that is executed at least once and
     then repeated as long as the condition remains true. */
} while(condition);
```

When a **do-while** statement is executed, the loop body is executed first and the value of condition is evaluated. If the value is false, the loop terminates. If it is true, the loop body is executed and then the value of the condition is tested again. If it becomes false, the loop terminates. If not, the loop body is executed again. This process repeats until the value of condition becomes false.

*A **do-while** statement should end with a semicolon.*

For example, the following program uses the **do-while** statement to display the integers from 1 to 10.

```
#include <stdio.h>
int main()
{
  int i = 1;
```

```
  do
  {
    printf("%d\n", i);
    i++;
  } while(i <= 10);

  return 0;
}
```

In the following exercises, we could use **for** or **while** loop statements. We use the **do-while** statement in order to show you how to use it.

Exercises

6.40 Write a program that reads an integer and displays the word "This" as many times as the value of the given integer.

```
#include <stdio.h>
int main()
{
  int i, num;

  printf("Enter number: ");
  scanf("%d", &num);
  i = 1;
  do
  {
    printf("This\n");
    i++;
  } while(i <= num);

  return 0;
}
```

6.41 Write a program that reads integers continuously and displays the square of the even numbers until the user enters a number in [10,20]. The program should display the number of positive and negative numbers and the number of those in [300,500] before it ends. Zero is counted as neither a positive or negative number.

```
#include <stdio.h>
int main()
{
  int i, pos, neg, cnt;

  pos = neg = cnt = 0;
  do
  {
    printf("Enter number: ");
    scanf("%d", &i);

    if((i & 1) == 0)
      printf("Num = %d\n", i*i);
```

```
    if(i > 0)
    {
      pos++;
      if(i >= 300 && i <= 500)
        cnt++;
    }
    else if(i < 0)
      neg++;
  } while(i < 10 || i > 20);

  printf("Pos = %d Neg = %d Cnt = %d\n", pos, neg, cnt);
  return 0;
}
```

6.42 Write a program that reads an integer N > 3 and calculates the result of the expression R1 = 1/1 − 1/2 + 1/3 − 1/4 +⋯ 1/N. The program should force the user to enter an integer greater than 3.

```
#include <stdio.h>
int main()
{
  int i, num;
  double a, val;

  do
  {
    printf("Enter number > 3: ");
    scanf("%d", &num);
  } while(num <= 3);
  val = 0;
  a = 1;
  for(i = 1; i <= num; i++)
  {
    val += a/i;
    a = -a;
  }
  printf("Val = %e\n", val);
  return 0;
}
```

6.43 Write a program that a teacher may use to check if the students know the multiplication table. The program should generate two random values that belong in [1,10] (i.e., a and b) and display a×b = (the smaller number should appear first). The student should fill the result and the program should display an informative message to indicate whether the answer is correct or wrong. If the user enters −1, the program should display the total number of correct and wrong answers and then terminate.

```
#include <stdio.h>
#include <stdlib.h>
#include <time.h>
int main()
{
  int i, j, num, fails, wins;
```

```
  fails = wins = 0;
  srand(time(NULL));
  do
  {
   i = rand()%10+1; /* The rand() function returns a random positive
      integer and the % operator constrains it in [0,9]. We add one
      to constrain it in [1,10]. */
   j = rand()%10+1;
   if(i < j)
     printf("\n%dx%d=", i, j);
   else
     printf("\n%dx%d=", j, i);

   scanf("%d", &num);
   if(num != -1)
   {
     if(num == i*j)
     {
       printf("Correct\n");
       wins++;
     }
     else
     {
       printf("Wrong(answer=%d)\n", i*j);
       fails++;
     }
   }
  } while (num != -1);

  printf("Fails = %d, Wins = %d\n", fails, wins);
  return 0;
}
```

Comments: Although we haven't discussed functions yet, you can think of them as building blocks that do something. As scanf() and printf() are used for data input and output, the srand(), rand(), and time() library functions are used to generate random positive integers. After reading Chapter 11, see Appendix C for a brief description of their purpose. The stdlib.h and time.h are included in the program to provide information for those functions.

goto Statement

The **goto** statement is used to transfer control to another statement within the same function, provided that this statement has a label. Its syntax is

```
goto location;
```

When the **goto** is executed, the program transfers control to the statement that follows the location label. The label's name must be unique in the function where the **goto** statement is used. The label is placed at the beginning of the target statement and its name must be followed by a colon (:).

Look at the following program. If the user enters –1, the `goto` statement will transfer the execution to the START label and the `for` loop will execute again.

```c
#include <stdio.h>
int main()
{
  int i, num;
START:
  for(i = 0; i < 5; i++)
  {
    printf("Enter number: ");
    scanf("%d", &num);
    if(num == -1)
      goto START;
  }
  return 0;
}
```

Usually, it is better to avoid the use of `goto` because the transition of the program's execution from one point to another and then to another and so on leads to obscure, hard-to-read and complex code that is hard to maintain. In fact, most programmers oppose its use, arguing that it has no place in a well-structured program.

However, there are cases where `goto` statement can be helpful, such as when exiting from nested `for` or `switch` statements, as in this example:

```c
for(i = 0; i < 10; i++)
  for(j = 0; j < 20; j++)
    for(k = 0; k < 30; k++)
    {
      if(condition)
        goto NEXT;
    }
NEXT:
...
```

We advise you to ignore statements like "never use `goto`", "`goto` is only for rookies," and "structural programming and `goto` do not go along" because it can be used to simplify some things.

And besides, `break`, `continue`, and `return` are nothing more than versions of the `goto` statement.

Unsolved Exercises

6.1   Write a program that reads integers continuously and calculates their sum until it exceeds `100`. Then, the program should display the sum and how many numbers were entered.

6.2   Write a program that reads an integer and, if it belongs in [10, 20], displays five times the square of the number. If the integer is outside the given set, it should display the word "`One`" ten times.

6.3   Write a program that displays all numbers from 111 to 999, except those beginning with 4 or ending with 6.

6.4   Write a program that reads the temperatures of July and displays how many days the temperature was equal, less, or more than 80°F.

6.5   Write a program that reads up to 100 integers. If an input number is greater than the last entered, the insertion of numbers should terminate and the program should display how many numbers were entered.

6.6   Write a program that reads three integers (i.e., a, b, and c) one after the other, not all together. The program should force the user to enter the numbers in descending order (i.e., a > b > c).

6.7   Write a program that reads integers continuously, until the user enters –1. Then, the program should display the largest positive and the minimum negative input numbers. If the user doesn't enter any positive or negative number, the program should display an informative message. (*Note*: zero is counted as neither a positive nor negative number.)

6.8   Write a program that must read 10 positive numbers. If an input number is negative, the program should prompt the user to enter another one. The program should display how many negative numbers were entered, before it terminates. Use a **for** loop. (*Note*: zero is counted as a positive number.)

6.9   Write a program that reads 10 integers and displays how many times the user entered successive values. For example, if the user enters -5, 10, 17, -31, -30, -29, 75, 76, 9, -4, the program should display 3 due to the pairs {-31, -30}, {-30, -29} and {75, 76}.

6.10   Write a program that reads an integer and displays the number of the bits set. For example, if the user enters 30 (in binary: 00000000000000000000000000011110), the program should display 4.

6.11   What is the output of the following program? Explain why.

```c
#include <stdio.h>
int main()
{
  int i, j, k = 100;

  for(i = 0; i < 2; i++)
  {
    printf("One ");
    for(j = 0; k; j++)
    {
      printf("Two ");
      k -= 50;
    }
  }
  return 0;
}
```

6.12   Write a program that produces the following output.

```
*****
 ****
  ***
   **
    *
```

6.13 Write a program that produces the following output.

```
*
*  *
*  *  *
*  *  *  *
*  *  *
*  *
*
```

6.14 Write a program that reads two integers (i.e., M, N) and produces an M×N grid. Each grid cell should be 3×2 characters wide. As an example, a 3×5 grid follows:

```
+--+--+--+--+--+
|  |  |  |  |  |  |
+--+--+--+--+--+
|  |  |  |  |  |  |
+--+--+--+--+--+
|  |  |  |  |  |  |
+--+--+--+--+--+
```

The three horizontal characters of each cell should be +-- and the two verticals +|

6.15 Write a program that reads 100 integers and displays the number which is entered the most times in sequence.

6.16 Write a program that reads an integer (i.e., N) and displays the result of:

$$\frac{1}{1\times3}+\frac{1}{3\times5}+\frac{1}{5\times7}+\ldots+\frac{1}{(N-2)\times N}$$

The program should force the user to enter an odd integer greater or equal to 3.

6.17 Write a program that reads a number (i.e., N) and calculates the result of $2^2 + 4^2 + 6^2 + \cdots + (2*N)^2$. The program should force the user to enter a positive number less than 20. Use one **for** loop.

7

Arrays

The variables we've used so far can store a single value. In this chapter, we'll talk about a new type of variable capable of storing a number of values. This type is called array. An array may be multidimensional. We'll focus on the simplest and most usual kinds; the one-dimensional and two-dimensional arrays. To introduce you to arrays, we'll show you how to use arrays of integers and floating-point numbers. We'll discuss other types of arrays, as well as their close relationship to pointers in later chapters.

Declaring Arrays

An array is a data structure that contains elements of the same data type. To declare a one-dimensional array, you must specify its name, the number of its elements, and their data type, like this:

```
data_type array_name[number_of_elements];
```

The elements of an array can be of any type, while their number must be enclosed in brackets. For example, the statement

```
int arr[1000];
```

declares the array `arr` with 1000 elements of type `int`.

After declaring an array, you can't change the number of its elements; it remains fixed.

The array's length is specified by an integer constant expression; however, it is a good practice to use a macro instead. If you ever need to change it, you only need to change the value of the constant. For example,

```
#define SIZE 150
float arr[SIZE]; /* The compiler replaces SIZE with 150 and creates an
  array of 150 floats. */
```

When declaring an array, the compiler allocates a memory block to store the values of its elements. These values are stored one after another in consecutive memory locations. This memory block is allocated in a region called stack, and it is automatically released when the function that declares the array terminates.

For example, with the statement `int arr[10];` the compiler allocates 40 bytes to store the values of the 10 integer elements.

The maximum memory size that can be allocated for an array depends on the available stack size.

For example, this program

```c
#include <stdio.h>
int main()
{
  double arr[300000];

  printf("Memory allocated\n");
  return 0;
}
```

may not run in your computer unless the available stack size is large enough to hold the values.

To avoid useless waste of memory, don't declare an array with more length than needed.

Accessing Array Elements

To access an array element, we write the array's name followed by the element's index enclosed in brackets. The index specifies the position of the element within the array, and it can be an integer constant, variable, or expression.

In an array of n elements, the first one is stored in position [0], the second one in position [1], and the last one in [n–1]. For example, the statement:

```c
float grd[1000];
```

declares the array grd with 1000 elements of type **float**, named grd[0], grd[1], ... grd[999].

An array element can be used in the same way as an ordinary variable.

For example, look at the following statements

```c
int i, j, arr[10];

arr[0] = 2; /* The value of the first element becomes 2. */
arr[9] = arr[0]; /* The value of the last element becomes equal to the
   value of the first element. */
i = j = arr[0]; /* The values of i and j become equal to the value of the
   first element, that is 2. */
arr[i+j] = 300; /* Since i+j = 2+2 = 4, the value of the fifth element
   becomes 300. */
```

Don't forget that the indexing of an array of n elements starts from 0 (not 1) up to n–1.

The following program declares an array of 5 integers, assigns the values 100, 101, 102, 103, and 104 to its elements, and displays them:

```c
#include <stdio.h>
int main()
{
  int i, arr[5];

  arr[0] = 100;
  arr[1] = 101;
  arr[2] = 102;
  arr[3] = 103;
  arr[4] = 104;

  for(i = 0; i < 5; i++)
    printf("%d\n", arr[i]);
  return 0;
}
```

C does not check if the index goes out of the array bounds. It is the programmer's responsibility to assure that this won't happen. If it does, the program may behave unpredictably.

For example, in the previous program, the valid values for indexing the array are from 0 to 4. However, if you write arr[5] = 20; the compiler won't raise an error message. It is your responsibility to avoid this kind of error; otherwise, the program may behave unpredictably. Consider the following program:

```c
#include <stdio.h>
int main()
{
  int i, j = 20, arr[3];

  for(i = 0; i < 4; i++)
    arr[i] = 100;

  printf("%d\n", j);
  return 0;
}
```

In the last iteration (i = 3), the statement arr[3] = 100; assigns a value to a nonexisting array element. In fact, the value 100 will overwrite the data in the memory following the arr[2] element. If this memory is reserved for j, its value will change and the program will display 100 instead of 20 !!!

Once more, be careful when you assign a value to an array element because exceeding the array bounds may cause abnormal program operation.

Array Initialization

Like ordinary variables, the elements of an array can be initialized when it is declared. In the most common form, the = operator follows the array's name and the values of its elements are enclosed in braces and separated with a comma. For example, with the declaration

```c
int arr[4] = {10, 20, 30, 40};
```

the values of `arr[0]`, `arr[1]`, `arr[2]`, and `arr[3]` become 10, 20, 30, and 40, respectively.

If the initialization list is shorter than the number of the elements, the remaining elements are set to 0. For example, with the declaration

```
int arr[10] = {10, 20};
```

the values of `arr[0]` and `arr[1]` become 10 and 20, respectively, and the rest elements `arr[2]` to `arr[9]` are set to 0.

If the array's length is omitted, the compiler will create an array with length equal to the number of the values in the list. For example, with the statement

```
int arr[] = {10, 20, 30, 40};
```

the compiler creates an array of four integers and assigns the values 10, 20, 30, and 40 to its elements.

To declare an array whose elements can't change during program execution, start its declaration with the keyword **const**. In that case, the array must be initialized. For example, with the declaration

```
const int arr[] = {10, 20, 30, 40};
```

if you attempt to change the value of an element, like `arr[0]` = 80, the compiler will raise an error message.

Exercises

7.1 Write a program that declares an array of 5 integers and assigns the values 10, 20, 30, 40, and 50 to its elements. Next, the program should display the elements greater than 20.

```
#include <stdio.h>
int main()
{
  int i, arr[] = {10, 20, 30, 40, 50};

  for(i = 0; i < 5; i++)
  { /* The braces are not necessary. */
    if(arr[i] > 20)
      printf("%d\n", arr[i]);
  }
  return 0;
}
```

Comments: What would be the output if we change the **for** condition from i<5 to i < **sizeof**(arr)/**sizeof**(arr[0])?

The expression **sizeof**(arr) calculates how many bytes allocates the array arr. This number divided by the element size, that is, **sizeof**(arr[0]), calculates the number of the arr elements.

Since each **int** reserves four bytes, the result of 20/4 is 5. Therefore, the **for** loop is executed five times and the program displays the arr elements greater than 20.

7.2   What would be the values of the array a in the following program?

```
#include <stdio.h>
int main()
{
  int i, a[3] = {4, 2, 0}, b[3] = {2, 3, 4};

  for(i = 0; i < 3; i++)
    a[b[i]-a[2-i]]++;
  return 0;
}
```

Answer: The statement **int** a[] = {4, 2, 0}; makes a[0] = 4, a[1] = 2, and a[2] = 0. Similarly, with the declaration of array b, we have b[0] = 2, b[1] = 3, and b[2] = 4. Let's trace the iterations:

First iteration (i = 0): a[b[0]-a[2]]++ = a[2-0]++ = a[2]++ = 1.
Second iteration (i = 1): a[b[1]-a[1]]++ = a[3-2]++ = a[1]++ = 3.
Third iteration (i = 2): a[b[2]-a[0]]++ = a[4-4]++ = a[0]++ = 5.

Therefore, the values of a[0], a[1], and a[2] become 5, 3, and 1, respectively.

7.3   What would be the values of the array arr in the following program?

```
#include <stdio.h>
int main()
{
  int i, arr[10] = {0};

  for(i = 0; i < 10; i++)
    arr[i++] = 20;
  return 0;
}
```

Answer: The elements of the array arr are initialized to 0. The statement arr[i++] = 20 first makes arr[i] equal to 20, then increases i by one. Then, the **for** statement increases i once more.

Therefore, the even elements of the array arr[0], arr[2], ... become 20, whereas the odd ones arr[1], arr[3], ... remain 0.

7.4   Write a program that declares an array of 5 elements and uses a loop to assign the values 1.1, 1.2, 1.3, 1.4, and 1.5 to them. Next, the program should display the array's elements in reverse order.

```
#include <stdio.h>
int main()
{
  int i;
  double arr[5];
```

```
    for(i = 0; i < 5; i++)
      arr[i] = 1.1 + (i*0.1);
    for(i = 4; i >= 0; i--)
      printf("%f\n", arr[i]);
    return 0;
}
```

Comments: Alternatively, we could replace the first loop with

```
arr[0] = 1.1;
for(i = 1; i < 5; i++)
  arr[i] = arr[i-1] + 0.1;
```

7.5   What is the output of the following program?

```
#include <stdio.h>
int main()
{
  unsigned char arr[5];
  int i;

  for(i = 0; i < 5; i++)
  {
    arr[i] = 256+i;
    printf("%d", arr[i]);
  }
  return 0;
}
```

Answer: Since the number 256 is encoded in nine bits (100000000) and the type of arr is **unsigned char**, only the lower eight bits of the expression 256+i can be stored.

For example, when i=1, only the lower eight bits of the number 257 (100000001) will be stored into arr[1]. Therefore, arr[1] becomes 1.

As a result, the program displays 0 1 2 3 4

7.6   Write a program that reads 10 integers and stores them in an array. Then, the program should check if the array is symmetric, that is, if the value of the first element is equal to the last one, the value of the second one equal to the value of the last but one, and so on.

```
#include <stdio.h>

#define SIZE 10

int main()
{
  int i, a[SIZE];

  for(i = 0; i < SIZE ; i++)
  {
    printf("Enter element a[%d]: ", i);
    scanf("%d", &a[i]);
  }
```

```
    for(i = 0; i < SIZE/2 ; i++)
      if(a[i] != a[SIZE-1-i])
      {
        printf("Non symmetric array\n");
        return 0; /* Since we found out that the array isn't symmetric
          the program terminates. */
      }
    printf("Symmetric array\n");
    return 0;
}
```

Comments: If the number of the elements is odd, would you change something in the code?

Since the middle element isn't compared with another element, it doesn't affect the array's symmetry. Therefore, this code works for both odd and even number of elements.

7.7   What is the output of the following program?

```
#include <stdio.h>
int main()
{
  int i, arr[] = {30, 20, 10, 0, -10, -20, -30};

  for(i = 0; arr[i]; i++)
    printf("%d\n", arr[i]);
  return 0;
}
```

Answer: The arr[i] expression in the **for** statement is equivalent to arr[i] != 0. Since the value of the fourth element is 0, the loop displays the values of the first three elements, that is, 30, 20, and 10, then it terminates.

7.8   Write a program that declares an array of 10 integers, assigns random values from 0 to 20 to its elements, and displays their average.

```
#include <stdio.h>
#include <stdlib.h>
#include <time.h>

#define SIZE 10

int main()
{
  int i, sum, arr[SIZE];

  sum = 0; /* Initialise with 0 the variable that calculates the sum
    of the array elements. */
  srand(time(NULL));
  for(i = 0; i < SIZE; i++)
  {
    arr[i] = rand() % 21; /* The rand() function returns a random
      positive integer and the % operator constrains it in [0,20]. */
    sum += arr[i];
  }
```

```
    printf("Avg = %f\n", (float)sum/SIZE);
    return 0;
}
```

7.9   What would be the values of the array `arr` in the following program?

```
#include <stdio.h>
int main()
{
  int i, arr[3] = {0, 1, 2};

  for(i = 1; i < 4; i++)
    arr[arr[arr[3-i]]] = i-1;
  return 0;
}
```

Answer: The statement **int** arr[3] = {0,1,2}; makes arr[0] = 0, arr[1] = 1, and arr[2] = 2. Let's trace the iterations:

> *First iteration* (i=1): arr[arr[arr[2]]] = arr[arr[2]] = arr[2] = i-1 = 1-1 = 0.
> *Second iteration* (i=2): arr[arr[arr[1]]] = arr[arr[1]] = arr[1] = i-1 = 2-1 = 1.
> *Third iteration* (i=3): arr[arr[arr[0]]] = arr[arr[0]] = arr[0] = i-1 = 3-1 = 2.

Therefore, the values of arr[0], arr[1], and arr[2] become 2, 1, and 0, respectively.

7.10   Write a program that reads the grades of 100 students (should be in [0–10]), stores them in an array, and displays the average, the maximum, and the minimum grade, as well as the serial numbers of the students who got them. (*Note*: If there are more than one student with the same maximum or minimum grade, the program should display the first found.)

```
#include <stdio.h>

#define SIZE 100

int main()
{
  int i, min_pos, max_pos;
  float sum, min_grd, max_grd, grd[SIZE];

  sum = max_grd = 0; /* Initialize max_grd with the minimum allowed
     value. */
  min_grd = 10; /* Initialization with the maximum allowed value. */
  for(i = 0; i < SIZE; i++)
  {
    printf("Enter grade of stud_%d in [0-10]: ", i+1);
    scanf("%f", &grd[i]);
    while(grd[i] < 0 || grd[i] > 10)
    {
      printf("Error - Enter new grade in [0-10]: ");
      scanf("%f", &grd[i]);
    }
```

```
    sum += grd[i];
    if(grd[i] > max_grd)
    {
      max_grd = grd[i];
      max_pos = i; /* Store the position of the student with the best
        grade. */
    }
    if(grd[i] < min_grd)
    {
      min_grd = grd[i];
      min_pos = i; /* Store the position of the student with the
        worst grade. */
    }
  }
  /* Since the first element of an array is always stored in
    position [0] we add one to the max_pos and min_pos variables to
    display the serial numbers. */
  printf("Avg: %.2f H(%d): %.2f L(%d): %.2f\n", sum/SIZE, max_pos+1,
    max_grd, min_pos+1, min_grd);
  return 0;
}
```

7.11 The following program reads 10 integers and stores in the array `freq` the number of occurrences of each input number. Is there a bug in this code?

```
#include <stdio.h>

#define SIZE 10

int main()
{
  int i, num, arr[SIZE], freq[SIZE];

  for(i = 0; i < SIZE; i++)
    scanf("%d", &arr[i]);

  for(i = 0; i < SIZE; i++)
  {
    num = arr[i];
    freq[num]++;
  }
  printf("\nNumber occurrences\n");
  for(i = 0; i < SIZE; i++)
    printf("Num %d appears %d times\n", arr[i], freq[i]);
  return 0;
}
```

Answer: The first bug is due to the fact that the elements of the array `freq` have not been initialized to 0. Therefore, the statement `freq[num]++` is meaningless because it increases the random value of `freq[num]` by one.

However, the most serious bug happens when the user enters a number out of the bounds of the array `freq`. For example, if the number 100 is stored in the array `arr`, then the value of `num` would become 100 and the statement `freq[100]++` would change the content of a memory location out of the array bounds.

7.12 Write a program that reads 10 integers and stores them in an array only if either one of the following conditions is true:

(a) If the current index position is even, such as 0,2,4,..., and the number is even

(b) If the current index position is odd, such as 1,3,5,..., and the number is odd

The program shouldn't accept the values 0 and –1. After the insertion of the numbers, the unassigned elements should be set to –1. The program should display the array elements before it ends.

```c
#include <stdio.h>

#define SIZE 10

int main()
{
  int i, num, arr[SIZE] = {0}; /* Since the value 0 is not an
    acceptable value, we are using it as a special value to indicate
    that an element is not assigned with a value. */

  for(i = 0; i < SIZE; i++)
  {
    do
    {
      printf("Enter number: ");
      scanf("%d",&num);
      if(num == 0 || num == -1)
        printf("Not valid input !!!\n");
    } while(num == 0 || num == -1);

    if(i & 1) /* Check if the number is odd. */
    {
      if(num & 1) /* Store the number only if both the current index
        position and the number are odd. */
        arr[i] = num;
    }
    else
    {
      if((num & 1) == 0) /* Store the number only if both the current
        index position and the number are even. */
        arr[i] = num;
    }
  }
  printf("\n*** Array elements ***\n");
  for(i = 0; i < SIZE; i++)
  {
    if(arr[i] == 0)
      arr[i] = -1;
    printf("%d\n",arr[i]);
  }
  return 0;
}
```

7.13 Write a program that reads the grades of 100 students and stores in successive positions of an array the grades within [5,10] and in a second array the grades within [0,5). If the user enters –1, the insertion of grades should end and the program should display the average of the grades stored in both arrays.

```c
#include <stdio.h>
int main()
{
  int i, k, m;
  float grd, sum_suc, sum_fail, arr1[100], arr2[100];

  sum_suc = sum_fail = 0;
  k = m = 0;
  for(i = 0; i < 100; i++)
  {
    printf("Enter grade: ");
    scanf("%f", &grd);
    if(grd == -1)
      break;

    if(grd >= 5 && grd <= 10)
    {
      sum_suc += grd;
      arr1[k] = grd;
      k++; /* The variable k indicates how many grades are stored in
        arr1. It is increased when a grade is stored. */
    }
    else if(grd >= 0 && grd < 5)
    {
      sum_fail += grd;
      arr2[m] = grd;
      m++; /* We could combine the two statements and write arr2[m++]
        = grd; */
    }
  }
  if(k != 0)
    printf("\nSuccess_Avg: %.2f\n", sum_suc/k);
  else
    printf("\nAll students failed\n");

  if(m != 0)
    printf("\nFail_Avg: %.2f\n", sum_fail/m);
  else
    printf("\nAll students passed\n");
  return 0;
}
```

7.14 Write a program that reads the temperatures of July, stores them in an array, and displays the two highest temperatures.

```c
#include <stdio.h>

#define SIZE 31

int main()
```

```
{
  int i;
  float max_1, max_2, temp[SIZE];

  printf("Enter temperature: ");
  scanf("%f", &temp[0]);

  max_1 = temp[0];
  for(i = 1; i < SIZE; i++)
  {
    printf("Enter temperature: ");
    scanf("%f", &temp[i]);

    /* Find the maximum value of the array. */
    if(temp[i] > max_1)
      max_1 = temp[i];
  }
  max_2 = max_1;
  /* If the array elements are different, max_2 becomes equal to a
     value other than max_1. If not, it remains equal to max_1. */
  for(i = 0; i < SIZE; i++)
  {
    if(max_1 != temp[i])
    {
      max_2 = temp[i];
      break;
    }
  }
  /* If max_1 is equal to max_2 implies that all array elements are
     the same and the loop is not executed. */
  if(max_1 != max_2)
    for(i = 0; i < SIZE; i++) /* Compare the array elements with
      max_2 and store into max_2 the second maximum value, after
      checking that it isn't equal to max_1. */
    {
      if((temp[i] > max_2) && (temp[i] != max_1))
        max_2 = temp[i];
    }
  printf("First_Max = %f and Sec_Max = %f\n", max_1, max_2);
  return 0;
}
```

7.15   Write a program that reads an integer and displays the appearances of each digit
       [0–9] in the number. For example, if the user enters 123, the program should display
       that the digits 1, 2, and 3 appear once and the rest digits none.

```
#include <stdio.h>
int main()
{
  int i, dig_times[10] = {0}; /* This array holds the appearances of
    each digit. For example, dig_times[0] indicates how many times
    the digit 0 appears. */

  printf("Enter number: ");
  scanf("%d", &i);
  if(i < 0) /* If the user enters a negative number make it positive. */
```

```
    i = -i;
  else if(i == 0) /* Check if 0 is entered. */
    dig_times[0] = 1;
  while(i > 0)
  {
    dig_times[i%10]++;
    i /= 10;
  }
  for(i = 0; i < 10; i++)
    printf("Digit %d appears %d times\n", i, dig_times[i]);
  return 0;
}
```

7.16 Write a program that reads a positive integer and displays it in binary.

```
#include <stdio.h>
int main()
{
  int i, j, num, bits[32]; /* This array holds the bits of the
    number, that is 0 or 1. Since the size of an integer is 4 bytes,
    the length of the array is declared as 32. */
  do
  {
    printf("Enter positive number: ");
    scanf("%d", &num);
  } while(num <= 0);

  i = 0;
  /* Successive divisions by 2 and store each last bit in the
    respective array position. */
  while(num > 0)
  {
    bits[i] = num % 2;
    num >>= 1; /* Equivalent to num /= 2, but most probably it is
      executed faster. */
    i++;
  }
  printf("Binary form: ");
  /* Display the number's bits from left to right. */
  for(j = i-1; j >= 0; j--)
    printf("%d", bits[j]);
  return 0;
}
```

7.17 What is the output of the following program?

```
#include <stdio.h>
int main()
{
  int i, a[] = {10, 20, 30, 40, 50};
  double b[] = {2.2, 1.94, 0.5, -1, -2};

  for(i = 0; a[i] = b[i]; i++)
    printf("%d ", a[i]);
  return 0;
}
```

Answer: The expression a[i] = b[i] is equivalent to (a[i] = b[i]) != 0, meaning that the elements of the array b would be copied to the respective elements of the array a as long as the value of a[i] does not become 0. If it does, the **for** loop terminates.

Since the type of the array a is **int**, only the integer parts of the b elements will be stored into a. Therefore, when the value 0.5 is copied, a[2] becomes 0 and the **for** loop terminates.

As a result, the program outputs 2 1.

7.18 Write a program that reads 20 integers and stores them in two arrays of 10 elements. The program should check if there are common elements in the two arrays, and, if it happens, the program should display the value of each common element and its position in both arrays. Otherwise, it should display a message that their elements are different.

```c
#include <stdio.h>

#define SIZE 10

int main()
{
  int i, j, cmn, arr1[SIZE], arr2[SIZE];

  for(i = 0; i < SIZE; i++)
  {
    printf("Enter number for the 1st array: ");
    scanf("%d", &arr1[i]);

    printf("Enter number for the 2nd array: ");
    scanf("%d", &arr2[i]);
  }
  cmn = 0; /* This variable counts the common elements. */
  for(i = 0; i < SIZE; i++)
  {
    for(j = 0; j < SIZE; j++) /* This for loop checks if an element
      of the first array exists in the second one. */
    {
      if(arr1[i] == arr2[j])
      {
        cmn++;
        printf("Cmn = %d (Pos_1 = %d Pos_2 = %d)\n", arr1[i], i, j);
      }
    }
  }
  printf("%d common elements were found\n", cmn);
  return 0;
}
```

7.19 Write a program that reads the populations of 100 cities and stores them in increase order in an array when entered. The program should display the elements of the array before it ends.

```c
#include <stdio.h>

#define SIZE 100
```

```
int main()
{
  int i, j, temp, pop[SIZE];

  for(i = 0; i < SIZE; i++)
  {
    printf("Enter population: ");
    scanf("%d", &pop[i]);
    /* Compare the input value with the stored elements. The
       comparison is performed up to the position of the last input
       element, indicated by i. */
    for(j = 0; j < i; j++)
    {
      if(pop[j] > pop[i]) /* Swap elements. */
      {
        temp = pop[j];
        pop[j] = pop[i];
        pop[i] = temp;
      }
    }
  }
  printf("\n*** Populations in increase order ***\n");
  for(i = 0; i < SIZE; i++)
    printf("%d ", pop[i]);
  return 0;
}
```

7.20 Write a program that reads the integer codes of 50 products and stores them in an array only if they have not already been stored. As such, the elements of the array must be different. The program should display them before it ends.

```
#include <stdio.h>

#define SIZE 50

int main()
{
  int i, j, num, found, code[SIZE];

  i = 0;
  while(i < SIZE)
  {
    printf("Enter code: ");
    scanf("%d", &num);

    found = 0;
    /* The variable i indicates how many codes have been stored in
       the array. The for loop checks if the input code is already
       stored. If it does, the variable found becomes 1 and the loop
       terminates. */
    for(j = 0; j < i; j++)
    {
      if(code[j] == num)
      {
        printf("Error: Code %d exists. ", num);
        found = 1;
```

```
        break;
      }
    }
    /* If the code is not stored, we store it and the index position
       is increased by one. */
    if(found == 0)
    {
      code[i] = num;
      i++;
    }
  }
  printf("\nCodes: ");
  for(i = 0; i < SIZE; i++)
    printf("%d ", code[i]);
  return 0;
}
```

Two-Dimensional Arrays

A two-dimensional array, like the matrix in math, consists of rows and columns and, like one-dimensional arrays, contains elements of the same data type.

Two-Dimensional Array Declaration

To declare a two-dimensional array, you must specify its name, the data type of its elements, and the number of its rows and columns.

```
data_type array_name[number_of_rows][number_of_columns];
```

The number of its elements is equal to the number of its rows multiplied by the number of its columns.

For example, the statement **double** arr[10][5]; declares the two-dimensional array arr with 50 elements of type **double**.

Accessing the Elements of a Two-Dimensional Array

To access an element, we write the name of the array followed by the element's row index and column index enclosed in double brackets [][]. Like one-dimensional arrays, the indexing of rows and columns starts from 0. For example, the statement

```
int a[3][4];
```

declares a two-dimensional array whose elements are the a[0][0], a[0][1], ..., a[2][3], as depicted in Figure 7.1.

Like one-dimensional arrays, when a two-dimensional array is declared, the compiler allocates a memory block from the stack to store the values of its elements. For example, with the statement **int** arr[10][5]; the compiler allocates a block of 200 bytes to store the values of its 50 elements.

	Column 0	Column 1	Column 2	Column 3
Row 0	a[0] [0]	a[0] [1]	a[0] [2]	a[0] [3]
Row 1	a[1] [0]	a[1] [1]	a[1] [2]	a[1] [3]
Row 2	a[2] [0]	a[2] [1]	a[2] [2]	a[2] [3]

Column index

Array name

Row index

FIGURE 7.1
Layout of a two-dimensional array with 3 rows and 4 columns.

The elements are stored in row order with the elements of row 0 first, followed by the elements of row 1, and so on. To access an element, we must specify its row index and its column index. For example,

```
int i = 2, j = 2, arr[3][4];
arr[0][0] = 100; /* The value of the first element becomes 100. */
arr[1][1] = 200; /* The value of the sixth element becomes 200. */
arr[2][3] = arr[0][0]; /* The value of the last element becomes equal
  with the value of the first element. */
arr[i-2][j-2] = 300; /* The value of the first element becomes 300. */
```

 Like one-dimensional arrays be careful not exceeding the bounds of any dimension.

Since the elements of a two-dimensional array are stored sequentially in memory, we can find the position of an element. In the general case of an array a with ROWS rows and COLS columns, the position of the a[i][j] element is calculated as follows:

```
position = (i × COLS)+ j+1
```

For example, the position of the element a[2][1] of an array with 3 rows and 4 columns is the tenth:

```
position = (i × COLS)+ j+1 = (2 × 4) + 1 + 1 = 10.
```

 To find an element's position, only the number of columns is needed.

Two-Dimensional Array Initialization

Like one-dimensional arrays, a two-dimensional array can be initialized when declared.
    A common initialization method is to use the = operator and put the values of the elements of each row inside braces {}. For example,

```
int arr[3][3] = {{10, 20, 30},
                 {40, 50, 60},
                 {70, 80, 90}};
```

The value of arr[0][0] becomes 10, arr[0][1] becomes 20, arr[0][2] becomes 30, and so on.

Alternatively, we can omit the inner braces and write

```
int arr[3][3] = {10, 20, 30, 40, 50, 60, 70, 80, 90};
```

Our preference is to use the inner braces to make clearer the initialization of each row.

If the initialization list is shorter than the number of the row's elements, the remaining elements are set to 0. For example,

```
int arr[3][3] = {{10, 20},
                 {40, 50},
                 {70}};
```

The values of `arr[0][2]`, `arr[1][2]`, `arr[2][1]`, and `arr[2][2]` are set to 0.

If we omit the initialization of a row, its elements are set to 0. For example,

```
int arr[3][3] = {{10, 20, 30}};
```

The elements of the second and third row are set to 0.

Like before, if we omit the inner braces and the initialization list is shorter than the number of the array elements, the remaining elements are set to 0. For example,

```
int arr[3][3] = {10, 20};
```

The value of `arr[0][0]` becomes 10, `arr[0][1]` becomes 20, and all the rest equal to 0.

The number of columns must always be present. However, the number of rows is optional. If you don't specify it, the compiler will create a two-dimensional array based on the initialization list. For example,

```
int arr[][3] = {10, 20, 30, 40, 50, 60};
```

Since the array `arr` has 3 columns and the initialization values are 6, the compiler creates a two-dimensional array of 2 rows and 3 columns.

Another usual initialization method is to use a pair of nested **for** loops. For example, the following program declares a two-dimensional array and assigns the value 1 to its elements:

```
#include <stdio.h>
int main()
{
  int i, j, arr[50][100];

  for(i = 0; i < 50; i++)
    for(j = 0; j < 100; j++)
      arr[i][j] = 1;
  return 0;
}
```

Exercises

7.21   Write a program that creates an identity 6×6 array and displays its elements as an identity 6×6 matrix in algebra form. (*Note*: In math, an identity matrix has 1's on the main diagonal's elements and 0's everywhere else.)

```
#include <stdio.h>

#define SIZE 6

int main()
{
  int i, j, arr[SIZE][SIZE] = {0}; /* Initialize the arr elements
    with 0. */
  for(i = 0; i < SIZE; i++)
  {
    for(j = 0; j < SIZE; j++)
    {
      if(i == j) /* Check if it is an element of the main diagonal.
        */
        arr[i][j] = 1;

      printf("%3d", arr[i][j]);
    }
    printf("\n"); /* Add it to separate the array rows. */
  }
  return 0;
}
```

7.22 Write a program that reads 8 integers, stores them in a 2×4 array, and displays the array elements in reverse order, from the lower-right element to the upper-left one.

```
#include <stdio.h>

#define ROWS 2
#define COLS 4

int main()
{
  int i, j, arr[ROWS][COLS];

  for(i = 0; i < ROWS; i++)
  {
    for(j = 0; j < COLS; j++)
    {
      printf("Enter arr[%d][%d]: ", i, j);
      scanf("%d", &arr[i][j]);
    }
  }
  printf("\nArray elements\n");
  printf("--------------\n");
  for(i = ROWS-1; i >= 0; i--)
  {
    for(j = COLS-1; j >= 0; j--)
      printf("arr[%d][%d] = %d\n", i, j, arr[i][j]);
  }
  return 0;
}
```

7.23 In linear algebra, a matrix is called a Toeplitz matrix when the elements of each diagonal parallel to the main diagonal are equal between each other. For example, the following 5×5 matrix demonstrates the generic form of a 5×5 Toeplitz matrix:

$$t = \begin{bmatrix} a & b & c & d & e \\ f & a & b & c & d \\ g & f & a & b & c \\ h & g & f & a & b \\ i & h & g & f & a \end{bmatrix}$$

Write a program that reads 5 integers and stores them in the first row and first column of a 5×5 array. Next, the program should create the Toeplitz matrix and display its elements.

```c
#include <stdio.h>

#define SIZE 5

int main()
{
  int i, j, num, t[SIZE][SIZE];

  for(i = 0; i < SIZE; i++)
  {
    printf("Enter number: ");
    scanf("%d", &num);

    t[0][i] = num; /* The elements of the first row become equal to
      the input numbers. */
    t[i][0] = num; /* The elements of the first column become equal
      to the input numbers. */
  }
  /* Create the Toeplitz matrix. */
  for(i = 0; i < SIZE-1; i++)
    for(j = 0; j < SIZE-1; j++)
      t[i+1][j+1] = t[i][j]; /* We traverse the array t and make each
        element equal to the upper left. */
  for(i = 0; i < SIZE; i++)
  {
    for(j = 0; j < SIZE; j++)
      printf("%3d", t[i][j]);
    printf("\n");
  }
  return 0;
}
```

7.24 Write a program that reads 8 integers, stores them in a 2×4 array, and displays the sum of all its elements, the sum of each row's elements, and the sum of each column's elements.

```c
#include <stdio.h>

#define ROWS 2
#define COLS 4

int main()
{
  int i, j, tot_sum, tmp, arr[ROWS][COLS];
```

```
  tot_sum = 0;
  for(i = 0; i < ROWS; i++)
  {
    /* Initialize with 0 when start calculating the sum of a row's
      elements. */
    tmp = 0;
    for(j = 0; j < COLS; j++)
    {
      printf("Enter the element arr[%d][%d]: ",i,j);
      scanf("%d", &arr[i][j]);

      tot_sum += arr[i][j];
      tmp += arr[i][j];
    }
    printf("Row_%d: Sum = %d\n", i+1, tmp);
  }
  printf("\n");
  for(i = 0; i < COLS; i++)
  {
    /* Initialize with 0 when start calculating the sum of a column's
      elements. */
    tmp = 0;
    for(j = 0; j < ROWS; j++)
      tmp += arr[j][i];

    printf("Col_%d: Sum = %d\n", i+1, tmp);
  }
  printf("\nTotal_Sum = %d\n", tot_sum);
  return 0;
}
```

7.25 Write a program that initializes with 0 the elements that are under the main diagonal of a 5×5 array, which, in linear algebra, is called an "upper triangular matrix." The program should set all the remainder elements with random values within [−3,3] and display the product of the main diagonal's elements, which, in linear algebra, is equal to the determinant of a triangular matrix.

```
#include <stdio.h>
#include <stdlib.h>
#include <time.h>

#define SIZE 5

int main()
{
  int i, j, determ, arr[SIZE][SIZE];

  determ = 1; /* Initialize with 1 the variable which calculates the
    product of the main diagonal's elements and is equal to the
    determinant of the matrix. */
  srand(time(NULL));
  for(i = 0; i < SIZE; i++)
  {
    for(j = 0; j < SIZE; j++)
    {
      if(i > j)
```

```
      arr[i][j] = 0; /* Initialize with 0 the elements under the
         main diagonal. We could have initialized arr with 0, but we
         do it here to make it clearer. */
    else
      arr[i][j] = rand()%7-3; /* The result of the expression
         rand()%7 is an integer within [0,6]. By subtracting 3, the
         integer is constrained in [-3,3]. */
    printf("%5d", arr[i][j]);
    if(i == j)
      determ *= arr[i][j];
    }
    printf("\n");
  }
  printf("\nThe determinant is: %d\n", determ);
  return 0;
}
```

7.26 Write a program that reads and stores the grades of 100 students in 10 lessons in a
100×10 array and displays the average, the maximum, and the minimum grade of
each student. The program should display the serial numbers of the students with
the best and worst average grade. If two or more students have the same best or
worst value, the program should display the first found. The program should force
the user to enter grades within [0, 10].

```c
#include <stdio.h>

#define STUDS    100
#define COURSES  10

int main()
{
  int i, j, min_pos, max_pos;
  float sum, min_grd, max_grd, avg_grd, min_avg_grd, max_avg_grd,
    grd[STUDS][COURSES];

  min_pos = max_pos = 0;
  max_avg_grd = 0; /* Initialization with the minimum allowed value. */
  min_avg_grd = 10; /* Initialization with the maximum allowed value. */
  for(i = 0; i < STUDS; i++)
  {
    sum = 0;
    max_grd = 0;
    min_grd = 10;
    for(j = 0; j < COURSES; j++)
    {
      do
      {
        printf("Enter grade of student_%d for lesson_%d: ", i+1, j+1);
        scanf("%f", &grd[i][j]);
      } while(grd[i][j] < 0 || grd[i][j] > 10);
      sum += grd[i][j];

      if(grd[i][j] >= max_grd)
        max_grd = grd[i][j];
      if(grd[i][j] <= min_grd)
```

```
      min_grd = grd[i][j];
    }
    avg_grd = sum/COURSES;
    if(avg_grd > max_avg_grd)
    {
      max_avg_grd = avg_grd;
      max_pos = i;
    }
    if(avg_grd < min_avg_grd)
    {
      min_avg_grd = avg_grd;
      min_pos = i;
    }
    printf("Student_%d: Avg = %.2f Max = %.2f Min = %.2f\n", i+1,
      avg_grd, max_grd, min_grd);
  }
  printf("\nStudent_%d has the higher average %.2f and student_%d
    has the lower average %.2f\n", max_pos+1, max_avg_grd, min_pos+1,
    min_avg_grd);
  return 0;
}
```

7.27 Write a program that reads 6 integers and stores them in a 2×3 array (i.e., a). Next, it reads another 6 integers and stores them in a second 3×2 array (i.e., b). The program should calculate and display the elements of a third 2×2 array (i.e., c), which is the product of the two matrices, that is, c = a×b.

```c
#include <stdio.h>

#define N 2
#define M 3

int main()
{
  int i, j, k, a[N][M], b[M][N], c[N][N] = {0};

  for(i = 0; i < N; i++)
  {
    for(j = 0; j < M; j++)
    {
      printf("Enter the element a[%d][%d]: ", i, j);
      scanf("%d", &a[i][j]);
    }
  }
  for(i = 0; i < M; i++)
  {
    for(j = 0; j < N; j++)
    {
      printf("Enter the element b[%d][%d]: ", i, j);
      scanf("%d", &b[i][j]);
    }
  }
  for(i = 0; i < N; i++)
    for(j = 0; j < N; j++)
      for(k = 0; k < M; k++)
```

```
       c[i][j] += a[i][k] * b[k][j];
  printf("\nArray c = a x b (%dx%d)\n", N, N);
  printf("------------------\n");
  for(i = 0; i < N; i++)
  {
    for(j = 0; j < N; j++)
      printf("%5d", c[i][j]);
    printf("\n");
  }
  return 0;
}
```

Comments: It is reminded from the linear algebra that the product of two matrices is produced by adding the products of the elements of each row of the first matrix with the corresponding elements of each column of the second matrix. Therefore, the outcome of a N×M matrix multiplied with a M×N matrix is a N×N matrix. For example, consider the following a(2×3) and b(3×2) matrices:

$$a = \begin{bmatrix} 1 & -1 & 1 \\ 0 & 2 & 1 \end{bmatrix} \text{ and } b = \begin{bmatrix} 1 & 0 \\ 2 & -2 \\ 2 & 3 \end{bmatrix}$$

The dimension of the c = a×b matrix would be 2×2, and its elements would have the following values:

$$c = \begin{bmatrix} 1\times1+(-1)\times2+1\times2 & 1\times0+(-1)\times(-2)+1\times3 \\ 0\times1+2\times2+1\times2 & 0\times0+2\times(-2)+1\times3 \end{bmatrix} = \begin{bmatrix} 1 & 5 \\ 6 & -1 \end{bmatrix}$$

Therefore, the value of each c_{ij} element is the outcome of the equation $c_{ij} = \sum_{k=1}^{M} a_{ik} \times b_{kj}$.

7.28 Write a program that simulates a cinema's ticket office. Suppose that the cinema has 30 rows with 20 seats each. The program should display a menu to perform the following operations:

1. *Buy a ticket*. The program should let the spectator to select the row and the seat. If he/she doesn't want to select a specific seat, the program should select a random seat. The ticket's price is $6.

2. *Ticket cancellation*. The program should read the row and the seat and cancel the reservation. The refund is $5.

3. *Display the box-office and a diagram to demonstrate the reserved and free seats.*

4. *Program termination.*

```
#include <stdio.h>
#include <stdlib.h>

#define ROWS 30
#define COLS 20
```

```c
int main()
{
  int i, j, sel, row, col, rsvd_seats, cost, seats[ROWS][COLS] =
    {0}; /* We use the array seats to manage the cinema's seats. If
    an element's value is 0, it implies that the seat is free. */

  rsvd_seats = cost = 0;
  while(1)
  {
    printf("\nMenu selections\n");
    printf("--------------\n");

    printf("1. Buy Ticket\n");
    printf("2. Ticket Refund\n");
    printf("3. Show Information\n");
    printf("4. Exit\n");

    printf("\nEnter choice: ");
    scanf("%d", &sel);

    switch(sel)
    {
      case 1:
        if(rsvd_seats == ROWS*COLS)
        {
          printf("\nSorry, no free seats\n");
          break;
        }
        printf("\nWould you like a specific seat (No: 0)? ");
        scanf("%d", &sel);
        if(sel == 0)
        {
          do
          {
            row = rand() % ROWS; /* Use rand() to select a random
              seat. */
            col = rand() % COLS;
          } while(seats[row][col] == 1);
        }
        else
        {
          do
          {
            printf("\nEnter row (0-%d): ", ROWS-1);
            scanf("%d", &row);
          } while(row < 0 || row > ROWS-1);

          do
          {
            printf("Enter seat (0-%d): ", COLS-1);
            scanf("%d", &col);
          } while(col < 0 || col > COLS-1);
        }
        if(seats[row][col] == 1)
          printf("\nSorry, seat in row_%d and column_%d is
            reserved\n", row, col);
```

```c
    else
    {
      seats[row][col] = 1;
      cost += 6;
      rsvd_seats++;
    }
  break;
  case 2:
    if(rsvd_seats == 0)
    {
      printf("\nAll seats are free\n");
      break;
    }
    do
    {
      printf("\nEnter row (0-%d): ", ROWS-1);
      scanf("%d", &row);
    } while(row < 0 || row > ROWS-1);

    do
    {
      printf("Enter seat (0-%d): ", COLS-1);
      scanf("%d", &col);
    } while(col < 0 || col > COLS-1);

    if(seats[row][col] != 1)
      printf("\nSeat in row_%d and column_%d is not reserved\n",
        row, col);

    else
    {
      seats[row][col] = 0;
      cost -= 5;
      rsvd_seats--;
    }
  break;
  case 3:
    printf("\nFree seats: %d, Income: %d\n\n", ROWS*COLS - rsvd_
      seats, cost);
    for(i = 0; i < ROWS; i++)
    {
      for(j = 0; j < COLS; j++)
      {
        if(seats[i][j] == 1)
          printf("%2s", "X");
        else
          printf("%2s", "#");
      }
      printf("\n");
    }
  break;
  case 4:
  return 0;
```

```
        default:
          printf("\nWrong choice\n");
          break;
      }
    }
    return 0;
}
```

Unsolved Exercises

7.1   Write a program that reads the grades of 100 students and stores them in an array. Then, the program should read two float numbers (i.e., a and b) and display how many students got a grade in [a, b]. (*Note*: the first input number should be equal or less than the second.)

7.2   Write a program that reads **double** numbers continuously and stores in an array of 100 places those with a value more than 5. If the user enters –1, the insertion of numbers should terminate. The program should display the minimum of the values stored in the array.

7.3   Write a program that reads 100 integers and stores them in an array. The program should display how many elements have a value greater than the value of the last element and how many elements have a value greater than the average.

7.4   Write a program that reads integers and stores them in an array of 100 places with the restriction that an input number is stored in the array only if it is less than the last entered.

7.5   Write a program that reads 100 integers and stores them in an array. Then, the program should rotate the elements one place to the right. For example, if the array were 1, -9, 5, 3, the rotated array would be 3, 1, -9, 5.

7.6   Write a program that reads 100 **double** numbers and stores them in an array. The program should calculate the distance between successive elements and display the minimum one. To calculate the distance of two elements, subtract their values and use the absolute value. For example, if the first four elements are 5.2, -3.2, 7.5, 12.22, the distances are $|-3.2-5.2| = 8.4$, $|7.5-(-3.2)| = 10.7$ and $|12.22-7.5| = 4.72$.

7.7   Write a program that reads 100 integers and stores them in an array. The program should display the number of the duplicated values. For example, if the array were {5, 5, 5, 5, 5} the program should display 4 (since number 5 is repeated four times) and if it were {-2, 3, -2, 50, 3} the program should display 2 (since numbers -2 and 3 are repeated once) and if it were {3, -1, 22, 13, 7} the program should display 0 (since no number is repeated).

7.8   Write a program that assigns random values to a 3×5 integer array with the restriction that the sum of each column should be equal to 100.

7.9   Write a program that assigns random values to a 5×5 integer array with the restriction that the values of the elements under the main diagonal should be less than those of the main diagonal and the elements over the main diagonal should be greater than those of the main diagonal.

7.10 Write a program that reads integers and stores them into a 3x5 array. The program should display the columns whose elements have different values. For example, if the array were:

$$\begin{bmatrix} 1 & -2 & 2 & 5 & 9 \\ 3 & 0 & 2 & 5 & 1 \\ 1 & 7 & 2 & -3 & 0 \end{bmatrix}$$

the program should display the elements of the second and fifth column.

8

Pointers

Pointers are the most important, but also the most difficult, part of C. This chapter mainly uses pointers to arithmetic variables to introduce you to the pointer concepts. It also describes the close relationship between pointers and arrays. You'll also learn how to use arrays of pointers and pointers to functions. Other uses of pointers, such as pointers as arguments in functions, as well as pointers to other type of data, will be gradually presented over the next chapters.

Pointers and Memory

The computer's RAM (random access memory) consists of millions of successive storage cells, called *bytes*. Each byte stores eight bits of information and is identified by a unique number, called memory address. For example, in a computer with n bytes, the memory address of each byte is a unique number from 0 to n–1, as shown in Figure 8.1.

When a variable is declared, the compiler reserves the required consecutive bytes to store its value. If a variable occupies more than one byte, the variable's address is the address of the first byte.

For example, with the declaration **int** a = 10; the compiler reserves four consecutive bytes (e.g., 5000-5003 as shown in Figure 8.1) and stores the value 10 (assuming that the less significant byte of the value is stored in the lower address byte).

The compiler associates the name a with its memory address. When the program uses the variable a, the compiler accesses its address. For example, with the statement a = 200; the compiler knows that the memory address of a is 5000 and sets its content to 200.

Declaring Pointers

A pointer variable is a variable that can hold the memory address of another variable. When we store the address of a variable in a pointer variable, we say that it "points to" the variable. To declare a pointer, we write

```
data_type *pointer_name;
```

The **data _ type** defines the type of the variable that the pointer points to. The operator * defines that the variable is a pointer. For example, with the declaration

```
int *ptr;
```

	Memory address	Memory content
	0	
	1	
	2	
	.	
	.	
	.	
	5000	10
	5001	0
	5002	0
	5003	0
	:	
	:	
	$n-1$	

FIGURE 8.1
Memory layout.

`ptr` is declared as a pointer variable, which can store the memory address of an **int** variable. Similarly, with the declaration

```
double *ptr;
```

the memory address of a **double** variable can be stored in `ptr`.

Pointer variables can be declared together with other variables of the same type. For example,

```
int *ptr, i, j, k;
```

As a matter of style, we prefer to declare the pointer variables before the ordinary variables. When a pointer variable is declared, the compiler allocates memory to store its value. For example, the following program uses the **sizeof** operator to find out how many bytes the `ptr` allocates:

```
#include <stdio.h>
int main()
{
  int *ptr;

  printf("Bytes: %d\n", sizeof(ptr));
  return 0;
}
```

A pointer variable allocates the same size, no matter the data type it points to. This value is platform dependent; it's typically four bytes. Therefore, if we declare the `ptr` as **char** `*ptr;` or **double** `*ptr;` the output would be the same.

Pointer Initialization

To find the memory address of a variable, we put the address operator & before its name. For example,

```
#include <stdio.h>
int main()
{
  int *ptr, a;

  ptr = &a; /* ptr "points" to the memory address of a. */
  printf("Address = %p\n", ptr); /* Display the memory address of a. */
  return 0;
}
```

With the statement ptr = &a; ptr becomes equal, or else "points" to the memory address of a. The %p specifier displays the memory address in hex.

A pointer variable can be initialized when declared, provided that the variable that it points to has already been declared.

```
int a, b, arr[100], *ptr = &a;
```

As a matter of style, we prefer to initialize the pointer variables in separate statements, not together with its declarations.

NULL Value

To make a pointer to point to nowhere (null pointer), we set its value to NULL. The NULL value is defined in several C header files and it equals zero. For example, the following program first displays the initial value of ptr and then 0:

```
#include <stdio.h>
int main()
{
  int *ptr;

  printf("Val = %p\n", ptr);
  ptr = NULL;
  printf("Val = %p\n", ptr);
  return 0;
}
```

To compare a pointer value against NULL, we write

```
if(ptr != NULL) /* Equivalent to if(ptr) */
if(ptr == NULL) /* Equivalent to if(!ptr) */
```

Use a Pointer

To access the content of a memory address referenced by a pointer variable, we use the indirection (dereferencing) * operator before its name. For example,

```c
#include <stdio.h>
int main()
{
  int *ptr, a;

  a = 10;
  ptr = &a;
  printf("Val = %d\n", *ptr); /* Display the content of the memory
    address that ptr points to. */
  return 0;
}
```

The value of *ptr is equal to the value of the variable that ptr points to. Since ptr points to the memory address of a, *ptr is equal to a. Therefore, the program displays 10.

A pointer variable must point to a valid memory address before being used within the program.

The following program may crash because ptr doesn't point to the address of a program variable before used:

```c
#include <stdio.h>
int main()
{
  int *ptr, a;

  a = *ptr; /* ptr does not point to a valid memory address. The program
    may crash. */
  printf("Val = %d\n", a);
  return 0;
}
```

Usually, in a Unix/Linux environment, this type of error is indicated with a "Segmentation fault" message.

The following program is normally executed because ptr points to the memory address of an existing variable before used in the statement i = *ptr; Since ptr points to the address of j, *ptr is equal to j, that is, 20. Therefore, with the statement i = *ptr; i becomes 20 and the program displays Val = 20.

```c
#include <stdio.h>
int main()
{
  int *ptr, i, j;

  j = 20;
  ptr = &j;

  i = *ptr;
  printf("Val = %d\n", i);
  return 0;
}
```

The * and & operators cancel each other when used together. For example, the following program displays three times the address of i:

```
#include <stdio.h>
int main()
{
  int *ptr, i;

  ptr = &i;
  printf("%p %p %p\n", &i, *&ptr, &*ptr);
  return 0;
}
```

Exercises

8.1 What is the output of the following program?

```
#include <stdio.h>
int main()
{
  int *ptr, i = 10;

  ptr = &i;
  i += 20;
  printf("Val = %d\n", *ptr);
  return 0;
}
```

Answer: Since ptr points to the address of i, *ptr is equal to i. With the statement i += 20; i becomes 30 and the program displays Val = 30.

8.2 Write a program that reads two integers, stores them in two variables, declares two pointers to them, and displays the memory addresses of both variables, the content of both pointers, as well as their memory addresses.

```
#include <stdio.h>
int main()
{
  int *ptr1, *ptr2, i, j;

  printf("Enter numbers: ");
  scanf("%d%d", &i, &j);

  ptr1 = &i;
  ptr2 = &j;

  printf("Num1 address = %p\n", ptr1);
  printf("Num2 address = %p\n", ptr2);

  printf("Ptr1 content = %d\n", *ptr1);
  printf("Ptr2 content = %d\n", *ptr2);

  printf("Ptr1 address = %p\n", &ptr1);
  printf("Ptr2 address = %p\n", &ptr2);
  return 0;
}
```

Comments: As with ordinary variables, we use the & operator to find the memory addresses of the pointer variables ptr1 and ptr2.

8.3 The following program uses a pointer to read and display a decimal number. Is there any programming bug?

```
#include <stdio.h>
int main()
{
  double *ptr, i;

  scanf("%lf", ptr);
  printf("Val = %f\n", *ptr);
  return 0;
}
```

Answer: The program won't work because ptr does not point to the address of i before used in scanf(). Had we added the statement ptr = &i; before scanf(), the program would be normally executed.

8.4 What is the output of the following program?

```
#include <stdio.h>
int main()
{
  int *ptr, i = 0;

  for(ptr = &i; i < 3; i++)
    printf("%d ", *ptr);
  return 0;
}
```

Answer: Since ptr points to the address of i, *ptr is equal to i. Therefore, in each loop iteration, the program displays the current value of i, that is, 0 1 2.

8.5 Write a program that uses a pointer to read a float number and display its absolute value.

```
#include <stdio.h>
int main()
{
  double *p, val;

  p = &val;
  printf("Enter number: ");
  scanf("%lf", p);

  if(*p >= 0)
    printf("%f\n", *p);
  else
    printf("%f\n", -*p);
  return 0;
}
```

8.6 What is the output of the following program?

```
#include <stdio.h>
int main()
{
  int i = 0, *ptr = &i;
```

```
    *ptr = *ptr ? 10 : 20;
    printf("Val = %d\n", i);
    return 0;
}
```

Answer: Since ptr points to the address of i, the expression:
*ptr = *ptr ? 10 : 20; is equivalent to i = i ? 10 : 20;
    Since i is 0 (false), the value of the expression is 20. Therefore, i becomes 20 and the program displays Val = 20.

8.7 What is the output of the following program?

```
#include <stdio.h>
int main()
{
    int *ptr1, *ptr2, *ptr3, i = 10, j = 20, k = 30;

    ptr1 = &i;
    ptr2 = &j;
    ptr3 = &k;

    *ptr1 = *ptr2 = *ptr3;
    k = i+j;

    printf("%d\n", *ptr3);
    return 0;
}
```

Answer: Since ptr1 points to the address of i, *ptr1 is equal to i. Similarly, *ptr2 is equal to j and *ptr3 is equal to k. Therefore, the statement *ptr1 = *ptr2 = *ptr3 is equivalent to i = j = k. Then, k becomes k = i+j = 30+30 = 60, and the program displays 60.

8.8 Write a program that uses two pointers to read the prices of two products and display the largest price.

```
#include <stdio.h>
int main()
{
    float *ptr1, *ptr2, i, j;

    /* The pointers should point to the addresses of the variables
       before calling scanf(). */
    ptr1 = &i;
    ptr2 = &j;

    printf("Enter prices: ");
    scanf("%f%f", ptr1, ptr2);
    if(*ptr1 > *ptr2)
        printf("%f\n", *ptr1);
    else
        printf("%f\n", *ptr2);
    return 0;
}
```

8.9   What is the output of the following program?

```
#include <stdio.h>
int main()
{
   int *ptr, i = 10;

   ptr = &i;
   (*ptr)++;
   ++*ptr;

   printf("Val = %d\n", i);
   return 0;
}
```

Answer: Since `ptr` points to the address of `i`, `*ptr` is equal to `i`. Therefore, the statement `(*ptr)++;` is equivalent to `i++;` and `i` becomes `11`.

The parentheses in the expression `(*ptr)++;` are necessary, because the postfix use of the `++` operator takes precedence over the `*` operator. As such, the expressions `(*ptr)++;` and `*ptr++;` operate differently.

Similarly, the statement `++*ptr;` is equivalent to `++i;` `i` is increased once more and the program displays `Val = 12`.

8.10   Write a program that uses two pointers to read two integers and swap their values.

```
#include <stdio.h>
int main()
{
   int *ptr1, *ptr2, i, j, tmp;

   ptr1 = &i;
   ptr2 = &j;

   printf("Enter numbers: ");
   scanf("%d%d", ptr1, ptr2);

   tmp = *ptr2;
   *ptr2 = *ptr1;
   *ptr1 = tmp;

   printf("Values: %d %d\n", i, j);
   return 0;
}
```

8.11   What is the output of the following program?

```
#include <stdio.h>
int main()
{
   int *ptr1, i = 10;
   double *ptr2, j = 1.234;

   ptr1 = &i;
   ptr2 = &j;
```

```
    *ptr1 = *ptr2;
    printf("%d %d %d\n", i, sizeof(ptr1), sizeof(ptr2));
    return 0;
}
```

Answer: Since ptr1 points to the address of i, *ptr1 is equal to i. Similarly, *ptr2 is equal to j. Therefore, the statement *ptr1 = *ptr2; is equivalent to i = j. Since ptr1 is a pointer to **int** variable, only the integer part of j is assigned to i. Therefore, i becomes 1.

As said, the pointer variables allocate the same size (usually, it is 4 bytes) no matter what they point to.

8.12 Use the pointer p and a **while** loop and complete the following program to display the integers from 1 to 10.

```
#include <stdio.h>
int main()
{
    int *p, i;
    ...
}
```

Answer:

```
#include <stdio.h>
int main()
{
    int *p, i;

    p = &i;
    *p = 1;
    while(*p <= 10)
    {
        printf("%d\n", *p);
        (*p)++;
    }
    return 0;
}
```

8.13 What is the output of the following program?

```
#include <stdio.h>
int main()
{
    int *ptr1, *ptr2, *ptr3, i = 10, j = 20, k = 30;

    ptr1 = &i;
    i = 100;

    ptr2 = &j;
    j = *ptr2 + *ptr1;

    ptr3 = &k;
    k = *ptr3 + *ptr2;
```

```
    printf("%d %d %d\n", *ptr1, *ptr2, *ptr3);
    return 0;
}
```

Answer: Since ptr1 points to the address of i, *ptr1 is equal to i. Similarly, *ptr2 is equal to j. Therefore, j = *ptr2 + *ptr1 = 20+100 = 120.

Since ptr3 points to the address of k, *ptr3 is equal to k. Therefore, k = *ptr3 + *ptr2 = 30+120 = 150.

Since the values of *ptr1, *ptr2, and *ptr3 are equal to i, j, and k, respectively, the program displays 100 120 150.

8.14 Use the pointer p2 and complete the following program to read the students' grades continuously until the user enters −1. Use the pointer p1 to display how many students got a grade within [5,10] and the pointer p3 to display the maximum grade.

```
#include <stdio.h>
int main()
{
    int *p1, sum;
    float *p2, *p3, grade, max;
    ...
}
```

Answer:

```
#include <stdio.h>
int main()
{
    int *p1, sum;
    float *p2, *p3, grade, max;

    p1 = &sum;
    *p1 = 0;

    p3 = &max;
    *p3 = 0;

    p2 = &grade;
    while(1)
    {
        printf("Enter grade: ");
        scanf("%f", p2);

        if(*p2 == -1)
            break;
        if(*p2 >=5 && *p2 <= 10)
        {
            (*p1)++;
            if(*p2 > *p3)
                *p3 = *p2;
        }
    }
    printf("%d students passed (max = %.2f)\n", *p1, *p3);
    return 0;
}
```

void* Pointer

A pointer variable of type **void*** is a generic pointer, in the sense that it can point to a variable of any data type. To access the content of a memory address pointed by a **void*** pointer, typecast is necessary. For example,

```c
#include <stdio.h>
int main()
{
  void *ptr;
  int i = 10;

  ptr = &i;
  *(int*)ptr += 20;
  printf("%d\n", i);
  return 0;
}
```

To access the value of `i`, we typecast `ptr` to the type of `i` and the program displays 30.

Use of const Keyword

To prohibit a pointer from changing the value of the variable it points to, add the **const** keyword before its type. For example, the following code won't compile because `ptr` isn't allowed to change the value of `i`. However, it is allowed to point to some other variable.

```c
int j, i = 10;
const int *ptr;
ptr = &i;
*ptr = 30; /* Not allowed action. */
ptr = &j; /* Allowed action. */
```

To prohibit a pointer from pointing to another variable, add the **const** keyword before its name. You have to initialize the pointer with an address when it is declared. For example, the following code won't compile because `ptr` isn't allowed to point to the address of `j`. However, it is allowed to change the value of `i`.

```c
int i, j;
int* const ptr = &i;
ptr = &j; /* Not allowed action. */
*ptr = 30; /* Allowed action. */
```

Pointer Arithmetic

Pointer arithmetic refers to the application of some arithmetic operations on pointers. The operators that can be used in pointer arithmetic are ++, --, +, and –, while the allowed operations are adding an integer to pointer, subtracting an integer from a pointer and sub-traction of pointers, which point to the same type of data.

Pointers and Integers

The addition of an integer n to a pointer variable in a statement like

```
ptr = ptr + n;
```

increases its value by n × size of the pointer's data type. For example, if ptr is declared as a pointer to

- **char**: ptr is increased by n since the size of **char** is one byte
- **int** or **float**: ptr is increased by n×4 since the size of both **int** and **float** is four bytes
- **double**: ptr is increased by n×8 since the size of **double** is eight bytes

In the following program, the statement ptr++; increases its value by four because it is declared as a pointer to **int**. In fact, this program displays two addresses and the second one is four bytes higher than the first one.

```
#include <stdio.h>
int main()
{
  int *ptr, i;

  ptr = &i;
  printf("Address = %p\n", ptr);
  ptr++;
  printf("Address = %p\n", ptr);
  return 0;
}
```

Similar to the addition, subtracting an integer n from a pointer variable decreases its value by n × size of the pointer's data type. For example, had we written ptr -= 10; instead of ptr++; the second address would be 40 bytes less.

Subtracting Pointers

The subtraction of two pointers is allowed only if both point to the same data type. The result of their subtraction is an integer number, which indicates the number of data items between them.

For example, suppose that the ptr1 and ptr2 are pointers to two integer variables stored in the addresses 1000 and 1040, respectively. The result of ptr2-ptr1 is not equal to 40 (i.e., 1040–1000), but equal to (1040–1000)/**sizeof**(**int**) = 40/4 = 10. This number indicates the number of integers between the two pointers.

If the type of both pointers and variables was **char**, the result of ptr2-ptr1 would be 40 because the size of the **char** type is one byte.

Comparing Pointers

The comparison of two pointers makes sense only if both point to members of the same data structure (such as an array). The operators ==, !=, >, <, >= and <= can be used to compare the pointers. For example, to check if ptr1 and ptr2 point to the same address or not, we can write **if**(ptr1 == ptr2) or **if**(ptr1 != ptr2), respectively.

Besides subtracting and adding integers to a pointer, comparing and subtracting pointers of the same type, no other arithmetic operation is allowed.

For example, the statements ptr *= 2, ptr += 7.5, ptr1+ptr2; are not allowed.

Exercises

8.15 What is the output of the following program?

```c
#include <stdio.h>
int main()
{
    int *ptr, i = 10, j = 20;

    ptr = &j;
    ptr++;
    printf("Val = %d\n", *ptr);
    return 0;
}
```

Answer: The program displays the value stored four bytes after the address of j. If it happens to be the address of i, the program would display 10, otherwise a random value.

8.16 Write a program that uses three pointers to read the grades of a student in three exercises. If all grades are greater or equal to 5, the program should display them in ascending order. Otherwise, the program should display their average.

```c
#include <stdio.h>
int main()
{
    float *ptr1, *ptr2, *ptr3, i, j, k;

    ptr1 = &i;
    ptr2 = &j;
    ptr3 = &k;

    printf("Enter grades: ");
    scanf("%f%f%f", ptr1, ptr2, ptr3);

    if((*ptr1 >= 5) && (*ptr2 >= 5) && (*ptr3 >= 5))
    {
        if(*ptr1 <= *ptr2 && *ptr1 <= *ptr3)
        {
```

```
    printf("%f ", *ptr1);
    if(*ptr2 < *ptr3)
      printf("%f %f\n", *ptr2, *ptr3);
    else
      printf("%f %f\n", *ptr3, *ptr2);
  }
  else if(*ptr2 <= *ptr1 && *ptr2 <= *ptr3)
  {
    printf("%f ", *ptr2);
    if(*ptr1 < *ptr3)
      printf("%f %f\n", *ptr1, *ptr3);
    else
      printf("%f %f\n", *ptr3, *ptr1);
  }
  else
  {
    printf("%f ", *ptr3);
    if(*ptr2 < *ptr1)
      printf("%f %f\n", *ptr2, *ptr1);
    else
      printf("%f %f\n", *ptr1, *ptr2);
  }
}
else
  printf("Avg = %.2f\n", (*ptr1 + *ptr2 + *ptr3)/3);
return 0;
}
```

8.17   What is the output of the following program?

```
#include <stdio.h>
int main()
{
  int *ptr, i = 10, j = 20, k = 30;

  ptr = &i;
  *ptr = 40;

  ptr = &j;
  *ptr += i;

  ptr = &k;
  *ptr += i + j ;

  printf("i = %d j = %d k = %d\n", i, j, k);
  return 0;
}
```

Answer: Since ptr points to the address of i, the statement *ptr = 40; is equivalent to i = 40; With the statement ptr = &j; ptr points to the address of j, so *ptr is equal to j. Therefore, the statement *ptr += i; is equivalent to j += i; that is, j = 20+40 = 60.

  With the statement ptr = &k; ptr points to the address of k, so *ptr is equal to k. Therefore, the statement *ptr += i+j; is equivalent to k += i+j, that is, k = k+i+j = 30+40+60 = 130.

  As a result, the program displays i = 40 j = 60 k = 130.

8.18 Use the pointers p1 and p2 and complete the following program to display the product of even numbers from 10 up to 20.

```
#include <stdio.h>
int main()
{
  int *p1, *p2, i, mul;
  ...
}
```

Answer:

```
#include <stdio.h>
int main()
{
  int *p1, *p2, i, mul;

  p1 = &i;
  p2 = &mul;
  for(*p1 = 10, *p2 = 1; *p1 <= 20; (*p1)+=2)
    *p2 = *p2 * *p1;

  printf("Mul = %d\n", *p2);
  return 0;
}
```

8.19 What is the output of the following program?

```
#include <stdio.h>
int main()
{
  int *ptr1, *ptr2, i = 10, j = 20;

  ptr1 = &i;
  *ptr1 = 150;

  ptr2 = &j;
  *ptr2 = 50;

  ptr2 = ptr1;
  *ptr2 = 250;

  ptr2 = &j;
  *ptr2 += *ptr1;

  printf("Val = %d\n", j);
  return 0;
}
```

Answer: Since ptr1 points to the address of i, the statement *ptr1 = 150; is equivalent to i = 150; Similarly, the statement *ptr2 = 50; is equivalent to j = 50;
    With the statement ptr2 = ptr1; ptr2 points to the same address that ptr1 points to, that is, the address of i. Therefore, the statement *ptr2 = 250; is equivalent to i = 250.
    With the statement ptr2 = &j; ptr2 points to the address of j, so *ptr2 is equal to j. Since ptr1 still points to the address of i, *ptr1 is equal to i.

The statement `*ptr2 += *ptr1;` is equivalent to `j += i`, that is, `j = j+i = 50+250 = 300`.

Therefore, the program displays `Val = 300`.

8.20 Use the pointers p1, p2, and p3 and complete the following program to read two integers and display the sum of the integers between them. For example, if the user enters 6 and 10, the program should display 24 (7+8+9). The program should force the user to enter numbers less than 100 and the first integer should be less than the second.

```
#include <stdio.h>
int main()
{
    int *p1, *p2, *p3, i, j, sum;
    ...
}
```

Answer:

```
#include <stdio.h>
int main()
{
    int *p1, *p2, *p3, i, j, sum;

    p1 = &i;
    p2 = &j;
    p3 = &sum;
    *p3 = 0;
    do
    {
        printf("Enter two numbers (a < b < 100): ");
        scanf("%d%d", p1, p2);
    } while(*p1 >= *p2 || *p2 > 100);

    (*p1)++;
    while(*p1 < *p2)
    {
        *p3 += *p1;
        (*p1)++;
    }
    printf("Sum = %d\n", *p3);
    return 0;
}
```

Pointers and Arrays

The elements of an array are stored in successive memory locations, with the first element stored at the lowest memory address. The array's type defines the distance of its elements.

In a character array (**char**), the memory address of each element is a byte higher from the previous one, while in an integer array (**int**) it is four bytes higher.

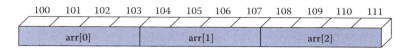

FIGURE 8.2
Array elements.

For example, suppose that with the declaration **int** arr[3]; the value of the first element is stored in the address bytes 100–103. Then, the value of the second element is stored in 104–107 and the value of the third one in 108–111, as shown in Figure 8.2.

The close relationship between pointers and arrays is based on the fact that the name of an array can be used as a pointer to its first element.

Similarly, arr+1 can be used as a pointer to the second element, arr+2 as a pointer to the third one, and so on. In general, the following expressions are equivalent:

```
arr == &arr[0]
arr + 1 == &arr[1]
arr + 2 == &arr[2]
...
arr + n == &arr[n]
```

Since the name of an array can be used as a pointer to its first element, its content is equal to the value of its first element. Therefore, *arr is equal to arr[0]. Similarly, since arr+1 is a pointer to the second element, *(arr+1) is equal to arr[1], and so on. In general, the following expressions are equivalent:

```
*arr == arr[0]
*(arr + 1) == arr[1]
*(arr + 2) == arr[2]
...
*(arr + n) == arr[n]
```

The parentheses are necessary because the * operator has higher precedence than the addition operator. Therefore, the expressions *(arr+n) and *arr+n operate differently. For example, consider the following program:

```c
#include <stdio.h>
int main()
{
  int *ptr, arr[5] = {10, 20, 30, 40, 50};

  ptr = arr;
  printf("Val1 = %d Val2 = %d\n", *ptr+2, *(ptr+2));
  return 0;
}
```

With the statement ptr = arr; ptr points to the address of arr[0], so *ptr is equal to arr[0], that is, 10. Since the * operator has higher precedence than the + operator, *ptr+2 = 10+2 = 12.

The expression `*(ptr+2)` is equal to the content of the address that `ptr` points to, increased by two integers' positions. Therefore, `*(ptr+2)` is equal to `arr[2]`.

The program displays `Val1 = 12 Val2 = 30`.

The following program uses array subscripting and pointer arithmetic to display the addresses and the values of all array elements.

```c
#include <stdio.h>
int main()
{
  int i, arr[5] = {10, 20, 30, 40, 50};

  printf("***** Using array index *****\n");
  for(i = 0; i < 5; i++)
    printf("Addr = %p Val = %d\n", &arr[i], arr[i]);

  printf("\n***** Using pointer arithmetic *****\n");
  for(i = 0; i < 5; i++)
    printf("Addr = %p Val = %d\n", arr+i, *(arr+i));
  return 0;
}
```

Our preference for array processing is to use array subscripting instead of pointer arithmetic to get a clearer code. For example, since the lack of parentheses introduces a bug, it is easier and safer to write `arr[i]` instead of `*(arr+i)`.

Therefore, even if the array processing using pointers might run a bit faster, the gain with today's fast processors and optimized compilers would be so small that it is not worth to make the code harder to read as well as prone to errors.

The following program uses another pointer variable to display the values and the addresses of all array elements:

```c
#include <stdio.h>
int main()
{
  int *ptr, i, arr[5] = {10, 20, 30, 40, 50};

  ptr = arr;
  for(i = 0; i < 5; i++)
  {
    printf("Addr = %p Val = %d\n", ptr, *ptr);
    ptr++; /* ptr becomes equal to the memory address of the next array
      element. Equivalently, we could write ptr = &arr[i]; */
  }
  return 0;
}
```

With the statement `ptr = arr;` `ptr` points to the first element, while in each loop iteration the statement `ptr++;` makes `ptr` to point to the next element.

*When the name of an array is used as a pointer, C treats it as a **const** pointer. Therefore, it is not allowed to change its value and make it point to some other variable.*

Therefore, had we written `arr++;` instead of `ptr++;` the compiler would raise an error message. However, we may copy its value in another pointer variable, as we did with the statement `ptr = arr;` and then use this variable to process the array elements.

Tip

Although a pointer variable isn't an array, it can be indexed like an array.

For example, the following program uses the pointer variable `ptr` like an array to display the values and the addresses of all array elements:

```c
#include <stdio.h>
int main()
{
  int *ptr, i, arr[5] = {10, 20, 30, 40, 50};

  ptr = arr;
  for(i = 0; i < 5; i++)
    printf("Addr = %p Val = %d\n", &ptr[i], ptr[i]); /* Using ptr as an
      array. */
  return 0;
}
```

Exercises

8.21 What is the output of the following program?

```c
#include <stdio.h>
int main()
{
   int *ptr, arr[] = {10, 20, 30, 40, 50};

   ptr = arr;
   *ptr = 3;

   ptr += 2;
   *ptr = 5;
   printf("Val = %d\n", arr[0]+arr[2]);
   return 0;
}
```

Answer: With the statement `ptr = arr;` `ptr` points to the address of `arr[0]`, so `*ptr` is equal to `arr[0]`. Therefore, the statement `*ptr = 3;` is equivalent to `arr[0] = 3`.
   The statement `ptr += 2;` makes `ptr` equal to the address of `arr[2]`, so `*ptr` is equal to `arr[2]`. Therefore, the statement `*ptr = 5;` is equivalent to `arr[2] = 5;`
   As a result, the program displays `Val = 8`.

8.22 What is the output of the following program?

```c
#include <stdio.h>
int main()
{
   int i = 10, *ptr = &i;

   ptr[0] = 50;
   printf("%d\n", i);
   return 0;
}
```

Answer: Since ptr points to the address of i, we can use it as an array of one element. Therefore, the statement ptr[0] = 50; is equivalent to i = 50; and the program displays 50.

What would have happened if we had written ptr[1] = 50; instead of ptr[0] = 50;?

Since ptr can be used as an array of one element, the statement ptr[1] = 50; attempts to change the value of an out-of-bound memory, which may cause a program crash.

8.23 Write a program that reads the grades of 10 students, stores them in an array, and displays the maximum and the minimum grade and their positions in the array. The program should check if the input grades are within [0, 10]. Use pointer arithmetic to process the array.

```c
#include <stdio.h>

#define SIZE 10

int main()
{
  int i, max_pos, min_pos;
  float max, min, arr[SIZE];

  max = 0;
  min = 10;
  max_pos = min_pos = 0;
  for(i = 0; i < SIZE; i++)
  {
    do
    {
      printf("Enter grade: ");
      scanf("%f", arr+i);
    } while(*(arr+i) > 10 || *(arr+i) < 0); /* Check if the grade is
      within [0,10]. */
    if(*(arr+i) > max)
    {
      max = *(arr+i);
      max_pos = i;
    }
    if(*(arr+i) < min)
    {
      min = *(arr+i);
      min_pos = i;
    }
  }
  printf("Max grade is %.2f in pos #%d\n", max, max_pos);
  printf("Min grade is %.2f in pos #%d\n", min, min_pos);
  return 0;
}
```

8.24 What is the output of the following program?

```c
#include <stdio.h>
int main()
{
  int *ptr1, *ptr2, arr[] = {10, 20, 30, 40, 50};
```

```
    ptr1 = &arr[0];
    ptr2 = &arr[3];
    printf("%d\n", ptr1[ptr2 - ptr1]);
    return 0;
}
```

Answer: The subtraction of two pointers that point to the same array calculates the number of the elements between them. Therefore, the result of `ptr2-ptr1` is equal to the difference of their subscripts, that is, `3-0 = 3`.

Since `ptr1` points to the first element, `ptr1[3]` is equal to `arr[3]` and the program displays 40.

8.25 Write a program that reads the daily temperatures of January and stores them in an array. Next, the program should read a number and display the first day number with a temperature less than this. Use pointer arithmetic to process the array.

```c
#include <stdio.h>

#define SIZE 31
int main()
{
  int i;
  double temp, arr[SIZE];
  for(i = 0; i < SIZE; i++)
  {
    printf("Enter temperatures: ");
    scanf("%lf", arr+i);
  }
  printf("Enter base temperature: ");
  scanf("%lf", &temp);
  for(i = 0; i < SIZE; i++)
  { /* The braces could be omitted. We put them to make the code
    more readable. */
    if(*(arr+i) < temp)
      break;
  }
  if(i == SIZE)
    printf("No temperature less than %.1f\n", temp);
  else
    printf("The first temperature less than %.1f was %.1f in day
      %d\n", temp, *(arr+i), i+1);
  return 0;
}
```

8.26 What is the output of the following program?

```c
#include <stdio.h>
int main()
{
  int *ptr, i, arr[5] = {10, 20, 30, 40, 50};
  ptr = arr+2;
  for(i = 0; i < 5; i++)
    printf("%d ", ptr[i]);
  return 0;
}
```

Answer: With the statement ptr = arr+2; ptr becomes equal to the address of arr[2]. Since we use ptr as an array, ptr[0] corresponds to arr[2], ptr[1] to arr[3], and ptr[2] to arr[4], respectively.

Since the array arr has five elements, what would be the values of ptr[3] and ptr[4]?

Their values are the random values that exist in the two memory blocks (four bytes each), following the address of arr[4].

Therefore, the program displays 30 40 50 and two random values.

8.27 Use the pointer ptr and complete the following program to read and store the grades of 50 students in the array arr and display the array's values in reverse order. Use pointer arithmetic to process the array.

```c
#include <stdio.h>

#define SIZE 50

int main()
{
  float *ptr, arr[SIZE];
  ...
}
```

Answer:

```c
#include <stdio.h>
#define SIZE 50

int main()
{
  float *ptr, arr[SIZE];

  ptr = arr;
  while((ptr - arr) != SIZE)
  {
    printf("Enter grade: ");
    scanf("%f", ptr);
    ptr++;
  }
  ptr--;
  while(ptr >= arr)
  {
    printf("%f\n", *ptr);
    ptr--;
  }
  return 0;
}
```

8.28 What is the output of the following program?

```c
#include <stdio.h>
int main()
{
  int *ptr1, *ptr2, i = 10, j = 20;

  ptr1 = &i;
  ptr2 = &j;
```

```
    ptr2 = ptr1;
    *ptr1 = *ptr1 + *ptr2;
    *ptr2 = 2*(*ptr2);
    printf("Val = %d\n", *ptr1 + *ptr2);
    return 0;
}
```

Answer: With the statement ptr2 = ptr1; ptr2 points to the same address with ptr1, that is, the address of i. Therefore, *ptr2 is equal to i.

Since both pointers point to the address of i, the statement *ptr1 = *ptr1 + *ptr2; is equivalent to i = i+i = 10+10 = 20.

Similarly, the statement *ptr2 = 2*(*ptr2); is equivalent to i = 2*i = 2*20 = 40. The program displays the value of the expression *ptr1 + *ptr2; that is, i+i = 40+40 = 80.

8.29 Write a program that declares two arrays of 10 integers each and assigns them random values. The program should use two pointers to copy the values of the second array to the first one and display their elements. Use pointer arithmetic to process the arrays.

```
#include <stdio.h>
#include <stdlib.h>
#include <time.h>

#define SIZE 10

int main()
{
  int *ptr1, *ptr2, i, arr1[SIZE], arr2[SIZE];

  srand((unsigned)time(NULL));
  for(i = 0; i < SIZE; i++)
  {
    *(arr1+i) = rand();
    *(arr2+i) = rand();
  }
  ptr1 = arr1;
  ptr2 = arr2;
  for(i = 0; i < SIZE; i++)
  {
    *ptr1 = *ptr2; /* Equivalent to arr1[i] = arr2[i]; */
    ptr1++;
    ptr2++;
  }
  ptr1 = arr1;
  ptr2 = arr2;
  for(i = 0; i < SIZE; i++)
    printf("arr1[%d] = %d, arr2[%d] = %d\n", i, *(ptr1+ i), i,
      *(ptr2+i));
  return 0;
}
```

Comments: The srand() and rand() functions are used for the generation of random numbers.

8.30 Does the following program contain any programming bug?

```c
#include <stdio.h>
int main()
{
  int i, arr[5] = {10, 20, 30, 40, 50};

  for(i = 0; i < 5; i++)
    printf("%d\n", i[arr]);

  printf("%d\n", 2[arr]-3[arr]);
  return 0;
}
```

Answer: Normally, you should answer that the expression i[arr] is wrong because arr and not i is declared as an array. For the same reason, the expressions 2[arr] and 3[arr] seem wrong too.

However, the compiler translates the expression i[arr] to *(i+arr), which is the same as *(arr+i), so equivalent to arr[i]. For example, 2[arr] is equivalent to arr[2].

Therefore, the program is compiled successfully and displays the values of the array's elements, as well as the difference of arr[2] and arr[3], that is, –10.

Obviously, we don't recommend using ever this reverse syntax; we just used this weird syntax to show you another example of the close relationship between arrays and pointers.

8.31 What is the output of the following program?

```c
#include <stdio.h>
int main()
{
  char *ptr, arr[4] = {6, 7, 8, 9};
  int i;

  ptr = arr;
  i = *(int*)ptr;
  printf("Val = %d\n", i);
  return 0;
}
```

Answer: With the typecast expression (int*)ptr we can temporarily handle ptr as a pointer to an int variable and get its value. Since the array arr reserves four bytes, the expression i = *(int*)ptr assigns to i the values of the array's elements. Their values in binary are

```
6 = 00000110
7 = 00000111
8 = 00001000
9 = 00001001
```

Therefore, i will be
    00001001000010000000011100000110 = 0x09080706 (hex) = 151521030
(decimal).

Notice that we could not use `ptr` and write `i = *(int*)arr` or `i = *(int*) &arr[0]`.

8.32 Complete the following program by using the variable `num` to read the codes of 100 products and the pointers `ptr1` and `ptr2` to store them in the array `arr`. The program should store a code in the array only if it is not already stored. Before it terminates, the program should display the products' codes. Use pointer arithmetic to process the array.

```
#include <stdio.h>

#define SIZE 100

int main()
{
  int *ptr1, *ptr2, num, arr[SIZE];

  ...
}
```

Answer:

```
#include <stdio.h>

#define SIZE 100

int main()
{
  int *ptr1, *ptr2, num, arr[SIZE];

  ptr1 = ptr2 = arr; /* ptr2 points to the position of the array
    where the last code is stored. */
  while((ptr2 - arr) != SIZE)
  {
    printf("Enter code: ");
    scanf("%d", &num);

    ptr1 = arr;
    while(ptr1 != ptr2) /* Starting from the beginning, we check if
      the input code is already stored. */
    {
      if(*ptr1 == num)
      {
        printf("Error: Code %d exists\n", *ptr1);
        break;
      }
      ptr1++;
    }
    /* If the code is not stored, it's stored and the pointer is
      increased. */
    if(*ptr1 != num)
    {
      *ptr2 = num;
      ptr2++;
    }
  }
  /* Display the codes. */
  ptr1 = arr;
```

```
    while(ptr1 < ptr2)
    {
      printf("C: %d\n", *ptr1);
      ptr1++;
    }
    return 0;
}
```

8.33 What are the values of `arr` elements in the following program?

```
#include <stdio.h>
int main()
{
  int *ptr, arr[5] = {20};

  for(ptr = arr+1; ptr <= arr+4; ptr++)
    *ptr = *(ptr-1) + *(ptr+1) + 1;
  return 0;
}
```

Answer: When the array `arr` is declared, `arr[0]` becomes 20 and the rest elements 0.

With the statement `ptr = arr+1;` `ptr` points to `arr[1]`. In each loop iteration, the statement `ptr++` makes it point to the next element. The **for** loop is executed until the `ptr` points to the address of the last element.

In each loop iteration, the value of the current element becomes equal to the value of the previous element, plus the value of the next element, plus one. For example, in the first iteration, the statement

`*ptr = *(ptr-1)+*(ptr+1)+1;` is equivalent to

`arr[1] = arr[0]+arr[2]+1 = 20+0+1 = 21;`

As a result, the values of `arr[0]`, ... `arr[3]` become equal from 20 to 23. What about the value of the last element?

`arr[4]` becomes equal to `arr[3]`, plus one, plus the random value that exists in the four-byte memory block following the address of `arr[4]`.

8.34 Use the pointer `ptr` and an iteration loop and complete the following program to decrease the elements of the array `arr` by one. Before it terminates, the program should display their sum. Use pointer arithmetic to process the array.

```
#include <stdio.h>
int main()
{
  int *ptr, sum, arr[5] = {11, 21, 31, 41, 51};

  ...
}
```

Answer:

```
#include <stdio.h>
int main()
```

```
{
  int *ptr, sum, arr[5] = {11, 21, 31, 41, 51};

  sum = 0;
  for(ptr = arr; ptr <= arr+4; ptr++)
  {
    --*ptr;
    sum += *ptr;
  }
  printf("Sum = %d\n", sum);
  return 0;
}
```

Comments: Let's analyze the first iteration. With the statement ptr = arr; ptr points to arr[0]. Therefore, *ptr is equal to arr[0], that is, 11. The statement --*ptr; decreases its value and arr[0] becomes 10. This value is added to sum.

Similarly, the next iterations decrease the rest elements by one and their values become 20, 30, 40, and 50, respectively. When the loop ends, sum will be equal to the sum of the array's elements, that is, 150.

8.35   What would be the values of the arr elements in the previous program, if we write --*ptr++ instead of --*ptr?

Answer: The statement --*ptr++; first decreases the content of the address that ptr points to by one and then increases ptr by one. Let's trace the iterations.

First iteration (ptr = arr). As explained, arr[0] becomes 10. Then, ptr is increased by one and points to arr[1].

Second iteration. With the statement ptr++; ptr points to arr[2]. Then, arr[2] becomes 30 and ptr is increased by one and points to arr[3].

Third iteration. With the statement ptr++; ptr points to arr[4]. Then, arr[4] becomes 50 and the next increase of ptr terminates the loop.

Therefore, the arr elements become 10, 21, 30, 41, 50.

8.36   What is the output of the following program?

```
#include <stdio.h>
int main()
{
  int a[] = {0, 0, 1, 2, 3}, b[] = {0, 0, 4, 5, 6};
  int *ptr1 = a, *ptr2 = b;

  while(!*ptr1++ && !*ptr2++);

  printf("%d %d\n", ptr1-a, ptr2-b);
  return 0;
}
```

Answer: This is a really tough one. Let's trace the iterations.

First iteration. Notice that in !*ptr1++; the ! operator is applied first and then ptr1 is increased. Since ptr1 points to a, *ptr1 is equal to a[0], that is, 0. The ! operator makes it 1. Then, ptr1 is increased and points to the next element. Similarly, the value of !*ptr2++; is 1. Since both terms are true, the loop continues.

Second iteration. Like before, the values of !*ptr1++; and !*ptr2++; are 1.

Third iteration. Since ptr1 points to a[2], the value of !*ptr1++; is 0 and ptr1 points to a[3]. Recall from Chapter 4 that if an operand is false the rest operands are not checked and the value of the expression becomes 0. Therefore, the loop terminates. Since the term !*ptr2++; is not checked, ptr2 is not increased.

As a result, since ptr1 points to a[3] and ptr2 points to a[2], the program displays 3 2.

Arrays of Pointers

An array of pointers is an array whose elements are pointers to the same data type. When declared, an * must prefix its name. For example, the statement

```
int *arr[3];
```

declares an array of three pointers to integers.

When you declare an array of pointers, don't enclose its name in parentheses.

For example, with the statement

```
int (*arr)[3];
```

the variable arr is declared as a pointer to an array of three integers and not as an array of three pointers.

The elements of an array of pointers are treated as the ordinary pointers. For example,

```
#include <stdio.h>
int main()
{
  int *arr[3], i = 100, j = 200, k = 300;

  arr[0] = &i;
  arr[1] = &j;
  arr[2] = &k;

  printf("%d %d %d\n", *arr[0], *arr[1], *arr[2]);
  return 0;
}
```

With the statement arr[0] = &i; arr[0] points to the address of i, therefore *arr[0] is equal to i. Similarly, *arr[1] is equal to j and *arr[2] is equal to k.

Therefore, the program displays 100 200 300.

Exercises

8.37 What is the output of the following program?

```c
#include <stdio.h>
int main()
{
  int *arr[3], i, p[3] = {10, 20, 30};

  for(i = 0; i < 3; i++)
  {
    arr[i] = &p[i];
    printf("%d ", *arr[i]);
  }
  return 0;
}
```

Answer: With the statement arr[i] = &p[i]; each element of arr points to the address of the corresponding element of p. Therefore, the program displays 10 20 30.

8.38 What is the output of the following program?

```c
#include <stdio.h>
int main()
{
  char *arr[3];
  int i;

  arr[0] = "This is";
  arr[1] = "a new";
  arr[2] = "message";
  for(i = 0; i < 3; i++)
    printf("Text: %s\tFirst char: %c\n", arr[i], *arr[i]);
  return 0;
}
```

Answer: With the statement **char** *arr[3]; the elements of arr are declared as pointers to **char**. As we'll see in Chapter 10, the compiler allocates memory to store the literals "This is", "a new", and "message".

With the statement arr[0] = "This is"; arr[0] points to the address of the first character of "This is", that is, 'T'. Therefore, *arr[0] is equal to 'T'.

Similarly, with the statement arr[1] = "a new"; arr[1] points to the address of the first character of "a new", that is, 'a'. Therefore, *arr[1] is equal to 'a'.

Finally, with the statement arr[2] = "message"; *arr[2] becomes equal to 'm'.

As a result, the program displays

```
Text: This is      First char: T
Text: a new        First char: a
Text: message      First char: m
```

8.39   What is the output of the following program?

```c
#include <stdio.h>
int main()
{
  int *arr[3], i, num;

  for(i = 0; i < 3; i++)
  {
    printf("Enter number: ");
    scanf("%d", &num);
    arr[i] = &num;
  }
  for(i = 0; i < 3; i++)
    printf("Num: %d\n", *arr[i]);
  return 0;
}
```

Answer: With the statement `arr[i] = &num;` each element points to the address of num. Since all three pointers point to the same address, their content would be equal to the last value of num.

Therefore, the second loop displays three times the last input value.

Pointer to Pointer

As with all variables, when a pointer variable is declared the compiler reserves memory to store its value. Therefore, we can declare another pointer variable to point to this address.

To declare a pointer to pointer variable, add an `*` twice. For example, the statement

```c
int **ptr;
```

declares `ptr` as a pointer to another pointer that points to an integer.

To use a pointer to pointer variable, the single `*` provides access to the address of the second pointer, while the double `**` provides access to the value of the variable that the second pointer points to. For example,

```c
#include <stdio.h>
int main()
{
  int *ptr1, **ptr, i = 20;

  ptr1 = &i;
  ptr = &ptr1;

  printf("Val = %d\n", **ptr);
  return 0;
}
```

With the statement `ptr = &ptr1;` `ptr` points to the address of `ptr1`, which points to the address of `i`.

Since `ptr` points to the address of `ptr1`, `*ptr` is equal to `ptr1`. Since `ptr1` points to the address of `i`, we have `*ptr = ptr1 = &i`. Therefore, `**ptr` is equal to `i` and the program displays `Val = 20`.

In general, it is allowed to declare pointers to pointers to other pointers and so on (such as `int ***ptr`), but, in practice, it is rarely needed to exceed a depth of two.

Exercises

8.40 What is the output of the following program?

```c
#include <stdio.h>
int main()
{
  int *ptr1, **ptr, i = 10, j = 20;

  ptr1 = &i;
  ptr = &ptr1;
  **ptr += 100;

  ptr1 = &j;
  **ptr += 100;
  printf("Val = %d\n", i+j);
  return 0;
}
```

Answer: As explained in the previous example, `*ptr = ptr1 = &i` and `**ptr` is equal to `i`. Therefore, the statement `**ptr += 100;` is equivalent to `i = i+100 = 10+100 = 110`.

With the statement `ptr1 = &j;` `ptr1` points to the address of `j`, so `**ptr` is equal to `j`. Therefore, the statement `**ptr += 100;` is equivalent to `j = j+100 = 20+100 = 120`.

As a result, the program displays `Val = 230`.

8.41 What are the values of `arr` elements in the following program?

```c
#include <stdio.h>
int main()
{
  int a = 0, b = 1, c = 2, d = 3, m, arr[3];
  int *ptr[4] = {&a, &b, &c, &d};

  for(m = 0; m < 3; m++)
    arr[*ptr[m]] = *ptr[m+1];
  return 0;
}
```

Answer: When the array `ptr` is declared, `ptr[0]` becomes equal to `&a`, `ptr[1]` equal to `&b`, `ptr[2]` equal to `&c`, and `ptr[3]` equal to `&d`.

Therefore, the values of `*ptr[0]`, `*ptr[1]`, `*ptr[2]`, and `*ptr[3]` are 0, 1, 2, and 3, respectively. Let's trace the iterations:

First iteration. `arr[*ptr[0]]` = `*ptr[1]`, so `arr[0]` = 1.

Second iteration. `arr[*ptr[1]]` = `*ptr[2]`, so `arr[1]` = 2.

Third iteration. `arr[*ptr[2]]` = `*ptr[3]`, so `arr[2]` = 3.

Therefore, the values of `arr` elements become 1, 2, and 3.

8.42 What would be the values of `arr` elements in the previous example if we had written `arr[*ptr[m]]` = `**(ptr+m)`; instead of `arr[*ptr[m]]` = `*ptr[m+1]`;

Answer: Since the name of an array can be used as a pointer to its first element, `ptr` points to `ptr[0]`, which points to the address of a. Therefore, we can treat `ptr` as a pointer to pointer. In that case, `ptr+m` is a pointer to the `ptr[m]` element. Let's trace the iterations.

First iteration. Since `ptr` points to `ptr[0]`, `*ptr` is equal to `ptr[0]`. Now, since `ptr[0]` points to the address of a, `**ptr` is equal to a. Therefore, `arr[0]` = 0.

Second iteration. Since `ptr+1` points to `ptr[1]`, `*(ptr+1)` is equal to `ptr[1]`. Now, since `ptr[1]` points to the address of b, `**(ptr+1)` is equal to b. Therefore, `arr[1]` = 1.

Third iteration. Similarly, `arr[2]` becomes 2.

Pointers and Two-Dimensional Arrays

The pointers are closely related to multidimensional arrays as to one-dimensional ones. In this section, we'll focus on the most common case of multidimensional arrays, that of two-dimensional arrays.

Recall from Chapter 7 that a statement like **int** `arr[2][3]`; declares a two-dimensional array of six elements. A graphical representation of such an array is presented in Figure 8.3.

The elements of a two-dimensional array are stored in successive memory locations, starting with the elements of the first row, followed by the elements of the second row, and so on. For example, the following program declares a two-dimensional array and displays the memory addresses of its elements. Their distance is the size of **int** type, that is, 4.

```
#include <stdio.h>
int main()
{
  int i, k, arr[2][3];

  for(i = 0; i < 2; i++)
    for(k = 0; k < 3; k++)
      printf("Address of [%d][%d] element is: %p\n", i, k, &arr[i][k]);
  return 0;
}
```

To process a two-dimensional array `arr[N][M]` through pointer arithmetic, we assume that its elements are the `arr[0]`, `arr[1]`, ..., `arr[N–1]`, and we treat each of them as a pointer to an array of M elements. For example, with the statement

```
int arr[2][3];
```

arr [0] [0]	arr [0] [1]	arr [0] [2]
arr [1] [0]	arr [1] [1]	arr [1] [2]

FIGURE 8.3
Two-dimensional array of two rows and three columns.

`arr[0]` can be used as a pointer to an array of three integers, which consists of the elements of the first row, `arr[0][0]`, `arr[0][1]`, and `arr[0][2]`. In particular, `arr[0]` points to the first element of the row, that is, `arr[0][0]`. Therefore, the value of `*arr[0]` is equal to `arr[0][0]`.

Since we can use `arr[0]` as a pointer to the first element of the row, `arr[0]+1` is a pointer to the second element, that is, `arr[0][1]`, and `arr[0]+2` points to the third element.

In the general case, `arr[0]+k` is a pointer to the `arr[0][k]` element of the first row, which means that

- `arr[0]+k` is equivalent to `&arr[0][k]`
- `*(arr[0]+k)` is equivalent to `arr[0][k]`

Similarly, `arr[1]+k` can be used as a pointer to the `arr[1][k]` element of the second row, which means that

- `arr[1]+k` is equivalent to `&arr[1][k]`
- `*(arr[1]+k)` is equivalent to `arr[1][k]`

To sum up, we consider that the elements of an array `arr[N][M]` are `arr[0]`, `arr[1]`, ..., `arr[N–1]`, which can be used as pointers to arrays that consist of the M elements of the corresponding row. The following program uses pointer arithmetic to display the elements of a two-dimensional array:

```
#include <stdio.h>
int main()
{
  int i, k, arr[2][3] = {10, 20, 30, 40, 50, 60};

  for(i = 0; i < 2; i++)
    for(k = 0; k < 3; k++)
      printf("Value of [%d][%d] element is: %d\n", i, k, *(arr[i]+k));
  return 0;
}
```

An alternative way to handle a two-dimensional array through pointers is to use its name as a pointer. For example, with the statement

```
int arr[2][3];
```

the name `arr` can be used as a pointer to the address of its first element. As explained, we consider that the first element of the array is `arr[0]`, which is a pointer to an array that consists of the three elements of the first row. As said, `arr[0]` points to `arr[0][0]`.

Since arr is a pointer to arr[0] and arr[0] is a pointer to arr[0][0], we can treat arr as a pointer to pointer variable. Since arr points to the address of arr[0][0], **arr is equal to arr[0][0].

Similarly, since arr+1 can be used as a pointer to arr[1] and arr[1] points to arr[1][0], **(arr+1) is equal to arr[1][0]. In the general case,

- arr+k is equivalent to &arr[k]
- *(arr+k) is equivalent to arr[k], consequently equivalent to &arr[k][0]
- **(arr+k) is equivalent to arr[k][0]

For example, the following program uses the name of a two-dimensional array as a pointer to pointer to display its elements:

```c
#include <stdio.h>
int main()
{
  int i, k, arr[2][3] = {10, 20, 30, 40, 50, 60};

  for(i = 0; i < 2; i++)
    for(k = 0; k < 3; k++)
      printf("Value of [%d][%d] element is: %d\n", i, k, *(*(arr+i)+k));
  return 0;
}
```

The expression *(*(arr+i)+k) is equivalent to *(arr[i]+k), applied in the previous program.

To sum up, the following program demonstrates three ways to display the elements of a two-dimensional array:

```c
#include <stdio.h>
int main()
{
  int i, k, arr[2][3] = {10, 20, 30, 40, 50, 60};

  for(i = 0; i < 2; i++)
    for(k = 0; k < 3; k++)
    {
      printf("Value of [%d][%d] element is: %d\n", i, k, arr[i][k]);
      printf("Value of [%d][%d] element is: %d\n", i, k, *(arr[i]+k));
      printf("Value of [%d][%d] element is: %d\n", i, k, *(*(arr+i)+k));
    }
  return 0;
}
```

It's needless to say which is the simplest one...

Before terminating a tough section, the equivalent pointer expressions to access an element of up to a four-dimensional array are

```c
arr[i]        ==  *(arr+i)
arr[i][j]     ==  *(*(arr+i)+j)
arr[i][j][k]  ==  *(*(*(arr+i)+j)+k)
arr[i][j][k][l] == *(*(*(*(arr+i)+j)+k)+l)
```

Exercises

8.43 What does the following program?

```c
#include <stdio.h>
int main()
{
    int i, arr[2][5] = {10, 20, 30, 40, 50, 60, 70, 80, 90, 100};

    for(i = 0; i < 2; i++)
        *(arr[i]+3) = 0;
    return 0;
}
```

Answer: In each iteration, `arr[i]` points to the first element of row `i`. The expression `arr[i]+3` is a pointer to the fourth element of row `i`. Therefore, `*(arr[i]+3)` is equal to `arr[i][3]`.

As a result, the program makes zero the elements of the fourth column, therefore `arr[0][3]` and `arr[1][3]` become 0.

8.44 What does the following program?

```c
#include <stdio.h>
int main()
{
    int *ptr, arr[2][5] = {10, 20, 30, 40, 50, 60, 70, 80, 90, 100};

    for(ptr = arr[1]; ptr < arr[1]+5; ptr++)
        *ptr = 0;
    return 0;
}
```

Answer: With the statement `ptr = arr[1];` `ptr` points to `arr[1]`, which points to the address of `arr[1][0]`. Since `*ptr` is equal to `arr[1][0]`, the statement `*ptr = 0;` is equivalent to `arr[1][0] = 0;`.

With the statement `ptr++;` `ptr` points to the next element of the indicated row. For example, when the pointer is first increased, it points to the address of `arr[1][1]`, then to the address of `arr[1][2]`, and so on, up to the address of `arr[1][4]`.

Therefore, this program makes zero all five elements of the second row.

Pointer to Function

Although we've not discussed functions yet, you can get an idea. As with variables, the compiler reserves memory to store the function's code. A function pointer points to the memory address, where the function's code is stored. The general form of its declaration is

```c
return_type (*pointer_name) (type_param_1 name_1, type_param_2 name_2, ...,
type_param_n name_n);
```

The **return _ type** defines the function's return type, while the variables name _ 1, name _ 2, ..., name _ n indicate the function's parameters, if any. Look at the following examples:

```
int (*ptr)(int arr[], int size); /* ptr is declared as a pointer to a
function, which takes as parameters an array of integers and an integer
and returns an integer. */
void (*ptr)(double *arr[]); /* ptr is declared as a pointer to a
function, which takes as parameter an array of pointers to doubles and
returns nothing. */
int test(void (*ptr)(int a)); /* The test() function returns an integer
value and takes as a parameter a pointer to another function, which takes
an integer parameter and returns nothing. */
```

The name of the function pointer must be enclosed in parentheses because the function's call operator () has a higher precedence than the * operator. For example, the statement

int *ptr(**int** a); instead of **int** (*ptr)(**int** a);

declares a function named ptr, which takes an integer parameter and returns a pointer to an integer.

To make a pointer to point to a function, the pointer's declaration must match the function's return type and its parameters. For example, consider this program:

```
#include <stdio.h>

void test(int a);

int main()
{
  void (*ptr)(int a); /* ptr is declared as a pointer to a function,
    which takes an integer parameter and returns nothing. */
  int i = 10;

  ptr = test; /* ptr points to the memory address of the test()
    function. */
  (*ptr)(10); /* Call the function that ptr points to. */
  return 0;
}

void test(int a)
{
  printf("Val = %d\n", 2*a);
}
```

As the name of an array can be used as a pointer, a function's name can be used as a pointer to its memory address.

Therefore, with the statement ptr = test; ptr points to the address of the test() function. The statement ptr = test; is allowed because the declaration of ptr matches the declaration of test().

The function's call through a pointer can be done either like an ordinary call or use the * operator. For example, both statements

ptr(10); and (*ptr)(10); call test() and the program displays: Val = 20.

As a matter of style, our preference is the second way, to be clear that the variable is a pointer and not a function.

Exercise

8.45 Write a function that takes as parameters the grades of two students and returns the greater one. Write a program that reads two grades and uses a function pointer to call the function and display the greater one.

```c
#include <stdio.h>

float test(float a, float b);

int main()
{
  float (*ptr)(float a, float b); /* ptr is declared as a pointer to
    a function, which takes two float parameters and returns a
    float. */
  float i, j, max;

  printf("Enter grades: ");
  scanf("%f%f", &i, &j);

  ptr = test;
  max = (*ptr)(i, j); /* Call the function that ptr points to. */
  printf("Max = %f\n", max);
  return 0;
}

float test(float a, float b)
{
  if(a > b)
    return a;
  else
    return b;
}
```

Comments: Without using the variable max we could write printf("Max = %f\n", (*ptr)(i,j));

Array of Pointers to Functions

An array's elements can be pointers to functions of the same prototype. For example, the statement

```c
void (*ptr[20])(int a);
```

declares an array of twenty pointers to functions, which take an integer parameter and return nothing.

In the following program, each element of the array ptr is a pointer to a function, which takes two integer parameters and returns an integer:

```
#include <stdio.h>

int test_1(int a, int b);
int test_2(int a, int b);
int test_3(int a, int b);

int main()
{
    int (*ptr[3])(int a, int b);
    int i, j, k;

    ptr[0] = test_1; /* ptr[0] points to the memory address of the test_1
        function. */
    ptr[1] = test_2;
    ptr[2] = test_3;

    printf("Enter numbers: ");
    scanf("%d%d", &i, &j);

    if(i > 0 && i < 10)
        k = ptr[0](i, j); /* Call the function that ptr[0] points to. */
    else if(i >= 10 && i < 20)
        k = ptr[1](i, j); /* Call the function that ptr[1] points to. */
    else
        k = ptr[2](i, j); /* Call the function that ptr[2] points to. */
    printf("Val = %d\n", k);
    return 0;
}
int test_1(int a, int b)
{
    return a+b;
}
int test_2(int a, int b)
{
    return a-b;
}
int test_3(int a, int b)
{
    return a*b;
}
```

The program reads two integers, checks the first one, and uses a function pointer to call the respective function. The program displays the return value.

As with an ordinary array, an array of function pointers can be initialized when declared. For example, we could write

```
int (*ptr[3])(int a, int b) = {test_1, test_2, test_3};
```

Let's suppose that the user enters 20 and 10. What would be the output of the following statement?

```
printf("%d\n", ptr[0](ptr[1](i,j), ptr[2](i,j)));
```

The expression ptr[1](i,j) calls test _ 2(), which returns i–j, that is, 10. Similarly, the expression ptr[2](i,j) calls test _ 3(), which returns i*j, that is, 200. Therefore, the expression is translated to `ptr[0](10,200)` and the program displays the return value of `test _ 1()`, that is, `210`.

Unsolved Exercises

8.1 Write a program that uses two pointer variables to read two **double** numbers and display the absolute value of their sum.

8.2 Write a program that uses a pointer variable to read a **double** number and display the sum of its integer part and its fractional part. For example, if the user enters -7.21, the program should display `0.21`.

8.3 Write a program that uses three pointer variables to read three integers and check if they are in successive increase order (i.e., –5, –4, –3). The program should force the user to enter negative numbers.

8.4 Write a program that uses three pointer variables to read three integers one after the other. The program should force the user to enter the three numbers in decrease order.

8.5 Complete the following program by using the pointer p1 to read 100 integers, the pointer p2 to display the minimum of the input numbers less than –5, and the pointer p3 to display the maximum of those greater than 10. If none value less than –5 or greater than 10 is entered, the program should display an informative message.

```
int main()
{
   int *p1, *p2, *p3, i, num, min, max;
   ...
}
```

8.6 Use the pointer `ptr` and complete the following program to store up to 100 integers in the array `arr`, with the restriction to store values greater than –20. If the value –1 is stored, the insertion of numbers should terminate and the program should display how many numbers were stored in the array.

```
int main()
{
   int *ptr, arr[100];
   ...
}
```

8.7 Use p1, p2 and temp only, to reverse the elements of the array `arr`. Then, use p1 to display the array elements.

```
#include <stdio.h>
int main()
{
   double arr[] = {1.3, -4.1, -3.8, 9.4, 2.5}, temp, *p1 = arr, *p2 =
      arr+4;
   ...
}
```

8.8   What would be the output of 8.36 (Exercise) if we replace the && operator with ||
and write **while**(!*ptr1++ || !*ptr2++);
Explain why.

8.9   Use `arr` and complete the following program to read three integers and display the
sum of the even numbers.

```
int main()
{
  int *arr[3], i, j, k, m, sum;
  ...
}
```

8.10   Write a program that reads 100 integers and stores them in an array. The program
should replace the duplicated values with -99. Use pointer arithmetic to process the
array. For example, if the array were {5, 5, 5, 5, 5} the program should make it
{5, -99, -99, -99, -99} and if it were {-2, 3, -2, 50, 3} it should make it {-2,
3, -99, 50, -99}.

8.11   Write a program that uses two pointer to pointer variables to read two integers and
swap their values.

8.12   Write a program that assigns random values to a 5×5 array and checks if the sum
of the elements in the main diagonal is equal to that of its secondary diagonal. Use
pointer arithmetic to process the array.

8.13   Write a program that reads integers and stores them in a 5×5 array and displays the
maximum sum stored in a column, and in a row as well. Use pointer arithmetic to
process the array.

9

Characters

Up to this point, we've mostly used variables of type **int**, **float**, and **double**. In this chapter, we'll focus on the **char** type. To show you how to work characters, we assume that the underlying character set relies on the most popular set, the 8-bit ASCII (American Standard Code for Information Interchange) code. As you can see in Appendix B, ordinary characters, such as letters and digits, are represented by integers from 0 to 255.

char Type

Since a character in the ASCII set is represented by an integer between 0 and 255, we can use the **char** type to store its value. When a character is stored into a variable, it is the character's ASCII value that is actually stored. In the following example,

```
char ch;
ch = 'c';
```

the value of ch becomes equal to the ASCII value of the character c. Therefore, the statement ch = 'c'; is equivalent to ch = 99.

Don't forget to enclose a character constant in single quotes.

For example, if you omit them and write ch = c; the compiler would treat c as an ordinary variable and its value would be assigned to ch. If the variable c isn't declared, the compiler will raise an error message.

The %c specifier is used to display a character, while the %d is used to display its ASCII value. For example,

```
#include <stdio.h>
int main()
{
  char ch;

  ch = 'a';
  printf("Char = %c and its ASCII code is %d\n", ch, ch);
  return 0;
}
```

When a character appears in an expression, C treats it as an integer and uses its integer value.

Since the characters are treated as integers, we can use them in numerical expressions. For example,

```
char ch = 'c';
int i;
ch++; /* ch becomes 'd'. */
ch = 68; /* ch becomes 'D'. */
i = ch-3; /* i becomes 'A', that is 65. */
```

Exercises

9.1 Write a program that displays the characters of the ASCII set and their values.

```
#include <stdio.h>
int main()
{
  int i;

  for(i = 0; i < 256; i++)
    printf("Char = %c and its ASCII code = %d\n", i, i);
  return 0;
}
```

9.2 Write a program that displays the upper- and lowercase letters and their ASCII values.

```
#include <stdio.h>
int main()
{
  int i;

  for(i = 0; i < 26; i++)
  {
    printf("%c (%d)\n", 'a'+i, 'a'+i);
    printf("%c (%d)\n", 'A'+i, 'A'+i);
  }
  return 0;
}
```

Comments: Since in ASCII set, the difference between any uppercase letter and the respective lowercase is 32, the second printf() could be replaced with

```
printf("%c (%d)\n", 'a'+i-32, 'a'+i-32);
```

9.3 Write a program that reads three characters and checks if they are consecutive in the ASCII character set.

```
#include <stdio.h>
int main()
{
  char ch1, ch2, ch3;

  printf("Enter characters: ");
  scanf("%c%c%c", &ch1, &ch2, &ch3);

  if((ch1+1 == ch2) && (ch2+1 == ch3))
    printf("Consecutive\n");
  else
```

```
      printf("Not Consecutive\n");
    return 0;
  }
```

Comments: When you run the program, don't insert a space between the characters because the space is a character as well.

9.4 Write a program that reads two characters and displays the characters between them in the ASCII set. For example, if the user enters af or fa, the program should display bcde.

```
#include <stdio.h>
int main()
{
  char ch1, ch2;

  printf("Enter characters: ");
  scanf("%c%c", &ch1, &ch2);

  if(ch1 < ch2)
  {
    ch1++;
    while(ch1 != ch2)
    {
      printf("%c", ch1);
      ch1++;
    }
  }
  else
  {
    ch2++;
    while(ch2 != ch1)
    {
      printf("%c", ch2);
      ch2++;
    }
  }
  return 0;
}
```

9.5 Suppose that the purpose of the following program is to read a character five times and print it. Does it work as expected?

```
#include <stdio.h>
int main()
{
  char ch;
  int i;

  for(i = 0; i < 5; i++)
  {
    scanf("%c", &ch);
    printf("Char = %c\n", ch);
  }
  return 0;
}
```

Answer: scanf() reads a character and stores it in ch. However, if the user presses Enter, the generated new line character '\n' is stored in stdin and it will be automatically assigned to ch when scanf() is called again. In that case, the program won't behave as expected.

A solution is to force scanf() to skip any white space before reading a character. To do that, add a space before %c, like scanf(" %c", &ch).

Notice that if the user enters the five characters one after another and then presses Enter, the program would display them because each scanf() reads one character and let the rest get read in next calls.

9.6 Write a program that reads a character and checks if it is a new line character, a space, a digit, or a letter.

```c
#include <stdio.h>
int main()
{
  char ch;

  printf("Enter character: ");
  scanf("%c", &ch);

  if(ch == ' ')
    printf("The character is a space\n");
  else if(ch == '\n')
    printf("The character is a new line\n");
  else if(ch >= '0' && ch <= '9')
    printf("The character is a number\n");
  else if(ch >= 'a' && ch <= 'z')
    printf("The character is a lower case letter\n");
  else if(ch >= 'A' && ch <= 'Z')
    printf("The character is an upper case letter\n");
  return 0;
}
```

9.7 Write a program that reads a character continuously, and if it is an uppercase letter it should display that character, otherwise it should display the respective lowercase letter. If the last two entered characters are ':' and 'q', the program should display how many 'w' and 'x' were entered and then terminate.

```c
#include <stdio.h>
int main()
{
  char ch, last_ch;
  int sum1 = 0, sum2 = 0;

  while(1)
  {
    printf("Enter character: ");
    scanf("%c", &ch);

    if(last_ch == ':' && ch == 'q') /* If the last entered character
      is ':' and the new one is 'q', the insertion of characters
      should terminate. */
      break;
    else if(ch >= 'a' && ch <= 'z')
```

```
      printf("Char = %c\n", ch-32); /* Print the upper case letter. */
    else
      printf("Char = %c\n", ch);

    last_ch = ch; /* The input character is stored in last_ch. */
    if(ch == 'w')
      sum1++;
    else if(ch == 'x')
      sum2++;

    getchar();
  }
  printf("%c: %d times, %c: %d times\n", 'w', sum1, 'x', sum2);
  return 0;
}
```

Comments: As we'll see later, we use getchar() to read the new line character '\n' left in stdin, when the user presses Enter. Alternatively, we could put a space in scanf(), as explained in 9.5 (Exercise).

9.8 Write a program that uses a **for** loop to display all lowercase letters in one line, all uppercase letters in a second line, and all characters that represent the digits 0-9 in a third line.

```
#include <stdio.h>
int main()
{
  char ch, end_ch;

  end_ch = 'z';
  for(ch = 'a'; ch <= end_ch; ch++)
  {
    printf("%c ", ch);
    if(ch == 'z')
    {
      ch = 'A'-1; /* Subtract 1, so that the ch++ statement in the
        next loop iteration will make it equal to 'A'. */
      end_ch = 'Z'; /* Change the end character, so that the loop
        displays all upper case letters. */
      printf("\n");
    }
    else if(ch == 'Z')
    {
      ch = '0'-1;
      end_ch = '9';
      printf("\n");
    }
  }
  return 0;
}
```

Comments: How about writing the **for** loop without using the end _ ch variable and any of the **if-else-if** statements, like this:

```
for(ch = 'a'; ch != '9'+1; ch++)
```

```
{
  printf("%c ", ch);
  ch = (ch == 'z') ? 'A'-1 : (ch == 'Z') ? '0'-1 : ch;
  (ch == 'A'-1 || ch == '0'-1) ? printf("\n") : 1;
}
```

A bit complicated, isn't it? We just wanted to show you an incomprehensible version of the same code. Don't forget our advice: write clear code for your own benefit and for the benefit of those who are going to read your code.

`getchar()` Function

The `getchar()` function reads a character from `stdin`. Its prototype is defined in `stdio.h`, like this:

```
int getchar();
```

`getchar()` starts reading characters when the user presses the *Enter* key. If it is executed successfully, it returns the character read. To indicate a read error or when the end of the input stream is met, `getchar()` returns a special constant value, called EOF.

To compare `getchar()` and `scanf()`, `getchar()` is executed faster because `scanf()` is a complex function designed to read many kinds of data and not only characters. Moreover, `getchar()` is usually implemented as a macro for additional speed.

Exercises

9.9 Write a program that uses `getchar()` to read characters. The program should display and count the input characters.

```
#include <stdio.h>
int main()
{
  int ch, sum;

  printf("Enter characters: ");
  sum = 0;
  ch = getchar();
  while(ch != '\n')
  {
    sum++;
    printf("%c", ch);
    ch = getchar();
  }
  printf("\nTotal number is = %d\n", sum);
  return 0;
}
```

Comments: getchar() returns the input characters one by one until the '\n' character is met. Notice that we could combine the two calls to getchar() into one, like this:

```
sum = 0;
while((ch = getchar()) != '\n') /* Notice the way that the
  parentheses are used. */
{
  sum++;
  printf("%c",ch);
}
```

This loop reads a character, stores it in ch, and checks if it is different from the new line character. However, if we want to be absolutely right, we should write

```
while((ch = getchar()) != '\n' && ch ! = EOF)
```

to check if the character was successfully read.

Notice that since the return type of getchar() is **int**, the type of ch should be **int** and not **char**.

9.10 Write a program that reads a number's digits and displays its value. If the user enters a sign, it should be taken into account. If the user inserts a nondigit character, the program should terminate.

```
#include <stdio.h>
int main()
{
  int ch, sign, val;

  val = 0; /* The value of val is 0 until the user enters the first
    digit. Then, if the user enters a non-digit character the program
    terminates. */
  printf("Enter number: ");
  while((ch = getchar()) != '\n' && ch != EOF)
  {
    if(ch == ' ' || ch == '\t')
    {
      if(val != 0)
      {
        printf("Error: Not spaces between digits\n");
        return 0;
      }
    }
    else if(ch == '+' || ch == '-')
    {
      if(val != 0)
      {
        printf("Error: Not signs between digits\n");
        return 0;
      }
      else
        sign = ch;
    }
    else if(ch >= '0' && ch <= '9')
```

```
      val = 10*val + (ch-'0'); /* The expression ch-'0' calculates
      the numerical value of the digit character. */
    else
    {
      printf("Error: Input isn't a digit\n");
      return 0;
    }
  }
  if(sign == '-')
    val = -val;
  printf("%d\n", val);
  return 0;
}
```

Comments: To find the numerical value of a digit character, we subtract the ASCII value of '0'. For example, if the user enters 4, ch becomes equal to '4', that is 52. To get the input digit, we write ch-'0'=52-48=4.

9.11 Write a program that reads an IP version 4 address (*IPv4*) and checks if it is valid. The form of a valid IPv4 address is x.x.x.x, where each x must be an integer within [0, 255].

```
#include <stdio.h>
int main()
{
  int ch, dots, bytes, temp;
  dots = bytes = temp = 0;
  printf("Enter IP address (x.x.x.x): ");

  while((ch = getchar()) != '\n' && ch != EOF)
  {
    if(ch < '0' || ch > '9')
    {
      if(ch == '.')
      {
        dots++;
        if(temp != -1)
        {
          if(temp > 255)
          {
            printf("Error: The value of each byte should be in
              [0, 255]\n");
            return 0;
          }
          bytes++;
          temp = -1; /* The code -1 means that the current IP byte is
            checked. */
        }
      }
      else
      {
        printf("Error: Acceptable chars are only digits and dots\n");

        return 0;
      }
    }
```

```
    else
    {
      if(temp == -1)
        temp = 0; /* Make it 0, to start checking the next IP byte. */
      temp = 10*temp + (ch-'0');
    }
  }
  if(temp != -1) /* Check the value of the last IP byte. */
  {
    if(temp > 255)
    {
      printf("Error: The value of each byte should be in [0, 255]\n");
      return 0;
    }
    bytes++;
  }
  if(dots != 3 || bytes != 4)
    printf("Error: The IP format should be x.x.x.x\n");
  else
    printf("The input address is a valid IPv4 address\n");

  return 0;
}
```

9.12 Every mobile phone operating in GSM (*2G*) and WCDMA (*3G*) wireless networks is characterized by a unique identifier of 15 digits, called IMEI (*International Mobile Equipment Identifier*). A method to check if the device is really made by the official manufacturer is to compare the IMEI's last digit, called *Luhn digit*, with a check digit. If it is equal, the device is most probably authentic. Otherwise, it isn't authentic for sure.

The check digit is calculated as follows: first, we calculate the sum of the first IMEI's 14 digits by adding

(a) The digits in the odd positions.

(b) The double of the digits in the even positions. But if a digit's doubling is a two-digit number, we add each digit separately. For example, suppose that the value of the checked digit is 8. Its double is 16, therefore we add to the sum the result of 1+6 = 7 and not 16.

If the last digit of the calculated sum is 0, that is, the check digit. If not, we subtract the last digit from 10 and that is the check digit.

For example, let's check the IMEI 357683036257378. The algorithm applied in the first 14 digits produces

```
3 + (2×5) + 7 + (2×6) + 8 + (2×3) + 0 + (2×3) + 6 + (2×2) + 5 +
(2×7) + 3 + (2×7) =
3 + (10) + 7 + (12) + 8 + (6) + 0 + (6) + 6 + (4) + 5 + (14) + 3 +
(14) =
3 + (1+0) + 7 + (1+2) + 8 + (6) + 0 + (6) + 6 + (4) + 5 + (1+4) + 3
+ (1+4) = 62
```

The check digit is 10−2 = 8 equal to the Luhn digit. Therefore, this IMEI is valid.

Write a program that reads the IMEI of a mobile phone (15 digits) and checks if it is authentic or not.

```c
#include <stdio.h>
int main()
{
  char ch, chk_dig;
  int i, sum, temp;

  sum = 0;
  printf("Enter IMEI (15 digits): ");
  for(i = 1; i < 15; i++) /* Read the first 14 IMEI's digits. */
  {
    ch = getchar();
    if((i & 1) == 1) /* Check if the digit's position is odd. */
      sum += ch-'0'; /* The expression ch-'0' calculates the
        numerical value of the digit character. */
    else
    {
      temp = 2*(ch-'0');
      if(temp >= 10)
        temp = (temp/10) + (temp%10); /* If the digit's doubling
          produces a two-digit number we calculate the sum of these
          digits. */
      sum += temp;
    }
  }
  ch = getchar(); /* Read the IMEI's last digit, that is the Luhn
    digit. */
  ch = ch-'0';

  chk_dig = sum % 10;
  if(chk_dig != 0)
    chk_dig = 10-chk_dig;

  if(ch == chk_dig)
    printf("*** Valid IMEI ***\n");
  else
    printf("*** Invalid IMEI ***\n");
  return 0;
}
```

10

Strings

Now that you've seen how to use single character data, it's time to learn how to use strings. A string in C is a series of characters that must end with a special character, the *null* character. This chapter will teach you how to read and write strings and covers some of the most important string-handling functions in the C library.

String Literals

A string literal is a sequence of characters enclosed in double quotes. C treats them as character arrays. In particular, when a C compiler encounters a string literal, it allocates memory to store the characters of the string plus one extra character, to mark the end of the string. This special character is called the null character, and it is represented by the first character in the ASCII set, which is the '\0'.

For example, if the compiler encounters the string literal `"message"`, it allocates eight bytes to store the seven characters of the string plus one for the null character.

Storing Strings

To store a string in a variable, we use an array of characters. Because of the C convention that a string should end with the null character, to store a string of up to N characters the size of the array should be N+1. For example, to declare an array capable of storing a string of up to 7 characters, we write

```c
char str[8];
```

An array can be initialized with a string, when it's declared. For example, with the declaration

```c
char str[8] = "message";
```

the compiler copies the characters of the `"message"` into the `str` array and adds the null character. In particular, the value of `str[0]` becomes `'m'`, `str[1]` becomes `'e'`, and the value of the last element `str[7]` becomes `'\0'`. In fact, this declaration is equivalent to

```c
char str[8] = {'m', 'e', 's', 's', 'a', 'g', 'e', '\0'};
```

Like an ordinary array, if the number of the characters is less than the size of the array, the remaining elements are initialized to 0, which is the same with '\0' because the ASCII code for the null character is 0. For example, with the declaration

```
char str[8] = "me";
```

str[0] becomes 'm', str[1] becomes 'e', and the values of the rest elements are initialized to 0, or equivalently to '\0'.

Similarly, with the declaration

```
char str[8] = {0};
```

all the elements of str are initialized to '\0'.

When declaring the array, we may omit its length and let the compiler compute it. For example, with the declaration

```
char str[] = "message";
```

the compiler calculates the length of "message", then allocates eight bytes for str to store the seven characters plus the null character. Leaving the compiler to compute the length, it is easier and safer since counting by hand can lead to a calculation error.

When declaring an array of characters to store a string, don't forget to reserve a place for the null character.

If the null character is missing and the program uses C library string functions to handle the array, this may cause unpredictable results because C string-handling functions assume that strings are null terminated.

Don't confuse the '\0' and '0' characters. The ASCII code of the null character is 0, whereas the ASCII code of the zero character is 48.

Exercises

10.1 What is the difference between "a" and 'a'?

Answer: The expression "a" is a string literal, which is stored in the memory as an array of two characters, the 'a' and '\0'. On the other hand, the expression 'a' is just the single 'a' character and it is represented by its ASCII code.

10.2 What is the output of the following program?

```
#include <stdio.h>
int main()
{
    char str[] = "This is the text";

    printf("%d\n", sizeof(str));
    return 0;
}
```

Answer: The expression **sizeof**(str) calculates how many bytes allocate the array str in memory. This number is the size of the text plus one byte for the null character. Therefore, the program displays 17. Had we written

```
printf("%d\n", sizeof("This is the text"));
```

the program would display 17 again.

10.3 Does the following program contain any bug?

```
#include <stdio.h>
int main()
{
  char str1[] = "abc";
  char str2[] = "efg";

  str2[4] = 'w';
  printf("%c\n", str1[0]);
  return 0;
}
```

Answer: With the statement **char** str1[] = "abc"; the compiler creates the str1 array with 4 places, to store the 'a', 'b', 'c', and '\0' characters.

Similarly, with the statement **char** str2[] = "efg"; the compiler creates the str2 array with 4 places, to store the 'e', 'f', 'g', and '\0' characters.

The attempt to store the 'w' character in a position that exceeds the length of str2 is wrong. In particular, the assignment str2[4] = 'w' overwrites the data out of the bounds of str2, which can make our program to behave unpredictably.

The program may display 'a', but it may also display 'w' if str1 is stored right after str2 in memory.

Writing Strings

To display a string, we can use either the printf() or puts() functions. printf() uses the %s conversion specification and a pointer to the string. For example, the following program uses the name of the array as a pointer to the first character of the string:

```
#include <stdio.h>
int main()
{
  char str[] = "This is text";

  printf("%s\n", str);
  return 0;
}
```

printf() displays the characters of the string beginning from the character that the pointer points to until it encounters the null character. Therefore, the program displays This is text.

Recall from Chapter 2 that if we want to specify the number of the characters to be displayed, we use the conversion specification %.ns, where n specifies the number of characters.

Another way to display a part of the string is to use pointer arithmetic. For example, in the aforementioned program, to display the part of the string beginning from the sixth character, that is, the is text, we write

`printf("%s\n", str+5);` or equivalently `printf("%s\n", &str[5]);`

If `printf()` doesn't encounter the null character, it continues writing characters until it finds the null character somewhere in the memory. For example,

```c
#include <stdio.h>
int main()
{
  char str[100];

  str[0] = 'a';
  str[1] = 'b';
  printf("%s\n", str);
  return 0;
}
```

Since `str` doesn't contain the null character, the program displays the characters `'a'` and `'b'` and then "nonsense" characters.

Had we declared the array as `char str[100] = {0};` then all `str` elements would be equal to 0, equivalently to `'\0'`, and the program would display ab.

As discussed, when `printf()` encounters the null character, it stops writing any more characters. For example,

```c
#include <stdio.h>
int main()
{
  char str[] = "SampleText";

  str[4] = '\0';
  printf("%s\n", str);
  return 0;
}
```

Since the null character is found in the fifth position, the program displays `Samp` and ignores all the rest.

Another function to write strings is the `puts()`. `puts()` takes only one argument; a pointer to the string to be displayed. Like `printf()`, `puts()` writes the characters of the string until the null character is met. After writing the string, `puts()` writes an additional new line character. For example,

```c
#include <stdio.h>
int main()
{
  char str[] = "This is text";

  puts(str);
  return 0;
}
```

Since `puts()` is designed only for writing strings, it tends to be faster than `printf()`.

Exercise

10.4 What is the output of the following program?

```
#include <stdio.h>
int main()
{
  char str[] = "Right'0'Wrong";

  printf("%s\n", str);
  return 0;
}
```

Answer: Did you answer `Right`? Wrong, we are sorry.

The `'0'` doesn't represent the null character `'\0'` as you might think, but these three characters: the `'` character, the zero character, and another `'` character.

Therefore, the program outputs `Right'0'Wrong`.

Had we written `Right'\0'Wrong`, the program would display `Right'`.

Pointers and String Literals

Since a string literal is stored as an array of characters, we can use it as a pointer of type `char*`. For example, the next "weird" program uses the string literal in two ways—as an array and as a pointer—and displays the fifth character of the string `"message"`.

```
#include <stdio.h>
int main()
{
  printf("%c %c\n","message"[4],*("message"+4));
  return 0;
}
```

An alternative way to handle a string literal is to declare a pointer variable and make it point to the first character of the string. For example,

```
#include <stdio.h>
int main()
{
  char *ptr;
  int i;

  ptr = "This is text";
  for(i = 0; ptr[i] != '\0'; i++)
    printf("%c %c\n", *(ptr+i), ptr[i]);
  return 0;
}
```

With the statement `ptr = "This is text";` the compiler allocates memory to store the string literal `"This is text"` and the null character. Then, `ptr` points to the first character of the string, as depicted in Figure 10.1.

FIGURE 10.1
Pointers and strings.

To access the characters of the string, we can use either pointer arithmetic or use the pointer as an array and use array subscripting. The **for** loop is executed until the null character is met.

The memory that is allocated to store a string literal is usually read-only; therefore, you might not be able to modify its content.

For example, the compiler may raise an error message during the execution of the following program:

```c
#include <stdio.h>
int main()
{
  char *ptr = "This is text";

  ptr[0] = 'a';
  return 0;
}
```

As discussed, to handle a string we can use an array of characters. However, the declarations

char ptr[] = "This is text"; and **char** *ptr = "This is text";

may look similar, but they have significant differences.

The first statement declares ptr as an array of 13 characters. We can store another string into ptr, provided that its length won't be more than 13 characters, otherwise the program may behave unpredictably.

The second statement declares ptr as a pointer variable. It's initialized to point to the memory that holds the string literal "This is text". The pointer ptr can point to another string during program execution, no matter what its length is. For example, consider the following program:

```c
#include <stdio.h>
int main()
{
  char *ptr = "First text";

  ptr = "This is a new text";
  printf("First char = %c\n", *ptr);
  return 0;
}
```

The pointer ptr initially points to the first character of the string literal "First text". Then, the statement ptr = "This is a new text"; makes ptr to point to the new memory allocated to store that literal string. Since ptr points to the first character, *ptr is equal to 'T' and the program displays T.

Before using a pointer variable, it must point to the address of a variable that already exists, like an array of characters. For example,

```
#include <stdio.h>
int main()
{
  char *ptr;

  ptr[0] = 'a';
  ptr[1] = 'b';
  ptr[2] = '\0';
  printf("%s", ptr);
  return 0;
}
```

Is the assignment ptr[0] = 'a' and the others correct? Certainly not. Since ptr doesn't point to a valid address, writing the characters 'a', 'b', and '\0' in some random addresses may cause the abnormal operation of the program.

Don't forget that using an uninitialized pointer variable is a serious error.

Had we declared an array of characters and make ptr to point to it, like this

```
char str[3], *ptr;
ptr = str;
```

the program would execute correctly.

Exercises

10.5 The following program stores two strings in two arrays, swaps them, and displays their new content. Is there any error?

```
#include <stdio.h>
int main()
{
  char temp[100];
  char str1[100] = "Let see";
  char str2[100] = "Is everything OK?";

  temp = str1;
  str1 = str2;
  str2 = temp;

  printf("%s %s", str1, str2);
  return 0;
}
```

Answer: Recall from Chapter 8 that the name of an array when used as a pointer is a constant pointer, meaning that it's not allowed to point to some other address. Therefore, the statements temp = str1; str1 = str2; str2 = temp; won't compile.

10.6 What is the output of the following program?

```
#include <stdio.h>
int main()
{
   char str1[] = "test", str2[] = "test";
   (str1 == str2) ? printf("One\n") : printf("Two\n");
   return 0;
}
```

Answer: The expression str1 == str2 compares str1 and str2 as pointers, not if they have the same content. Since str1 and str2 are stored in different memory addresses, the program displays Two.

What would be the output if we write

```
(*str1 == *str2) ? printf("One\n") : printf("Two\n");
```

Since str1 can be used as a pointer to its first element, *str1 is equal to 't'. Similarly, *str2 is equal to 't'. Therefore, the program would display One.

Read Strings

Like printf(), scanf() uses the %s conversion specification to read a string. By default, scanf() reads characters until it encounters a white space character (i.e., space, tab, or new line character). Then, it appends a null character at the end of the string. In the following program, suppose that the user enters the string this is the text:

```
#include <stdio.h>
int main()
{
  char str[100];

  printf("Enter text: ");
  scanf("%s", str);
  printf("%s\n", str);
  return 0;
}
```

Since scanf() stops reading once it encounters a space character, only the word this is stored into str. Therefore, the program displays this. Besides %s, the conversion specification of scanf() can take many forms, which are beyond the scope of this book. For example, to force scanf() read multiple words write scanf("%[^\n]",str);

gets() is another function to read strings. It is declared in stdio.h, like this:

```
char *gets(char *str);
```

Like scanf(), gets() reads characters from stdin and stores them in the memory pointed to by str. gets() discards the new line character and appends a null character at the end of the string. In contrast to scanf(), gets() stops reading once it encounters a new line character, not any white space character. Upon success, gets() returns the str pointer, NULL otherwise. For example, the following program displays the input string, even if it consists of many words.

```c
#include <stdio.h>
int main()
{
  char str[100];

  printf("Enter text: ");
  gets(str);
  printf("%s\n", str);
  return 0;
}
```

Like puts() and printf(), gets() tends to execute faster than scanf() since gets() is designed exclusively to read characters.

In the aforementioned programs, gets() and scanf() take as an argument an array of characters that will hold the input string. Since the name of the array is used as a pointer, there is no need to put the & operator in front of its name.

As you guess, if you pass an uninitialized pointer to gets() and scanf(), the program may crash. For example,

```c
#include <stdio.h>
int main()
{
  char *ptr;

  printf("Enter text: ");
  gets(ptr);
  printf("%s\n", ptr);
  return 0;
}
```

Since ptr doesn't point to an allocated memory, the program may crash.

As scanf() *and* gets() *read and store characters into an array, they may pass the end of the array, causing the unpredictable behavior of the program. Therefore, be careful when using them to read strings. They aren't safe.*

For example, suppose that in the following program the user enters a string with more than five characters.

```c
#include <stdio.h>
int main()
{
  char str[5];
  int i = 20;

  printf("Enter text: ");
  gets(str);
  printf("%s %d\n", str, i);
  return 0;
}
```

The first five characters will be stored into str, but the rest will be stored into a memory out of the bounds of the array, causing the abnormal operation of the program. For example, if the value of i is stored into this memory, this value would change.

For a Safe Reading of Strings

To read safely a string and avoid the case of a memory overflow, we suggest the following:

1. Use fgets() instead of gets(). As we'll see in Chapter 15, fgets() is safer than gets() because it specifies the maximum number of the characters that can be read from the input stream.

2. When using scanf() to read strings, use the %ns conversion specification, where n specifies the maximum number of characters to be stored.

3. If you know the maximum possible size of the input string, use a function to allocate dynamically this memory (i.e., see malloc() in Chapter 14) and after the user enters the string use realloc() to shrink its size and make it equal to the size of the string. For example, in the following program, we assume that the maximum input size is less than 5000 characters:

```c
#include <stdio.h>
#include <string.h>
#include <stdlib.h>
int main()
{
  char *str;
  int len;

  str = (char *)malloc(5000);
  if(str == NULL)
  {
    printf("Error: Not available memory\n");
    exit(1);
  }
  printf("Enter text: ");
  gets(str);

  len = strlen(str);
  str[len] = '\0';

  realloc(str, len+1);
  printf("%s\n", str);

  free(str);
  return 0;
}
```

You'll get a better understanding of the aforementioned program after reading Chapter 14.

4. Make a loop and use getchar() to read characters one by one until the new line character is met or EOF is returned. Store the characters into a dynamically allocated memory (i.e., initial size of 500 bytes). When the memory becomes full,

use `realloc()` to increase its size by adding its initial length. For example, the first time it becomes full add 500 to make its size 1000 bytes, the second time add another 500 bytes to make it 1500, and so on. Once all the characters are read, make a last call to `realloc()` to shrink the size of the allocated memory and make it equal to the length of the input string.

That's enough about safe reading; we think that something from these will cover your needs.

For the sake of brevity and simplicity, we are going to use `gets()` *to read strings, assuming that the maximum length of the input string would be up to a reasonable number, for example, up to 100 characters. However, in your applications, never ever use* `gets()` *unless it is guaranteed that the input string would fit into the array.*

Exercises

10.7 Suppose that the user enters an integer and then presses the Enter key. What would be the output of the following program?

```c
#include <stdio.h>
int main()
{
    char str[100];
    int num;

    printf("Enter number: ");
    scanf("%d", &num);

    printf("Enter text: ");
    gets(str);
    printf("\n%d %s\n",num,str);
    return 0;
}
```

Answer: `scanf()` reads the number and stores it into num variable. The new line character that is generated when the Enter key is pressed is left in `stdin` and it will be the first character read in next call to `gets()`.

Since `gets()` stops reading when it encounters the '\n' character, the user won't be able to enter any other character. Therefore, the program displays only the input number.

A solution is to use `getchar()` before calling `gets()` in order to read the '\n' character.

10.8 Write a program that reads a string of up to 100 characters and displays the number of its characters, the number of 'b' occurrences, and the input string after replacing the space character with the new line character and the 'a' appearances with 'p'.

```c
#include <stdio.h>
int main()
{
    char str[100];
```

```
    int i, cnt;

    printf("Enter text: ");
    gets(str);

    cnt = 0;
    for(i = 0; str[i] != '\0'; i++)
    {
      if(str[i] == ' ')
        str[i] = '\n';
      else if(str[i] == 'a')
        str[i] = 'p';
      else if(str[i] == 'b')
        cnt++;
    }
    printf("Len = %d Times = %d\nText = %s\n", i, cnt, str);
    return 0;
}
```

Comments: The loop is executed until the end of the string is met, that is, once `str[i]` becomes equal to `'\0'`. After the loop ends, the value of `i` declares the length of the string.

10.9 Use a pointer variable to replace the **for** loop of the aforementioned program with a **while** loop. Use this pointer to make the replacements and calculate the length of the input string, as well.

```
#include <stdio.h>
int main()
{
  char *ptr, str[100];
  int cnt;

  printf("Enter text: ");
  gets(str);

  cnt = 0;
  ptr = str;
  while(*ptr != '\0')
  {
    if(*ptr == ' ')
      *ptr = '\n';
    else if(*ptr == 'a')
      *ptr = 'p';
    else if(*ptr == 'b')
      cnt++;
    ptr++;
  }
  printf("Len = %d Times = %d\nText = %s\n", ptr-str, cnt, str);
  return 0;
}
```

10.10 Write a program that first stores the 26 lower letters of the English alphabet into an array, then the upper letters, and before it terminates it displays the content of the array.

```
#include <stdio.h>
int main()
```

```
{
  char str[53];
  int i;

  for(i = 0; i < 26; i++)
  {
    str[i] = 'a'+i;
    str[26+i] = 'A'+i;
  }
  str[52] = '\0';
  printf("%s\n", str);
  return 0;
}
```

Comments: Each loop iteration stores the ASCII code of the respective character into the array `str`. For example, in the first iteration (`i = 0`), we have `str[0] = 'a'+0 = 97` and `str[26] = 'A'+0 = 65`.

10.11 What is the output of the following program?

```
#include <stdio.h>
int main()
{
  char *ptr, str[] = "another";

  ptr = str;
  printf("%d %c\n", *ptr+3, *(ptr+3));
  return 0;
}
```

Answer: Since `ptr` points to the address of the first element of `str`, `*ptr` is equal to `str[0]`, that is equal to `'a'`. Since the `*` operator has higher precedence than the `+` operator, we have `*ptr+3 = 'a'+3`.

Therefore, the program displays the ASCII code of the character three places after `'a'`. This is `'d'`, and the program displays `100`.

Since the expression `*(ptr+3)` is equivalent to `str[3]`, the program displays `'t'`.

10.12 Write a program that reads strings of up to 100 characters continuously and displays them, after replacing the lower letters with upper letters and vice versa. The program should display the number of the lower and upper letters in the new string. The reading of strings should terminate when the user enters the string "end".

```
#include <stdio.h>
int main()
{
  char str[100];
  int i, small_let, big_let;

  while(1)
  {
    printf("Enter text: ");
    gets(str);

    if(str[0] == 'e' && str[1] == 'n' && str[2] == 'd')
      break;
```

```
    i = small_let = big_let = 0;
    while(str[i] != '\0')
    {
      if(str[i] >= 'a' && str[i] <= 'z')
      {
          str[i] -= 32; /* In ASCII code, the difference between
            an upper case letter and the respective lower case
            letter is 32. */
          big_let++;
      }
      else if(str[i] >= 'A' && str[i] <= 'Z')
      {
          str[i] += 32;
          small_let++;
      }
      i++;
    }
    printf("%s contains %d lower case and %d upper case letters\n",
      str, small_let, big_let);
  }
  return 0;
}
```

10.13 What is the output of the following program?

```
#include <stdio.h>
int main()
{
  char *arr[3], str[100];
  int i;

  for(i = 0; i < 3; i++)
  {
    printf("Enter text: ");
    gets(str);
    arr[i] = str;
  }
  for(i = 0; i < 3; i++)
    printf("Text: %s\n", arr[i]);
  return 0;
}
```

Answer: The arr variable is declared as an array of three pointers to character.

In each iteration, the statement arr[i] = str; makes all pointers to point to the first character of the str array.

Since all pointers point to the same address, the second loop displays three times the last input string.

10.14 What is the output of the following program?

```
#include <stdio.h>
int main()
{
  char *str = "this";
```

```
    for(; *str; printf("%s ", str++));
    return 0;
}
```

Answer: The expression *str is equivalent to *str != '\0'. Therefore, the loop is executed until the null character is met.

In the first iteration, str points to the first character of "this". Therefore, printf() displays this and str is advanced to point to the next character. The next call to printf() displays his and so on. Therefore, the program outputs

```
this his is s
```

10.15 What is the output of the following program?

```
#include <stdio.h>
int main()
{
    char *str = "Example";
    int *ptr = (int*)str;

    ptr++;
    printf("%s\n", (char*)ptr+3);
    return 0;
}
```

Answer: Since ptr is declared as a pointer to **int**, the statement ptr++; makes it point to the fifth character of the string. In printf(), since we typecast the type of ptr to **char***, the expression (char*)ptr+3 displays the part of the string from the eighth character and on. Since the eighth character is the null character, the program displays nothing.

String Functions

This section presents some of the most common string-handling functions in the C library. We won't cover every aspect of them, but we'll give you enough of what you need in order to use them in your programs.

Although we haven't discussed functions yet, you should be able to get a sense of what these functions are doing.

strlen() Function

The strlen() function is declared in string.h, like this:

```
size_t strlen(const char *str);
```

The type size _ t is defined in C library as an unsigned integer type (usually as **unsigned int**). strlen() returns the number of the characters in the string pointed by str, not counting the null character. The pointer is declared **const**, so that strlen() can't modify the content of the string.

In Chapter 11, we'll implement the `str _ len()` function, which is an implementation of the `strlen()`.

Here is an example of how to use `strlen()`. The following program reads a string of up to 100 characters and uses `strlen()` to display its length:

```c
#include <stdio.h>
#include <string.h>
int main()
{
  char str[100];
  int len;

  printf("Enter text: ");
  gets(str);

  len = strlen(str);
  printf("Text has %d characters\n", len); /* There is no need to declare
    the variable len. Instead, we could write: printf("Text has %d
    characters\n", strlen(text)); */
  return 0;
}
```

Exercises

10.16 What is the output of the following program?

```c
#include <stdio.h>
#include <string.h>
int main()
{
  char str[] = "Text";

  printf("%d %d\n", strlen(str+4), strlen("Text"+1));
  return 0;
}
```

Answer: Since the name of an array can be used as a pointer to its first element, str+4 is a pointer to the fifth character of `str`, that is, the null character. Therefore, the first `strlen()` returns 0.

Since the string literal can be used as a pointer, the second `strlen()` returns the number of the characters from the second character and so on. Therefore, the second `strlen()` returns 3.

As a result, the program outputs 0 3.

10.17 Write a program that reads a string of up to 100 characters, and if it ends with 'aa', the program should display it in reverse order.

```c
#include <stdio.h>
#include <string.h>
int main()
{
  char str[100];
  int i, len;
```

```
    printf("Enter text: ");
    gets(str);

    len = strlen(str);
    if(len > 1 && str[len-1] == 'a' && str[len-2] == 'a')
    {
      printf("Reversed text: ");
      for(i = len-1; i >= 0; i--)
        printf("%c", str[i]);
    }
    return 0;
}
```

Comments: Since array indexing starts from [0] position, the last two characters are stored in positions [len-1] and [len-2].

10.18 Write a program that reads a string of up to 100 characters, copies it in a second string variable after replacing each single 'a' with a double 'a', and displays the second string.

```
#include <stdio.h>
#include <string.h>
int main()
{
  char str1[100], str2[200]; /* The new string will be stored into
    str2. Its size is declared as the double of the size of the str1
    array, just for the case that the input string contains only
    'a's. */
  int i, j;

  printf("Enter text: ");
  gets(str1);

  j = 0;
  for(i = 0; i < strlen(str1); i++)
  {
    str2[j] = str1[i]; /* Copy each character of the input string in
      the position indicated by j. */
    if(str1[i] == 'a')
    {
      j++; /* Increase j to store another 'a'. */
      str2[j] = 'a';
    }
    j++; /* Increase j to store the next character. */
  }
  str2[j] = '\0'; /* Add the null character. */
  printf("Text: %s\n", str2);
  return 0;
}
```

10.19 Write a program that reads a string of up to 100 characters, and if its length is less than three characters, it should force the user to enter a new one. Next, the program should read a character and check if the string contains the input character three times in a row. The program should display the position of the first triad found.

```
#include <stdio.h>
#include <string.h>
```

```
int main()
{
  char ch, str[100];
  int i, len;

  do
  {
    printf("Enter text (more than 2 chars): ");
    gets(str);
    len = strlen(str);
  } while(len < 3);
  printf("Enter character: ");
  ch = getchar();

  for(i = 0; i <= len-3; i++)
    if(str[i] == ch && str[i+1] == ch && str[i+2] == ch)
    {
      printf("There are three successive '%c's in position %d\n",
        ch, i+1);
      return 0;
    }
  printf("There aren't three successive '%c's\n", ch);
  return 0;
}
```

10.20 Write a program that reads a string of up to 100 characters and displays it after replacing all 'a' characters that exist at the beginning and at the end of the string with the space character (' '). For example, if the user enters "aaccadaa", the program should display "ccad".

```
#include <stdio.h>
#include <string.h>
int main()
{
  char str[100];
  int i, len;

  printf("Enter text: ");
  gets(str);

  len = strlen(str);
  for(i = 0; i < len; i++)
  {
    if(str[i] == 'a')
      str[i] = ' ';
    else
      break;
  }
  for(i = len-1; i >= 0; i-- )
  {
    if(str[i] == 'a')
      str[i] = ' ';
    else
      break;
  }
```

```
    printf("Text: %s\n", str);
    return 0;
}
```

Comments: The first **for** loop starts from the beginning of the string and compares its characters with 'a'. If it is an 'a', it is replaced with the space character, otherwise the **break** statement terminates the loop. Similarly, the second **for** loop replaces all 'a' characters at the end of the string with the space character.

10.21 What is the output of the following program?

```
#include <stdio.h>
#include <string.h>
int main()
{
    char *ptr, str[] = "csfbl";

    for(ptr = &str[0]; ptr < str+5; ptr++)
        --*ptr;

    printf("%s\n", ptr-strlen(str));
    return 0;
}
```

Answer: The statement --*ptr; decreases the content of the address that ptr points to.

For example, in the first iteration, ptr points to the first character, that is, 'a'. The statement --*ptr; changes the value of str[0] and makes it equal to the next character in the ASCII code. Therefore, str[0] becomes equal to 'b'. Then, the statement ptr++; makes ptr to point to the next element of str array.

Following the same logic, the next iterations make the values of str[1], str[2], str[3], and str[4] equal to 'r', 'e', 'a', and 'k', respectively.

The **for** loop terminates once the value of ptr becomes equal to str+5, that is, once the null character is met.

Since strlen() returns the length of the string, the expression ptr-strlen(str) is equivalent to ptr-strlen(str) = str+5-5 = str.

Therefore, the program displays the new string stored into str, which is break.

Indeed, have a break before moving on to the next exercise.

10.22 What is the output of the following program?

```
#include <stdio.h>
#include <string.h>
int main()
{
    char *ptr, str[] = "Text";
    int i;

    ptr = str;
    for(i = 0; i < strlen(str); i++)
    {
        printf("%c", ptr[i]);
        ptr++;
    }
    return 0;
}
```

Answer: Since `strlen()` returns 4, the **for** loop will be executed four times. Let's trace the iterations:

First iteration (`i = 0`). The value of `ptr[0]` is displayed, that is, `'T'`.

Second iteration (`i++ = 1`). Since `ptr` has been increased by one, `ptr` points to the second character of the string, that is, `'e'`. Since we handle `ptr` as an array, `ptr[0]` is `'e'` and `ptr[1]` is `'x'`. Therefore, the program displays `'x'`.

Third iteration (`i++ = 2`). Now, `ptr` points to the third character of the string, that is, `'x'`. Therefore, `ptr[0]` is `'x'`, `ptr[1]` is `'t'`, and `ptr[2]` is equal to `'\0'`. As a result, the program displays nothing.

Fourth iteration (`i++ = 3`). Now, `ptr` points to the fourth character of the string, that is, `'t'`. Therefore, `ptr[0]` is `'t'`, `ptr[1]` is `'\0'`, and `ptr[2]` and `ptr[3]` are equal to the values that exist in the address past `str+4`. Therefore, the program displays a random character.

To sum up, the program displays `Tx(space)(random character)`.

10.23 Write a program that forces the user to enter a string with more than 5 characters and less than 100 characters and displays it. Don't use `gets()`.

```c
#include <stdio.h>
#include <string.h>
int main()
{
  char str[100];
  int i, ch;

  printf("Enter text (> 5 && < 100): ");
  while(1)
  {
    i = 0;
    while((ch = getchar()) != '\n' && ch != EOF)
    {
      if(i < 99)
      {
        str[i] = ch;
        i++;
      }
    }
    str[i] = '\0';
    if(strlen(str) > 5)
      break;
    else
      printf("Enter text (> 5 && < 100): ");
  }
  printf("%s\n", str);
  return 0;
}
```

10.24 What is the output of the following program?

```c
#include <stdio.h>
#include <string.h>
```

```
int main()
{
  char str[] = "example";

  if((str[strlen(str+5)] == 'm') || (*(str+2)+1) == 'b')
    printf("One\n");
  else
    printf("Two\n");
  return 0;
}
```

Answer: `strlen()` returns the length of the part of the string after the fifth character. Since this part contains the `'l'` and `'e'` characters, `strlen()` returns 2.

Since the expression `*(str+2)` is equivalent to `str[2]`, that is, `'a'`, the value of the expression `*(str+2)+1` is equal to `str[2]+1`, that is, `'a'+1 = 'b'`.

Therefore, the `if` statement is equivalent to

```
if((str[2] == 'm') || ('b' == 'b'))
```

and the program displays `One`.

10.25 A Universal Product Code (UPC) barcode consists of 12 digits. The last digit is a check digit used for error detection. To calculate its value, we use the first 11 digits, as follows:

1. Add the digits in the odd positions and multiply the result by three.
2. Add the digits in the even positions to the previous result.
3. Divide the result with 10. Subtract the remainder from 10, and that is the check digit. If the subtraction gives 10 (meaning that the remainder is 0), use 0 as check digit.

For example, the check bit for the barcode 12345678901 is calculated as follows:

1. $1 + 3 + 5 + 7 + 9 + 1 = 26$. Multiplied by 3 gives $26*3 = 78$.
2. $2 + 4 + 6 + 8 + 0 = 20$. Added to the previous result gives 98.
3. Check digit $= 10 - (98\%10) = 10 - 8 = 2$.

Write a program that reads a UPC and verifies if the check bit is correct. The program should force the user to enter a valid UPC, meaning that the length of the string should be 12 and it must contain digits only.

```
#include <stdio.h>
#include <string.h>
int main()
{
  char upc[13];
  int i, flag, chk_dig, sum;

  while(1)
  {
    printf("Enter UPC (12 digits): ");
    gets(upc);

    if(strlen(upc) != 12)
```

```
    {
      printf("Error: wrong length\n");
      continue;
    }
    flag = 1;
    for(i = 0; i < 12; i++)
    {
      if(upc[i] < '0' || upc[i] > '9')
      {
        printf("Error: only digits allowed\n");
        flag = 0;
        break;
      }
    }
    if(flag == 1)
      break;
  }
  sum = 0;
  for(i = 0; i < 11; i+= 2)
    sum += upc[i] - '0'; /* Subtract '0' to get the numerical value
      of the digit character. */
  sum *= 3;
  for(i = 1; i < 11; i+= 2)
    sum += upc[i] - '0';

  chk_dig = 10-(sum%10);
  if(chk_dig == 10)
    chk_dig = 0;

  if(chk_dig == (upc[11] - '0'))
    printf("Valid check digit\n");
  else
    printf("Wrong check digit. The correct is %d.\n", chk_dig);

  return 0;
}
```

strcpy() Function

The strcpy() function is declared in string.h, like this:

```
char *strcpy(char *dest, const char *src);
```

strcpy() copies the string pointed to by src to the memory location pointed to by dest. Once the null character is copied, strcpy() terminates and returns the dest pointer. Since src is declared **const**, strcpy() can't modify the string.

For example, the following strcpy() copies the string "something" into str.

```
char str[10];
strcpy(str, "something");
```

The following program reads a string of up to 100 characters and uses strcpy() to copy it into a second array:

```
#include <stdio.h>
#include <string.h>
```

```
int main()
{
  char str1[100], str2[100];

  printf("Enter text: ");
  gets(str2);

  strcpy(str1, str2);
  printf("Copied text: %s\n", str1);
  return 0;
}
```

Because strcpy() *doesn't check if the string pointed to by* src *fits into the memory pointed to by* dest, *it is the programmer's responsibility to assure that the destination memory is large enough to hold all characters. Otherwise, the memory past the end of* dest *will be overwritten causing the unpredictable behavior of the program.*

For example, consider the following program:

```
#include <stdio.h>
#include <string.h>
int main()
{
  char c = 'a', str[10];

  strcpy(str, "Longer text. The program may crash");
  printf("%s %c\n", str, c);
  return 0;
}
```

Since the size of str array isn't large enough to hold the characters of the string, the data past the end of str will be overwritten.

And yes, it's an error to pass an uninitialized pointer to strcpy(). For example,

```
#include <stdio.h>
#include <string.h>
int main()
{
  char *str;

  strcpy(str, "something");
  printf("%s\n", str);
  return 0;
}
```

Since str doesn't point to an allocated memory, the program may crash.

Exercises

10.26 What is the output of the following program?

```
#include <stdio.h>
#include <string.h>
```

```
int main()
{
  char str1[] = "abcd";
  char str2[] = {'e', 'f', 'g'};

  strcpy(str1, str2);
  printf("%s\n", str1);
  return 0;
}
```

Answer: Since `str2` array doesn't contain the null character, the copy operation won't perform successfully and the program won't display `efg` as you might expect.

10.27 What is the output of the following program?

```
#include <stdio.h>
#include <string.h>
int main()
{
  char str1[5], str2[5];

  printf("%c\n", strcpy(str1, strcpy(str2, "test"))[0]);
  return 0;
}
```

Answer: The inner `strcpy()` copies the string `test` into `str2` and returns the `str2` pointer. The outer `strcpy()` copies the string pointed to by `str2` into `str1`. Therefore, the string `test` is copied into `str1`.

Since the outer `strcpy()` returns the `str1` pointer, `printf()` translates to `printf("%c\n",str1[0]);` and the program displays `t`.

10.28 What is the output of the following program?

```
#include <stdio.h>
#include <string.h>
int main()
{
  char str[10];

  printf("%c\n", *(str+strlen(strcpy(str, "example"))/2));
  return 0;
}
```

Answer: The inner `strcpy()` copies the string `"example"` into `str` and returns `str`. Therefore, `strlen()` returns the length of the string stored into `str`, that is, 7.

As a result, `printf()` translates to `printf("%c\n", *(str+7/2));` and the program displays the character stored in position `str+7/2 = str+3`, that is, `m`.

10.29 What is the output of the following program?

```
#include <stdio.h>
#include <string.h>
int main()
{
  char str[10] = "test";

  printf("%d %s\n", *strcpy(str, "n")**strcpy(str+2, "xt"), str);
  return 0;
}
```

Answer: The first `strcpy()` copies the characters of the string n, that is, the characters 'n' and '\0' to `str[0]` and `str[1]`, respectively, and returns the `str` pointer. Therefore, the expression `*strcpy(str,"n")` can be replaced by `*str`, that is, 'n'.

The second `strcpy()` copies the string "xt" in the third position of `str` and returns the `str+2` pointer. Therefore, the expression `*strcpy(str+2,"xt")` can be replaced by `*(str+2)`, that is, 'x'.

As a result, the program displays the product of the ASCII codes of 'n' and 'x', that is, 13200. However, it doesn't display next as you might expect, but only n, because the first `strcpy()` replaced 'e' with '\0'.

`strncpy()` Function

The `strncpy()` function is declared in `string.h`, like this:

```
char *strncpy(char *dest, const char *src, size_t count);
```

`strncpy()` is similar to `strcpy()`, with the difference that only the first count characters of the string pointed to by `src` will be copied into `dest`.

If the value of count is less than or equal to the length of the string that `src` points to, a null character won't be appended to the memory pointed to by `dest`. If it is greater, null characters are appended, up to the value of count.

> To avoid the case of a memory overwrite, use `strncpy()` instead of `strcpy()` because `strncpy()` specifies the maximum number of the copied characters.

Here's an example of how to use `strncpy()`.

```c
#include <stdio.h>
#include <string.h>
int main()
{
  char str1[] = "Old text";
  char str2[] = "New";
  char str3[] = "Get";

  strncpy(str1, str2, 3);
  printf("%s\n", str1);

  strncpy(str1, str3, 5);
  printf("%s\n", str1);
  return 0;
}
```

Since the number 3 is less than the length of the string stored into `str2`, the first `strncpy()` copies the first 3 characters into `str1` and doesn't append a null character. Therefore, the program displays New text.

Since the number 5 is greater than the length of the string stored into `str3`, the second `strncpy()` copies three characters from `str3` into `str1` and appends two null characters to reach the value 5. Therefore, the program displays Get.

`strcat()` Function

The `strcat()` function is declared in `string.h`, like this:

```
char *strcat(char *dest, const char *src);
```

strcat() appends the string pointed to by src to the end of the string pointed to by dest. strcat() appends the null character and returns dest, which points to the resulting string.

For example, the following program reads two strings of up to 100 characters and uses strcat() to merge them and store the resulting string into an array:

```c
#include <stdio.h>
#include <string.h>
int main()
{
  char str1[100], str2[100];
  char str3[200] = {0}; /* Initially, store the null character. */

  printf("Enter first text: ");
  gets(str1);

  printf("Enter second text: ");
  gets(str2);

  strcat(str3, str1); /* Since str3 contains only the null character,
    strcat() stores the string pointed to by str1 into str3. */
  strcat(str3, str2); /* Append the string stored into str2. */
  printf("The merged text is: %s\n", str3);
  return 0;
}
```

It is the programmer's responsibility to ensure that there is enough room in the memory pointed to by dest to add the characters of the string pointed to by src. Otherwise, the memory past the end of dest will be overwritten, with unpredictable results on the program's operation.

Consider the following program:

```c
#include <stdio.h>
#include <string.h>
int main()
{
  char str[20] = "example";

  strcat(str, "not available memory");
  printf("The merged text is: %s\n", str);
  return 0;
}
```

Since the size of str isn't large enough to hold the characters of both strings, the data after the end of str will be overwritten.

strcmp() and strncmp() Functions

The strcmp() function is declared in string.h, like this:

```c
int strcmp(const char *str1, const char *str2);
```

strcmp() compares the string pointed to by str1 with the string pointed to by str2. If the strings are identical, strcmp() returns 0. If the first string is less than the second, strcmp() returns a negative value, whereas if it is greater it returns a positive value. A string is considered less than another if either of the following conditions is true:

(a) The first n characters of the strings match, but the value of the next character in the first string is less than the value of the respective character in the second string.

(b) All characters match, but the first string is shorter than the second.

For example, assuming that the comparison of the characters is based on their ASCII codes, the statement strcmp("onE", "one") returns a negative value because the ASCII code of the first nonmatching character 'E' is less than the ASCII code of 'e'.

On the other hand, the statement strcmp("w", "many") returns a positive value because the ASCII code of the first nonmatching character 'w' is greater than the ASCII code of 'm'.

In another example, the statement strcmp("some", "something") returns a negative value because the first four characters match, but the first string is smaller than the second.

In Chapter 11, we'll implement the str _ cmp() function, which is an implementation of the strcmp().

strncmp() is similar to strcmp(), with the difference that it compares a specific number of characters. It is declared in string.h, like this:

```
int strncmp(const char *str1, const char *str2, int count);
```

The parameter count specifies the number of the compared characters.

Exercises

10.30 Write a program that reads two strings of up to 100 characters and uses strcmp() to compare them. If the strings are different, the program should use strncmp() to compare their first three characters and display a relative message.

```
#include <stdio.h>
#include <string.h>
int main()
{
  char str1[100], str2[100];
  int ret;

  printf("Enter first text: ");
  gets(str1);

  printf("Enter second text: ");
  gets(str2);

  ret = strcmp(str1, str2);
  /* Without using the variable ret we could write
    if(strcmp(str1,str2) == 0) */
```

```
  if(ret == 0)
    printf("Same texts\n");
  else
  {
    printf("Different texts\n");
    if(strncmp(str1, str2, 3) == 0)
      printf("But the first 3 chars are the same\n");
  }
  return 0;
}
```

10.31 What is the output of the following program?

```
#include <stdio.h>
#include <string.h>
int main()
{
  char str[5];

  str[0] = 't';
  str[1] = 'e';
  str[2] = 's';
  str[3] = 't';

  if(strcmp(str, "test") == 0)
    printf("One\n");
  else
    printf("Two\n");
  return 0;
}
```

Answer: Although the obvious answer is One, the program displays Two. Why is that the case?

The str array contains the 't','e','s', and 't' characters, but not the null character. On the other hand, the string literal "test" ends with the null character. Since they are different, strcmp() returns a nonzero value and the program displays Two.

What would be displayed if we wrote **char** str[5] = {'t','e','s','t'};

Since the noninitialized elements are set to 0, str[4] becomes '\0' and the program would display One.

10.32 Write a program that reads continuously strings up to 100 characters and displays the "smallest" and "largest" strings. If the input string begins with "end", the insertion of strings should terminate. Don't compare this string with the others.

```
#include <stdio.h>
#include <string.h>
int main()
{
  char str[100], min_str[100], max_str[100];

  printf("Enter text: ");
  gets(str);

  /* Use the first string as a base to compare the rest. */
  strcpy(min_str, str);
```

```
    strcpy(max_str, str);
    if(strncmp(str, "end", 3) == 0)
    {
      printf("\nMax = %s Min = %s\n", max_str, min_str);
      return 0;
    }
    while(1)
    {
      printf("Enter text: ");
      gets(str);

      if(strncmp(str, "end", 3) == 0)
        break;

      if(strcmp(str, min_str) < 0)
        strcpy(min_str, str);

      if(strcmp(str, max_str) > 0)
        strcpy(max_str, str);
    }
    printf("\nMax = %s Min = %s\n", max_str, min_str);
    return 0;
}
```

10.33 What is the output of the following program?

```
#include <stdio.h>
#include <string.h>
int main()
{
  char str1[10], str2[] = "engine";

  printf("%c\n", str1[strcmp(strcpy(str1, "ine"), str2+3)]);
  return 0;
}
```

Answer: strcpy() copies the string "ine" into str1 and returns the str1 pointer. strcmp() compares the string pointed to by the return value of strcpy(), that is, str1, with the part of the string stored in str2+3 address, that is, "ine". Therefore, since both pointers point to identical strings, strcmp() returns 0 and the program displays the value of str1[0], that is, 'i'.

10.34 Write a program that reads a string of up to 100 characters and displays it, after reversing its characters (e.g., if the user enters "code", the program should display "edoc").

```
#include <stdio.h>
#include <string.h>
int main()
{
  char temp, str[100];
  int i, len;

  printf("Enter text: ");
  gets(str);

  len = strlen(str);
  for(i = 0; i < len/2; i++)
```

```
    {
        temp = str[i];
        str[i] = str[len-i-1];
        str[len-i-1] = temp;
    }
    printf("Reversed text:%s\n", str);
    return 0;
}
```

Comments: The **for** loop swaps the first half characters with the other half. That's why the loop is executed from 0 up to len/2.

For example, if the user enters code, str[0] becomes equal to 'c', str[1] equal to 'o', str[2] equal to 'd', and str[3] equal to 'e'.

Since the length of the string is 4, len is 4. The last character of the string is stored in str[len-1] element. The characters are swapped, like this:

First iteration (i = 0):

```
temp = str[0] = 'c'
str[0] = str[len-i-1] = str[4-0-1] = str[3] = 'e'
str[len-i-1] = str[4-0-1] = str[3] = temp = 'c'
```

Therefore, in the first iteration, the first character was swapped with the last one, that is, str[0] became 'e' and str[3] became 'c'.

Second iteration (i++ = 1):

```
temp = str[1] = 'o'
str[1] = str[len-i-1] = str[4-1-1] = str[2] = 'd'
str[len-i-1] = str[4-1-1] = str[2] = temp = 'o'
```

Therefore, in the second iteration, the second character was swapped with the one before the last, that is, str[1] became 'd' and str[2] became 'o'.

The loop terminates and the program displays edoc.

10.35 Write a program that reads continuously strings up to 100 characters and displays a message to indicate if the input string is a palindrome, which means if it can be read in the same way in either direction (e.g., the string "level" is a palindrome since it's read in the same way in both directions). If the input string is "exit", the insertion of strings should terminate.

```
#include <stdio.h>
#include <string.h>
int main()
{
    char str[100];
    int i, diff, len;

    while(1)
    {
        printf("Enter text: ");
        gets(str);

        if(strcmp(str, "exit") == 0)
            break;
```

```
        len = strlen(str);
        diff = 0;
        for(i = 0; i < len/2; i++)
        {
          if(str[i] != str[len-1-i])/* If two characters are not the
            same, then the loop terminates. */
          {
            diff = 1;
            break;
          }
        }
        if(diff == 1)
          printf("%s is not a palindrome\n", str);
        else
          printf("%s is a palindrome\n", str);
      }
    return 0;
}
```

Comments: For an explanation of the **for** loop, read the comments of the previous exercise.

10.36 Write a program that reads a string of up to 100 characters and displays which character appears the most times and the number of its appearances.

```
#include <stdio.h>
#include <string.h>
int main()
{
  char ch, max_ch, str[100];
  int i, max_times, occurs[256] = {0}; /* To declare the size of the
    array we assume that the ASCII set is used. */

  printf("Enter text: ");
  gets(str);

  max_ch = max_times = 0;
  for(i = 0; str[i] != '\0'; i++)
  {
    ch = str[i];
    occurs[ch]++; /* This array contains the number of appearances of
      each character. For example, occurs[97] holds the number of
      appearances of character 'a' within the string. */
    if(occurs[ch] > max_times)
    {
      max_times = occurs[ch];
      max_ch = ch;
    }
  }
  printf("'%c' appears %d times\n", max_ch, max_times);
  return 0;
}
```

Comments: If the string contains more than one character, which appears the same most times, the program displays the one found first. For example, if the user enters "exit1", the output would be: `'e' appears  1  times`, because all characters appear once and the character `'e'` is the first one.

10.37 The data compression algorithm RLE (Run Length Encoding) is based on the fact that a symbol within the data stream may be repeated many times in a row. This repetitive sequence can be replaced by

(a) An integer that declares the number of the repetitions

(b) The symbol itself

Write a program that reads a string of up to 100 characters and uses the RLE algorithm to compress it. Don't compress digits and characters that appear once.

For example, the string `fffmmmm1234jjjjjjjjjjx` should be compressed to `3f4m123410jx`

```c
#include <stdio.h>
#include <string.h>
int main()
{
  char str[100];
  int i, cnt;

  printf("Original text : ");
  gets(str);

  printf("Compressed text: ");
  i = 0;
  while(i < strlen(str))
  {
    cnt = 1;
    if(str[i] < '0' || str[i] > '9')/* Digits are not compressed. */
    {
      while(str[i+cnt] == str[i])/* Check if the current character,
        that is str[i], is repeated in the next places. */
        cnt++;

      if(cnt == 1)
        printf("%c", str[i]);
      else
        printf("%d%c", cnt, str[i]);
    }
    else
      printf("%c", str[i]);

    i += cnt;
  }
  return 0;
}
```

10.38 Write a program that reads a string of up to 100 characters and displays the number of appearances of its lowercase letters and its digits.

```c
#include <stdio.h>
int main()
```

```
{
  char ch, str[100];
  int i, low_let[26] = {0}; /* The size of the array is equal to the
    number of low-case letters. This array holds the number of the
    appearances of each letter. For example, low_let[0] holds the
    appearances of 'a' and low_let[25] the appearances of 'z'. */
  int dig[10] = {0}; /* Similarly, dig[0] holds the appearances of
    digit 0 and dig[9] the appearances of digit 9. */
  printf("Enter text: ");
  gets(str);
  for(i = 0; str[i] != '\0'; i++)
  {
    ch = str[i];
    if((ch >= 'a') && (ch <= 'z'))
      low_let[ch - 'a']++; /* For example, if the read character is
        'a', the value of low_let['a'-'a'] = low_let[0], which holds
        the appearances of 'a' will be increased by one. */
    else if((ch >= '0') && (ch <= '9'))
      dig[ch - '0']++;
  }
  printf("***** Lower case letters appearances\n");
  for(i = 0; i < 26; i++)
    if(low_let[i] != 0)/* Check if the character appears once at
      least. */
      printf("Letter %c appeared %d times\n", 'a'+i, low_let[i]);

  printf("***** Digits appearances\n");
  for(i = 0; i < 10; i++)
    if(dig[i] != 0)
      printf("Digit %d appeared %d times\n", i, dig[i]);
  return 0;
}
```

10.39 Write a program that reads a string of up to 100 characters and displays the words that it consists of and their number (consider that a word is a sequence of characters that doesn't contain the space character). For example, if the user enters "how   many   words ?" (notice that more than one space may be included between the words), the program should display

```
how
many
words
?
The text contains 4 words
```

```
#include <stdio.h>
int main()
{
  char str[100];
  int i, words;

  i = words = 0;
  printf("Enter text: ");
  gets(str);
```

```
if(str[0] != ' ' && str[0] != '\0')/* If the first character is
    other than the space character means that a word begins, so the
    value of words is increased by one. */
    words++;

while(str[i] != '\0')
{
    if(str[i] == ' ')
    {
        /* Since more than one space characters may be included between
            words, we check if the next character, that is str[i+1], is
            the space character. If it isn't, it means that a new word
            begins, so the value of words is increased by one. */
        if(str[i+1] != ' ' && str[i+1] != '\0')
        {
            words++;
            printf("\n");
        }
    }
    else
        printf("%c", str[i]);
    i++;
}
printf("\nThe text contains %d words\n", words);
return 0;
}
```

Two-Dimensional Arrays and Strings

The most convenient way to store multiple strings is to use a two-dimensional array. For example, the statement

```
char str[10][40];
```

declares the str array with 10 rows. Each row can store a string of up to 40 characters.

We can store string literals in a two-dimensional array together with its declaration. For example,

```
char str[3][40] = {"One", "Two", "Three"};
```

As shown in Figure 10.2, the characters of "One" are stored in the first row of str, the characters of "Two" in the second row, and the characters of "Three" in the third row, respectively. Since the strings were not long enough to fill the rows, null characters are padded.

Recall from Chapter 8 that we can treat each of the elements str[0], str[1],..., str[N–1] of a two-dimensional array str[N][M] as a pointer to an array of M elements.

Therefore, str[0] can be used as a pointer to an array of 40 characters, which holds the string "One". Similarly, str[1] and str[2] can also be used as pointers to the other two strings.

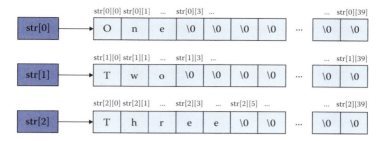

FIGURE 10.2
Two-dimensional arrays and strings.

Exercises

10.40 A simple algorithm to encrypt data is the algorithm of the single transformation. Let's see how it works. In one line, we write the letters of the used alphabet. In a second line, we write the same letters in a different order. That second line is called the cryptography key. For example, have a look at Figure 10.3.

Each letter of the original text is substituted with the respective key letter. For example, based on Figure 10.3, if the user enters "test", the encrypted text is "binb".

Write a program that reads a string up to 100 characters, the key string up to 26 characters, and encrypts the lowercase characters of the input string. Then, the encrypted string should be decrypted and the program should display the original string. Notice that the key characters must appear only once.

```
#include <stdio.h>
#include <string.h>

#define LETTERS 26

int main()
{
  char str[100], key[LETTERS];
  int i, j, len, found;

  printf("Enter original text: ");
  gets(str);

  do
  {
    printf("Enter key (%d different characters): ", LETTERS);
    gets(key);
    found = 0;
```

a	b	c	d	e	f	g	h	i	j	k	l	m	n	o	p	q	r	s	t	u	v	w	x	y	z
d	y	p	r	i	a	j	u	h	t	q	w	e	s	f	o	v	c	n	b	l	x	m	k	z	g

FIGURE 10.3
Single transformation encryption example.

```
  for(i = 0; i < LETTERS; i++)
  {
    for(j = i+1; j < LETTERS; j++)
    {
      if(key[i] == key[j])
      {
        found = 1;
        printf("Key characters should be different\n");
        break;
      }
    }
    if(found == 1)
      break;
  }
}
while(found != 0);

len = strlen(str);
for(i = 0; i < len; i++)
{
  if(str[i] >= 'a' && str[i] <= 'z')
    str[i] = key[str[i]-'a'];
}
printf("Encrypted text: %s\n", str);
for(i = 0; i < len; i++)
{
  for(j = 0; j < LETTERS; j++)
  {
    if(str[i] == key[j])
    {
      str[i] = 'a'+j;
      break;
    }
  }
}
printf("Original text:%s\n", str);
return 0;
}
```

10.41 What is the output of the following program?

```
#include <stdio.h>
int main()
{
  char arr[7][10] = {"Monday", "Tuesday", "Wednesday", "Thursday",
  "Friday", "Saturday", "Sunday"};
  int i;

  for(i = 0; i < 7; i++)
    if(arr[i][2] == 'n' && arr[i][3] == 'd' && *(arr[i]+4) == 'a')
      printf("%s is No.%d week day\n", arr[i], i+1);
  return 0;
}
```

Answer: The characters of "Monday" are stored in the first row of arr, the characters of "Tuesday" in the second row, and so on. The **for** loop checks each row of arr and displays the strings whose third, fourth, and fifth characters are 'n', 'd', and 'a', respectively. Notice that the expression *(arr[i]+4) is equivalent to arr[i][4]. Therefore, the program displays

```
Monday is No.1 week day
Sunday is No.7 week day
```

10.42 Write a program that reads 10 names of up to 40 characters, stores them in an array, and displays those that begin with 'a' and end with 's'.

```c
#include <stdio.h>
#include <string.h>

#define NUM 10

int main()
{
  char str[NUM][40];
  int i, len;

  for(i = 0; i < NUM; i++)
  {
    printf("Enter name: ");
    gets(str[i]);

    len = strlen(str[i]);
    if(len && str[i][0] == 'a' && str[i][len-1] == 's')
      printf("%s\n", str[i]);
  }
  return 0;
}
```

10.43 Write a program that reads 10 names of up to 40 characters, stores them in an array, and uses an array of pointers to display them in alphabetical order.

```c
#include <stdio.h>
#include <string.h>

#define NUM 10

int main()
{
  char *ptr[NUM], *temp, str[NUM][40];
  int i, j;

  for(i = 0; i < NUM; i++)
  {
    printf("Enter name: ");
    gets(str[i]);
    ptr[i] = str[i]; /* The elements of the array point to the input
      strings. */
  }
  for(i = 0; i < NUM; i++)
```

```
{
  for(j = i+1; j < NUM; j++)
  {
    /* If the string pointed to by ptr[j] is less than the string
       pointed to by ptr[i], swap the respective pointers. */
    if(strcmp(ptr[j], ptr[i]) < 0)
    {
      temp = ptr[j];
      ptr[j] = ptr[i];
      ptr[i] = temp;
    }
  }
}
for(i = 0; i < NUM; i++)
  printf("%s\n", ptr[i]);
return 0;
}
```

Unsolved Exercises

10.1 Write a program that reads characters until the sum of their ASCII values exceeds 500 or the user enters 'q'. The program should display how many characters were read.

10.2 Write a program that reads characters and displays how many characters between the first two consecutive '*' are (a) letters (b) digits, and (c) other than letters and digits. If there are no two '*', the program should display an informative message. For example, if the user enters: 1abc*D2Efg _ #!*345Higkl*mn+op*qr the program should display: Between first two stars (letters:4, digits:1, other:3).

10.3 Write a program that reads two strings up to 100 characters and displays how many times the second string is contained in the first one. The length of the second string should be less or equal to the first one.

10.4 Modify 10.39 (Exercise) to display the words in reverse order. For example, if the user enters "imagine the case", the program should display "case the imagine".

10.5 Write a program that reads three strings up to 100 characters and stores them in three arrays (i.e., str1, str2, and str3). Then, the program should copy their contents one place right, meaning that the content of the str3 should be copied to str1, the content of str1 to str2, and that of str2 to str3.

10.6 Write a program that reads characters and stores them in an array of 100 places with the restriction that none duplicated character is stored. If the user enters 'q', the insertion of characters should terminate.

10.7 Write a program that reads two strings up to 100 characters and removes every appearance of the second string inside the first one. After each removal, the remaining part of the first string should be shifted to the left, a number of places equal to the characters of the second string. The program should display the first string, before it ends. For example, if the first string is "this that" and the second is "th", the program should display "is at".

10.8 Write a program that reads two strings up to 100 characters and displays the longest part in the first string that doesn't contain any character of the second string. For example, if the first string is "example" and the second one is "day", the program should display "mple" because that part doesn't contain any character of the "day".

10.9 Write a program that reads an integer and converts it to a string. For example, if the user enters 12345, the program should store the characters '1', '2', '3', '4' and '5' into an array.

10.10 Write a program that reads two strings up to 100 characters and displays the largest part of the first string that contains exclusively characters from the second string. For example, if the first string is "programming" and the second string is "im", the program should display "mmi". Assume that the second string contains different characters.

10.11 Write a program that reads an integer in the form of a string and converts the string into that number. For example, if the user enters "12345", the program should convert that string to the number 12345 and assign it to a variable.

10.12 Write a program that reads a string up to 8 characters that represents a hexadecimal number (accepted characters 0-9, a-f, A-F) and displays the corresponding decimal value. For example, if the user enters "1AF", the program should display 431.

10.13 Write a program that reads 20 strings up to 100 characters and stores them in a two-dimensional array. Then, the program should read a string, and if it is found in the array it should be removed. To remove the string, move the strings below one row up. In the place of the last row moving up, insert the null character. For example, if the array were

```
one
two
three
four
...
```

and the user enters two, the array would be

```
one
three
four
...
```

11

Functions

In C, a function is an independent block of code that performs a specific task when called, and it may return a value to the calling program. Each function is essentially a small program, with its own variables and statements. A program is divided into smaller parts using functions, making it easier to understand, modify, and control. Another benefit of functions is that they are reusable. For example, the printf() and scanf() functions can be used in every C program.

Up to this point, we've written just one function, the main(). In this chapter, you'll learn how to declare and use your own function within your programs.

Function Declaration

A function declaration or function prototype specifies the name of the function, its return type, and a list of parameters. The general form of function declaration is

```
return_type function_name(parameter_list);
```

Try to choose descriptive names for your functions. It's much easier to read the code of a function when its name indicates its role. For example, if you write a function that calculates the sum of some numbers, name it something like sum rather than an arbitrary name like func, test, or lala.

Usually, the declarations of functions are put in a separate file. For example, the prototypes of C library functions reside in several header files. For each library function used, the program should use the **#include** directive to add the file that contains the declaration of the function. For example, the declarations of printf() and scanf() functions reside in stdio.h, which is included in the program with **#include** <stdio.h> line.

For the sake of simplicity, we'll declare our own functions in the same file with main().

Return Type

A function may return one value at most. The **return_type** specifies the type of the value that the function returns. A function may return any type of variable, except arrays. If the return type is missing, the function is presumed to return a value of type **int**.

To indicate that a function doesn't return any value, the return type should be set to **void**.

Function Parameters

A function may take a list of parameters, separated by commas. Each parameter is preceded by its type. As we'll see later, a function parameter is essentially a variable of the function, which will be assigned with a value when the function is called.

If the function has no parameters, we use either the word **void** or an empty pair of parentheses. Although we'll use an empty pair in our examples—this is a habit coming from an older version of C—if you want to be in full accordance with the C standard you should use the word **void** between the parentheses. The reason is that a pair of empty parentheses means that the function takes an unknown number of parameters and not any.

Let's see some examples of function declarations:

```
void show(char ch); /* Declare a function with name show, which takes a
  char parameter and returns nothing. */
double show(int a, float b); /* Declare a function with name show, which
  takes an integer and a float parameter and returns a value of type
  double. */
int *show(int *ptr1, double a); /* Declare a function with name show,
  which takes a pointer to integer parameter and a double parameter and
  returns a pointer to integer. */
```

Function Definition

The general form of the function definition is

```
return_type function_name(parameter_list)
{
  /* Function body */
}
```

The first line of the function's definition must resemble its declaration, with the difference that no semicolon is added at the end. The code or else the body of the function contains declarations and statements enclosed in braces.

The function's body is executed only if the function is called somewhere in the program. The execution of a function terminates if either an exit statement (i.e., **return**) is called or its last statement is executed.

Now that we've seen how to define a function, let's see an example of using a function:

```
#include <stdio.h>

void test(); /* Function declaration. */

int main()
{
  test(); /* Function call. */
  return 0;
}
```

```
void test()/* Function definition. */
{
  /* Function body. */
  printf("In\n");
}
```

Although C doesn't require to put function definitions after main(), defining functions in that order makes it easier for the reader to locate the starting point of the program. However, if a function definition is put before the first call of the function (i.e., in the aforementioned program before main()), it's not necessary to declare the function. Just remember that the compiler should have seen the prototype of a function before calling it in order to know explicitly how to call it.

return Statement

The **return** statement is used to terminate immediately the execution of a function. The execution of the program continues from the point where the function call was made.

In some programs so far, we've used the **return** statement to terminate the main() function, that is, the program itself. For example, the following program terminates if the user enters the value 2, otherwise it prints the input value:

```
#include <stdio.h>
int main()
{
  int num;

  while(1)
  {
    printf("Enter number: ");
    scanf("%d", &num);

    if(num == 2)
      return 0; /* Program termination. */
    else
      printf("Num = %d\n", num);
  }
  return 0; /* The code doesn't reach here. */
}
```

Notice that the last **return** will never be executed because the first **return** terminates the program.

The value returned by main() indicates the termination status of the program. To indicate normal termination main() should return 0, whereas a value other than 0 indicates abnormal termination.

If the function returns nothing, just write **return**. The **return** statement at the end of a **void** function is unnecessary because the function will return automatically.

In the following example, the avg() function compares the values of the two parameters, and if they are different, it displays their average. If they are equal, the function terminates.

```
void avg(int a, int b)
{
  /* Function body. */
  if(a == b)
    return;

  printf("%f\n", (a+b)/2.0);
  /* The return statement is unnecessary. */
}
```

If the function is declared to return a value, then the **return** statement should be followed by a returned value. This value is returned to the point at which the function was called. For example, we modified avg() to return an integer value.

```
int avg(int a, int b)
{
  if(a == b)
    return 0;

  printf("%f\n", (a+b)/2.0);
  return 1; /* Now, we must use the return statement to return a value. */
}
```

The type of the returned value should match the function's return type. If it doesn't match, the compiler will try to convert the returned value to the return type. For example,

```
int test()
{
  return 4.9;
}
```

Since test() is declared to return an **int** value, the returned value is implicitly converted to **int** and the function returns 4.

Function Call

When a function is called, the execution of the program continues with the execution of the function's code. When it terminates, the program returns to the point at which the function was called. A function can be called as many times as it is needed.

When calling a function, the compiler allocates memory to store the function's parameters and the variables that are declared within its body. This memory is reserved from a specific part of the memory, called *stack*. Once the function terminates, this memory is automatically deallocated.

Function Call without Parameters

A call to a function that doesn't take any parameters is made by writing the function name followed by a pair of empty parentheses. The calling program doesn't pass any data to the called function.

In the following program, the calling program, that is, the `main()` function, calls twice the `test()` function:

```
#include <stdio.h>

void test();

int main()
{
  printf("Call_1 ");
  test(); /*Function call. The parentheses are empty, because the
    function doesn't take any parameters. */
  printf("Call_2 ");
  test(); /* Second function call. */
  return 0;
}
void test()/* Function definition. */
{
  /* Function body. */
  int i;
  for(i = 0; i < 2; i++)
    printf("In ");
}
```

At the first call of `test()`, the program continues with the execution of the function body. When `test()` terminates, the execution of the program returns to the calling point and continues with the execution of the next statement. Therefore, the main program displays `Call_2` and calls `test()` again. As a result, the program displays `Call_1 In In Call_2 In In`.

In the following program, `test()` returns an integer value:

```
#include <stdio.h>

int test();

int main()
{
  int sum;

  sum = test(); /* Function call. The returned value is stored in sum. */
  printf("Sum = %d\n", sum);
  return 0;
}

int test()
{
  int i = 10, j = 20;
  return i+j;
}
```

`test()` declares two integer variables with values 10 and 20 and returns their sum, that is, 30. This value is stored in the `sum` variable and the program displays `Sum = 30`.

Notice that it is not needed to declare the variable `sum`. We could write `printf("Sum = %d\n", test());` In this way, `test()` is first executed and then `printf()` prints the return value.

Function Call with Parameters

A call to a function that takes parameters is made by writing the function name followed by a list of arguments, enclosed in parentheses. The difference between *parameters* and *arguments* is that the term *parameter* refers to the variables that appear in the definition of the function while the term *argument* refers to the expressions that appear in the function call. For example, consider the following program:

```c
#include <stdio.h>

int test(int x, int y);

int main()
{
  int sum, a = 10, b = 20;

  sum = test(a, b); /* The variables a and b become the function's
    arguments. */
  printf("Sum = %d\n", sum);
  return 0;
}
int test(int x, int y)/* The variables x and y are called parameters. */
{
  return x+y;
}
```

The argument can be any valid expression, such as constant, variable, math, or logical expression, even another function with a return value.

When a function is called, the number of the arguments and their types should match the number and the types of the corresponding parameters in the function definition. If the arguments are less, the compiler will raise an error message.

If the types of the arguments don't match the types of the parameters, the compiler will try to convert implicitly the types of the mismatch arguments to the types of the corresponding parameters. If it succeeds, the compiler may display a warning message to inform the programmer for the type conversion. If not, it will raise an error message.

For example, suppose that the test() function of the aforementioned program is called like this: test(10.9, b).

Since the type of the first parameter is **int**, the compiler will pass the value 10 to test() and the program would be compiled.

Let's see in more detail what happens when test() is called. The arguments are evaluated and its values are assigned one-to-one to the corresponding function's parameters. Essentially, each parameter is a variable that is initialized with the value of the corresponding argument.

Therefore, the values of x and y are initialized with the values of a and b (10 and 20), respectively. Figure 11.1 assumes some arbitrary values to show you what happens in the memory when test() is called.

When the program is executed, the compiler reserves eight bytes (i.e., 100-107) to store the values of the integers a and b. When test() is called, the compiler reserves another eight bytes (i.e., 2000-2007) to store the values of the integers x and y. Then, it copies the values of the arguments a and b in the corresponding locations of the parameters x and y.

Since the memory locations of x and y are different from those of a and b, any changes in the values of x and y don't affect the values of a and b. When test() terminates, the memory reserved for x and y is deallocated automatically.

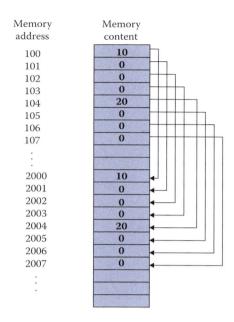

FIGURE 11.1
Passing arguments.

Passing Values

There are two ways to pass values to a function: *by value* and *by reference*.

In the aforementioned program, the arguments are passed *by value*. As discussed, when the arguments are passed *by value* any change in the values of the function's parameters doesn't affect the values of the arguments. Here's another example:

```
#include <stdio.h>

void test(int x);

int main()
{
  int a = 10;

  test(a);
  printf("Val = %d\n", a);
  return 0;
}

void test(int x)
{
  x = 20;
}
```

Since the variables a and x are stored in different memory locations, the change of x doesn't affect the value of a and the program displays 10.

At that point, a good question would be: What would the program output if we change the name of x to a?

As we'll see later, the variables of a function belong exclusively to that function and they are not related with the variables of other functions, even if they are named the same. Therefore, the output would be the same.

If we want the function to be able to modify the value of an argument, we should pass it *by reference*, meaning that we should pass its address. For example,

```c
#include <stdio.h>

void test(int *ptr1);

int main()
{
  int *ptr, i = 10;

  ptr = &i;
  test(ptr); /* Without using the variable ptr we could write test(&i); */
  printf("Val = %d\n", i);
  return 0;
}

void test(int *ptr1)
{
  *ptr1 = 20;
}
```

When test() is called, we have ptr1 = ptr = &i. Since ptr1 points to the address of i, the function may change the value of i. Therefore, the statement *ptr1 = 20; modifies the value of i and the program displays Val = 20.

> *Since a function can't return more than one value, passing arguments by reference is the most flexible way to change the values of the arguments.*

When a function is called, it's allowed to pass some arguments *by value* and some others *by reference*. For example,

```c
#include <stdio.h>

void test(int *ptr1, int a);

int main()
{
  int i = 100, j = 200;

  test(&i, j);
  printf("%d %d\n", i, j);
  return 0;
}

void test(int *ptr1, int a)
{
  *ptr1 = 300;
  a = 400;
}
```

In this program, j passes *by value* and i by reference. Since the function has access to the address of i, it changes its value from 100 to 300. Therefore, the program displays 300 and 200.

Exercises

11.1 In the following program, which one of the scanf() and printf() functions may change the value of a?

```
#include <stdio.h>
int main()
{
  int a;

  scanf("%d", &a);
  printf("Value:%d\n", a);
  return 0;
}
```

Answer: scanf() may change the value a because the address of a is passed to it. On the other hand, printf() can't change the value of a because it is passed by value.

11.2 Write a function that takes as parameter the radius of a circle and returns its area. Write a program that reads the length of the radius, calls the function, and displays the return value.

```
#include <stdio.h>

double area(double radius);

int main()
{
  double len;

  do
  {
    printf("Enter radius: ");
    scanf("%lf", &len);
  } while(len <= 0);

  printf("Circle area is %f\n", area(len));
  return 0;
}

double area(double radius)
{
  return 3.14*radius*radius;
}
```

11.3 Write two functions that take an integer parameter and return the square and the cube of this number, respectively. Write a program that reads an integer and uses the functions to display the sum of the number's square and cube.

```
#include <stdio.h>

int square(int a);
int cube(int a);

int main()
{
  int i, j, k;

  printf("Enter number: ");
  scanf("%d", &i);

  j = square(i);
  k = cube(i);
  printf("sum = %d\n", j+k); /* Without using the variables j and
    k, we could write printf("sum = %d\n",square(i) + cube(i)); */
  return 0;
}

int square(int a)
{
  return a*a;
}

int cube(int a)
{
  return a*a*a;
}
```

11.4 Write a function that takes as parameters three values and displays their minimum. Write a program that reads three grades and uses the function to display their minimum.

```
#include <stdio.h>

float min(float a, float b, float c);

int main()
{
  float i, j, k;

  printf("Enter grades: ");
  scanf("%f%f%f", &i, &j, &k);

  printf("Min grade = %f\n", min(i, j, k));
  return 0;
}

float min(float a, float b, float c)
{
  if(a <= b && b <= c)
    return a;
  else if(b < a && b < c)
    return b;
  else
    return c;
}
```

11.5 Write a function that takes as parameters an integer and a character and displays the character as many times as the value of the integer. Write a program that reads an integer and a character and uses the function to display the character.

```c
#include <stdio.h>

void show_char(int num, char ch);

int main()
{
  char ch;
  int i;

  printf("Enter character: ");
  scanf("%c", &ch);

  printf("Enter number: ");
  scanf("%d", &i);

  show_char(i, ch);
  return 0;
}
void show_char(int num, char ch)
{
  int i;

  for(i = 0; i < num; i++)
    printf("%c", ch);
}
```

Comments: The type of the show_char() is declared **void** because it doesn't return any value.

11.6 What is the output of the following program?

```c
#include <stdio.h>

int f(int a);

int main()
{
  int i = 10;
  printf("%d\n", f(f(f(i))));
  return 0;
}

int f(int a)
{
  return a+1;
}
```

Answer: Each time f() is called it returns the value of its argument incremented by one. The calls are executed from the inner to the outer. Therefore, the first call returns 11, which becomes the value of the argument in the second call. The second call returns 12, which becomes the value of the argument in the third call. Therefore, the program displays 13.

11.7   Write a function that takes as parameter a character and uses the `switch` statement to return the same character if it's 'a', 'b', or 'c', otherwise the character 'z'. Write a program that reads a character, calls the function, and displays the return character.

```c
#include <stdio.h>

char check(char ch);

int main()
{
  char ch;

  printf("Enter character: ");
  scanf("%c", &ch);

  printf("%c\n", check(ch));
  return 0;
}
char check(char ch)
{
  switch(ch)
  {
    case 'a':
    case 'b':
    case 'c':
    return ch;

    default:
    return 'z';
  }
}
```

11.8   Write a function that takes as parameters three values and returns the average of those within [1,2]. Write a program that reads the prices of three items, calls the function, and displays the return value.

```c
#include <stdio.h>

double avg(double a, double b, double c);

int main()
{
  double i, j, k, ret;

  printf("Enter prices: ");
  scanf("%lf%lf%lf", &i, &j, &k);

  ret = avg(i, j, k);
  if(ret == -1)
    printf("No item with price in [1, 2]\n");
  else
    printf("Avg = %f\n", ret);
  return 0;
}
```

```
double avg(double a, double b, double c)
{
  int k = 0;
  double sum = 0;

  if(a >= 1 && a <= 2)
  {
    sum += a;
    k++;
  }
  if(b >= 1 && b <= 2)
  {
    sum += b;
    k++;
  }
  if(c >= 1 && c <= 2)
  {
    sum += c;
    k++;
  }
  if(k != 0)
    return sum/k;
  else
    return -1;
}
```

11.9   What is the output of the following program?

```
#include <stdio.h>

void test(int *ptr1, int *ptr2);

int main()
{
  int i = 100, j = 200;

  test(&i, &j);
  printf("%d %d\n", i, j);
  return 0;
}

void test(int *ptr1, int *ptr2)
{
  int sum, diff;

  sum = *ptr1 + *ptr2;
  diff = *ptr1 - *ptr2;

  *ptr1 = sum;
  *ptr2 = diff;
}
```

Answer: Since the addresses of i and j are passed to test(), it may change their values.

When test() is called, we have ptr1 = &i and ptr2 = &j. Therefore, the values of *ptr1 and *ptr2 are 100 and 200, respectively.

With the statement `sum = *ptr1+*ptr2 = 100+200 = 300`, `sum` becomes 300. Since `ptr1` points to the address of `i`, the statement `*ptr1 = sum` makes `i` equal to 300.

Similarly, with the statement `diff = *ptr1-*ptr2 = 100-200 = -100`, `diff` becomes -100. Since `ptr2` points to the address of `j`, the statement `*ptr2 = diff` makes `j` equal to -100.

As a result, the program displays `300  -100`.

11.10 Write a function that takes an integer parameter (i.e., n) and returns the result of $1^3 + 2^3 + 3^3 + \cdots + n^3$. Write a program that reads a positive integer up to 1000 and uses the function to display the result of the expression.

```c
#include <stdio.h>

double sum_cube(int num);

int main()
{
  int i;

  do
  {
    printf("Enter number: ");
    scanf("%d", &i);
  } while(i < 0 || i > 1000);
  printf("Result = %.0f\n", sum_cube(i));
  return 0;
}

double sum_cube(int num)
{
  int i;
  double sum; /* It's declared as double in order to store larger
    numbers. */
  sum = 0;
  for(i = 1; i <= num; i++)
    sum += i*i*i;

  return sum;
}
```

11.11 What is the output of the following program? Remind that `printf()` returns the number of the displayed characters.

```c
#include <stdio.h>
int main()
{
  int i, sum;
  float j = 1.2345;

  sum = 0;
  for(i = 0; i < 3; i++)
    sum = sum + printf("%.2f\n", j);

  printf("Val = %d\n", sum);
  return 0;
}
```

Answer: Each call to printf() displays 1.23 and adds the new line character ('\n'). Since each printf() returns 5 (three characters for the digits, one for the dot, and one for the new line character), the value of sum becomes 15. Therefore, the program outputs

```
1.23
1.23
1.23
Val = 15
```

11.12 Write a function that takes as parameters two pointers to floats and swaps their content. Write a program that reads two floats and uses the function to swap their content.

```c
#include <stdio.h>

void swap(float *ptr1, float *ptr2);

int main()
{
  float i, j;

  printf("Enter numbers: ");
  scanf("%f%f", &i, &j);

  swap(&i, &j);
  printf("i = %f j = %f\n", i, j);
  return 0;
}
void swap(float* ptr1, float* ptr2)
{
  float m;

  m = *ptr1; /* Equivalent to m = i; */
  *ptr1 = *ptr2; /* Equivalent to i = j; */
  *ptr2 = m; /* Equivalent to j = m = i; */
}
```

11.13 Write a program that generates a random number within [0, 1] with two decimal digits and prints it.

```c
#include <stdio.h>
#include <stdlib.h>
#include <time.h>

int main()
{
  int i;
  double var;

  srand((unsigned)time(NULL)); /* This srand() call sets a starting
    point related to the current time for generating random positive
    integers, each time the program runs. */
  i = rand()% 101; /* rand() returns a random integer which is
    constrained in [0,100]. */
```

```
        var = i/100.0; /* This division constrains the value within [0,1]
          with two decimal digits. */
        printf("%.2f\n", var);
        return 0;
      }
```

Variables Scope

The scope of a variable is the part of the program in which the variable can be accessed. As we'll see, a variable gets a scope when it is declared.

Local Variables

A variable declared within the body of a function is called *local*. The scope of a local variable is constrained in the function to which it belongs, meaning that it's not visible to other functions.

Since a local variable isn't visible outside the function in which it is declared, we can use the same name to declare local variables in other functions.

For example, in the following program, the local variable i declared in main() is different from the variable i declared in test(), although they have the same name:

```
#include <stdio.h>

void test();

int main()
{
  int i = 10;

  test();
  printf("I_main = %d\n", i);
  return 0;
}

void test()
{
  int i = 200;
  printf("I_test = %d\n", i);
}
```

Since they are different variables, the program displays

```
I_test = 200
I_main = 10.
```

Note that if we delete the declaration int i; from test(), the compiler would raise an error that the variable i in test() is undeclared.

The parameters of the function are also considered local variables of the function. For example,

```
#include <stdio.h>

void test(int i, int j);
```

```
int main()
{
  int i = 100, j = 100;

  test(i, j);
  printf("%d %d\n", i, j);
  return 0;
}

void test(int i, int j)
{
  int a = 2000; /* The local variables of test() are a, i and j. */
  i = j = a;
}
```

Since the parameters of a function are treated as local variables, the variables i and j of test() are different from those of main(). Therefore, the program displays 100 100.

When a function is called, the compiler allocates memory to store the values of its local variables. This memory is automatically deallocated when the function terminates. As we'll see, an exception to that rule is the local variables declared as **static**.

Since the memory of a local variable is deallocated, a function should not return the address of a local variable. For example,

```
#include <stdio.h>

int *test();

int main()
{
  int *ptr, j;

  ptr = test();
  printf("%d\n", *ptr);

  j = *ptr;
  printf("%d\n", j);
  return 0;
}
int *test()
{
  int i = 10;
  return &i;
}
```

When test() is called, the compiler allocates memory for the local variable i. Then, it returns the address of i.

Since this memory is deallocated, the value 10 may be overwritten. Therefore, this program may display 10 and a random value and not the values 10 and 10, as you'd expect.

Don't return the address of a local variable unless it is declared as **static**.

Global Variables

A variable that is declared outside of the body of any function is called *global*. The scope of a global variable is extended from the point of its declaration to the end of the file in which it is declared. As a result, it is visible to all functions that follow its declaration. Typically, when the same variable is used in many functions, many programmers choose to declare it as global instead of passing it as an argument in function calls.

When naming a global variable, choose a descriptive name that indicates its role. Don't use meaningless names that are often used for local variables, like i. The default value for an uninitialized global variable is 0.

The following program declares the global variable glob and assigns the value 10 to it.

```c
#include <stdio.h>

void add();
void sub();

int glob = 10;

int main()
{
  add();
  printf("Val = %d\n", glob);
  sub();
  printf("Val = %d\n", glob);
  return 0;
}

void add()
{
  glob++;
}

void sub()
{
  glob-- ;
}
```

Since glob is visible to all functions, they may access it. Therefore, the program displays

```
Val = 11
Val = 10.
```

A local variable is different from a global variable even if they are named the same. For example,

```c
#include <stdio.h>

void test();

int a = 100;

int main()
{
  test();
  printf("Val = %d\n", a);
  return 0;
}
```

```
void test()
{
  int a;
  a = 2000;
}
```

Since the variable a declared in test() is different from the global variable a, any changes in the value of a inside test() don't affect the value of the global a and vice versa.

Therefore, the program displays Val = 100.

Typically, the source code of a large program is split in several files. In that case, if a global variable is declared in one file and another source file needs to access it, declare it as **extern** in that second file. For example, the statement

```
extern int size;
```

informs the compiler that the variable size isn't declared in that file but in another file. Since the compiler is informed that the variable is declared elsewhere, it doesn't allocate extra memory to store it. In effect, the word **extern** enables several source files to share the same variable.

Static Variables

As discussed, the memory that is allocated to store the local variables is deallocated when the function terminates. Therefore, there is no guarantee that a local variable would still have its old value when the function is called again.

To force a local variable to retain its value, declare it as **static**. Unlike ordinary local variables, a **static** variable resides at the same memory location throughout program execution and that memory is not deallocated when the function returns. Therefore, a **static** variable retains its last value.

A **static** variable is initialized only once, the first time that the function is called. If it is not initialized with a specific value, it is automatically assigned the value 0. In next calls of the function, it retains its last value and it isn't initialized again.

Consider the following example:

```
#include <stdio.h>

void test();

int main()
{
  test();
  test();
  test();
  return 0;
}

void test()
{
  static int i = 100;
  int j = 0;

  i++;
  j++;
  printf("%d %d\n", i, j);
}
```

When test() is first called, the value of i becomes 101. Since i is declared as **static**, it retains its value and the next call makes it 102. Similarly, the next call of test() makes i equal to 103.

On the other hand, j doesn't retain its last value because it hasn't been declared as **static**. Therefore, the program displays

```
101 1
102 1
103 1
```

Arrays as Arguments

When a parameter of a function is a one-dimensional array, we write the name of the array followed by a pair of brackets. The length of the array can be omitted; in fact, this is the common practice. For example,

```
void test(int arr[]);
```

When passing an array to a function, don't put brackets after its name. For example,

```
test(arr);
```

When an array name is passed to a function, it is always used as a pointer. Essentially, the passing argument is the memory address of its first element and not a copy of the array itself. Since no copy of the array is made, the time required to pass an array to a function doesn't depend on the size of the array.

Since the array name is treated as a pointer, the function may access its elements. For example, consider the following program:

```
#include <stdio.h>

void test(int arr[]);

int main()
{
  int i, p[5] = {10, 20, 30, 40, 50};

  test(p);
  for(i = 0; i < 5; i++)
    printf("%d ", p[i]);
  return 0;
}

void test(int arr[])
{
  arr[0] = arr[1] = 0;
}
```

When test() is called, we have arr = p = &p[0]. Therefore, the statements arr[0] = 0; and arr[1] = 0; change the values of the first two elements and the program displays 0 0 30 40 50.

Alternatively, we could use pointer arithmetic to access the array elements, as shown here:

```
void test(int arr[])
{
  *arr = 0;
  arr++;
  *arr = 0;
}
```

Note that we could use the name p instead of `arr` since you know by now that local variables of different functions are not related even if they are named the same.

An array parameter can be declared as a pointer, as well. For example, the declarations

```
void test(int arr[]); and void test(int *arr);
```

are equivalent. The compiler treats both of them the same.

Our suggestion is to use the array notation in order to make it clear that the passing argument is an array.

Tip

To prevent a function from changing the values of the array elements, use the word **const** *in its declaration.*

For example, with the declaration

```
void test(const int arr[]);
```

`test()` can't modify the value of any `arr` element.

When passing an array to a function, we can pass a part of it. For example,

```
#include <stdio.h>

void test(int *ptr);

int main()
{
  int i, arr[6] = {1, 2, 3, 4, 5, 6};

  test(&arr[3]);
  for(i = 0; i < 6; i++)
    printf("%d ", arr[i]);
  return 0;
}

void test(int *ptr)
{
  int i, tmp[3] = {10, 20, 30};

  for(i = 0; i < 3; i++)
    ptr[i] = tmp[i];

  *ptr = *(ptr-1);
}
```

When `test()` is called, we pass the address of `arr[3]`. Since we use `ptr` as array, `ptr[0]` corresponds to `arr[3]`, `ptr[1]` corresponds to `arr[4]`, and `ptr[2]` to `arr[5]`.

Therefore, the **for** loop makes the values of `arr[3]`, `arr[4]`, and `arr[5]` equal to 10, 20, and 30, respectively.

Since ptr points to arr[3], the statement *ptr = *(ptr-1); is equivalent to arr[3] = arr[2].

As a result, the program displays 1 2 3 3 20 30.

 *Although we can use the **sizeof** operator to find the size of an array variable, we can't use it in a function to determine the size of an array parameter.*

For example, consider the following program:

```
#include <stdio.h>

void test(int arr[]);

int main()
{
  int p[10] = {1, 2, 3, 4, 5, 6, 7, 8, 9, 10};

  test(p);
  return 0;
}

void test(int arr[])
{
  printf("Size = %d bytes\n", sizeof(arr));
}
```

Since an array parameter is treated as a pointer, the **sizeof** operator calculates the size of a pointer variable, that is, 4 bytes, and not the actual size of the array.

Therefore, the program displays Size = 4 bytes.

An easy way to make known to the function the length of an array is to pass it as an additional argument. For example, we could declare test() like

```
void test(int arr[], int size);
```

and write test(p, 10); to call it.

Alternatively, we can define a constant (i.e., **#define** SIZE 10) to hold the length of the array and use that constant instead of passing an additional argument.

Exercises

11.14 Write the power(**int** base, **int** exp) function that raises the integer base to the positive exponent exp and returns the result. Write a program that reads an integer and a positive integer exponent, calls the function, and displays the return value.

```
#include <stdio.h>

int power(int base, int exp);

int main()
{
  int i, j;

  do
```

```
  {
    printf("Enter an integer and a positive integer (exponent): ");
    scanf("%d%d", &i, &j);
  }
  while(j < 0);

  printf("Result = %d\n", power(i, j));
  return 0;
}

int power(int base, int exp)
{
  int k, out;

  out = 1; /* Necessary initialization. */
  for(k = 0; k < exp; k++)
    out = out * base;
  return out;
}
```

11.15 Write a function that takes as parameter a string and a character and returns the appearances of the character in the string. Write a program that reads a character and a string (up to 100 characters) continuously, calls the function, and displays the return value. If the user enters the string "end", the insertion of strings should terminate.

```
#include <stdio.h>
#include <string.h>

int str_chr(char str[], char ch);
int main()
{
  char ch, str[100];

  while(1)
  {
    printf("Enter text: ");
    gets(str);
    if(strcmp(str, "end") == 0)
      break;

    printf("Enter character to search: ");
    scanf("%c", &ch);

    printf("'%c' is contained %d times in '%s'\n\n", ch,
      str_chr(str, ch), str);
    getchar();
  }
  return 0;
}

int str_chr(char str[], char ch)
{
  int i, times;

  i = times = 0;
```

```
  while(str[i] != '\0')
  {
    if(str[i] == ch)
      times++;
    i++;
  }
  return times;
}
```

11.16   Write a function that takes as parameters an array containing the students' grades in a test and two grades (i.e., A and B) and returns the average of the grades within [A, B]. Write a program that reads the grades of 50 students and the two grades A and B and uses the function to display the average. The program should force the user the value of A to be less or equal to B.

```c
#include <stdio.h>

#define SIZE 50

float avg_arr(float arr[], int min, int max);

int main()
{
  int i;
  float a, b, ret, arr[SIZE];

  for(i = 0; i < SIZE; i++)
  {
    printf("Enter grade: ");
    scanf("%f", &arr[i]);
  }
  do
  {
    printf("Enter min and max grades: ");
    scanf("%f%f", &a, &b);
  } while(a > b);

  ret = avg_arr(arr, a, b);
  if(ret == -1)
    printf("None grade in [%f,%f]\n", a, b);
  else
    printf("Avg = %.2f\n", ret);
  return 0;
}

float avg_arr(float arr[], int min, int max)
{
  int i, cnt = 0;
  float sum = 0;

  for(i = 0; i < SIZE; i++)
  {
    if(arr[i] >= min && arr[i] <= max)
    {
      cnt++;
      sum += arr[i];
    }
  }
```

```
    if (cnt == 0)
      return -1;
    else
      return sum/cnt;
}
```

11.17 Write a function that takes as parameter a string and returns its length (don't use `strlen()`). Write a program that reads a string up to 100 characters and uses the function to display its length.

```
#include <stdio.h>

unsigned int str_len(const char *str);

int main()
{
  char arr[100];

  printf("Enter text: ");
  gets(arr);
  printf("Length = %d\n", str_len(arr));
  return 0;
}

unsigned int str_len(const char *str)
{
  unsigned int i = 0;
  while (str[i] != '\0') /* Count the characters up to the null
    character. */
    i++;
  return i;
}
```

Comments: The implementation of the `str_len()` function is similar to the implementation of the `strlen()` library function.

11.18 What does the following function?

```
unsigned int test(const char *str)
{
  const char *ptr = str;

  while (*str++);  /* Equivalent to while (*str++ != '\0'); */
  return str-ptr-1;
}
```

Answer: That's a tough one. Let's explain it.

Remember that the semicolon (;) at the end of the **while** loop means that it doesn't contain any statements.

In each iteration, the **while** loop compares the value of `*str` with 0 (equivalent to '\0') and then `str` is increased, to point to the next array element. For example, in the first iteration, `*str` is equal to `str[0]`, then it becomes equal to `str[1]`, and so on. Once `*str` becomes equal to the null character, the loop terminates.

Recall from Chapter 8 and pointer arithmetic that the result of the subtraction of two pointers that point to the same array is the number of elements, in our case the number of characters, between them.

`ptr` points to the first element, while `str` after its last increase points to the next character after the null character. That's why we put –1 to subtract this place.

So, what really does this function? It returns the length of the string stored into `str`.

In fact, it does the same thing as the `str_len()` function of the previous exercise. The reason we added this exercise is to show you that a problem may be solved in several ways, others simpler and others more complex.

Even if `test()` is executed a bit faster than `str_len()`, the reader of `test()` needs much more time to realize what exactly this function does. Once more, try to write simple and clear code for your own benefit and for those who are going to read your code.

In another example of complex coding, the following listing uses a **for** loop instead of a **while** loop to produce the same result.

```c
unsigned int test(const char *str)
{
  const char *ptr = str;

  for(; *str; str++);
  return str-ptr;
}
```

11.19 Write a function that takes as parameters two strings and uses them as pointers to copy the second one into the first (don't use `strcpy()`). Write a program that reads two strings up to 100 characters and uses the function to swap them and display their content.

```c
#include <stdio.h>

void copy(char s1[], char s2[]);

int main()
{
  char str1[100], str2[100], tmp[100];

  printf("Enter first text: ");
  gets(str1);

  printf("Enter second text: ");
  gets(str2);

  copy(tmp, str1); /* Copy the first string into tmp. */
  copy(str1, str2); /* Copy the second string into str1. */
  copy(str2, tmp); /* Copy the first string into str2. */
  printf("\nFirst text:%s\n", str1);
  printf("Second text:%s\n", str2);
  return 0;
}

void copy(char s1[], char s2[])
{
  while(*s2 != '\0')
```

```
    {
      *s1 = *s2;
      s1++;
      s2++;
    }
    *s1 = '\0';
}
```

11.20 Write a function that takes as parameters two strings and returns 1 if the second string is contained at the end of the first one. Otherwise, it should return 0. Write a program that reads two strings up to 100 characters and uses the function to check whether the second string is contained at the end of the first one or not.

```c
#include <stdio.h>
#include <string.h>

int str_end(char str1[], char str2[]);

int main()
{
  char str1[100], str2[100];
  int ret;

  printf("Enter first text: ");
  gets(str1);

  printf("Enter second text: ");
  gets(str2);

  ret = str_end(str1, str2);
  if(ret == 0)
    printf("%s is not at the end of %s\n", str2, str1);
  else
    printf("%s is at the end of %s\n", str2, str1);
  return 0;
}

int str_end(char str1[], char str2[])
{
  int i, len1, len2;

  len1 = strlen(str1);
  len2 = strlen(str2);

  if(len1 < len2)/* If the length of the second string is bigger,
    the function returns. */
    return 0;

  for(i = 1; i <= len2; i++)
    if(str1[len1-i] != str2[len2-i])/* If two characters are not the
      same, it's not needed to compare the rest and the function
      returns. */
      return 0;

  /* If this point is reached, it means that all compared
    characters were the same, and the function returns 1. */
  return 1;
}
```

11.21 What is the output of the following program?

```
#include <stdio.h>

void test(int *ptr1, int *ptr2);

int main()
{
   int i = 10, j = 20;

   test(&i, &j);
   printf("i = %d j = %d\n", i, j);
   return 0;
}
void test(int *ptr1, int *ptr2)
{
   int m, *tmp;

   tmp = ptr1;
   ptr1 = &m;

   *ptr1 = 100;
   *ptr2 += m;

   ptr2 = tmp;
   *ptr2 = 100;
}
```

Anwser: When test() is called, we have ptr1 = &i and ptr2 = &j. With the statements ptr1 = &m; and *ptr1 = 100, m becomes 100. Since ptr2 points to the address of j, *ptr2 is equal to 20. Therefore, the statement *ptr2 += m is equivalent to j = j+m and j becomes 120.

Since tmp points to the address of i, the statement ptr2 = tmp is equivalent to ptr2 = &i. Therefore, the statement *ptr2 = 100 changes the value of i to 100. As a result, the program displays i = 100 j = 120.

11.22 Write a function that takes as parameters two pointers to floats and uses them to return a pointer to the float with the greater value. Write a program that reads two floats and uses the function to display the greater.

```
#include <stdio.h>

double *max(double *ptr1, double *ptr2);

int main()
{
   double *ptr, i, j;

   printf("Enter numbers: ");
   scanf("%lf%lf", &i, &j);

   ptr = max(&i, &j);
   printf("The max of %f and %f is %f\n", i, j, *ptr);
   return 0;
}
double *max(double *ptr1, double *ptr2)
{
   if(*ptr1 > *ptr2)
      return ptr1;
```

```
    else
        return ptr2;
}
```

Comments: max() compares the two numbers and returns the pointer to the greater one. This pointer is copied to ptr and printf() displays the greater number. Note that without declaring ptr, we could write

```
printf("The max value of %f and %f is %f\n", i, j, *max(&i, &j));
```

11.23 What is the output of the following program?

```
#include <stdio.h>

void test(int *ptr1, int *ptr2, int *ptr3);

int main()
{
    int i = 10, j = 20, k = 30;

    test(&i, &j, &k);
    printf("i = %d, j = %d, k = %d\n", i, j, k);
    return 0;
}

void test(int *ptr1, int *ptr2, int *ptr3)
{
    ptr1 = ptr2 = ptr3;

    *ptr1 = 100;
    *ptr2 = 200;
    *ptr3 = 0;
}
```

Answer: When test() is called, we have ptr1 = &i, ptr2 = &j, and ptr3 = &k. Then, the statement ptr1 = ptr2 = ptr3; makes ptr1 = &k and ptr2 = &k.

Therefore, the statement *ptr1 = 100; changes the value of k to 100. Since ptr2 points to the address of k, the statement *ptr2 = 200; makes k equal to 200. Similarly, the statement *ptr3 = 0; makes it 0.

The values of i and j remain the same and the program displays i = 10, j = 20, k = 0.

11.24 What is the output of the following program?

```
#include <stdio.h>
#include <string.h>

void test(char ch, char *ptr);

int main()
{
    char str[20] = "bacdefghij";

    test(*str-1, &str[5]);
    printf("%s\n", str);
    return 0;
}
```

```
void test(char ch, char *ptr)
{
  strcpy(ptr, "12345");
  *ptr = ch;
}
```

Answer: When `test()` is called, we have `ch = *str+1 = str[0]-1 = 'b'-1`, so `ch` becomes equal to the character before `'b'`, which is `'a'`.

We also have `ptr = &str[5]`. Since `ptr` points to the sixth element of `str`, the statement `strcpy(ptr, "12345");` changes the content of `str` to `"bacde12345"`.

The statement `*ptr = ch;` is equivalent to `str[5] = ch = 'a'`.

Therefore, the program displays `bacdea2345`.

11.25 Write a function that takes as parameters two strings and returns a pointer to the larger string. If both strings have the same number of characters, it should return `NULL`. Write a program that reads two strings up to 100 characters and uses the function to display the larger one.

```
#include <stdio.h>
#include <string.h>

char *max_str(char str1[], char str2[]);

int main()
{
  char *ptr, str1[100], str2[100];

  printf("Enter first text: ");
  gets(str1);

  printf("Enter second text: ");
  gets(str2);

  ptr = max_str(str1, str2);
  if(ptr == NULL)
    printf("Same number of characters.\n");
  else
    printf("Result:%s\n", ptr);

  return 0;
}

char *max_str(char str1[], char str2[])
{
  int i, j;

  i = strlen(str1);
  j = strlen(str2);

  if(i > j)
    return str1;
  else if(i < j)
    return str2;
  else
    return NULL;
}
```

11.26 What is the output of the following program?

```c
#include <stdio.h>

void test(int *ptr1, int *ptr2, int a);

int main()
{
  int i = 1, j = 2, k = 3;

  test(&i, &j, k);
  printf("%d %d %d\n", i, j, k);
  return 0;
}
void test(int *ptr1, int *ptr2, int a)
{
  ptr1 = ptr2;

  *ptr1 = 100;
  *ptr2 = 200;
  a = *ptr1 + *ptr2;
  printf("%d\n", a);
}
```

Answer: When test() is called, we have ptr1 = &i, ptr2 = &j, and a = k. With the statement ptr1 = ptr2; ptr1 and ptr2 point to the same address, that is, the address of j. Therefore, the statement *ptr1 = 100; makes j equal to 100.

Since ptr2 points to the address of j, the statement *ptr2 = 200; changes the value of j to 200.

Since the statement a = *ptr1 + *ptr2; is equivalent to a = j+j = 200+200 = 400, test() displays 400.

Since any changes in the value of a don't affect k, k remains the same, that is, 3. As a result, the program displays

```
400
1 200 3
```

11.27 Write a function that takes as parameters two strings, uses them as pointers, and returns 0 if they are identical or the ASCII difference of their first two characters that are not the same. Write a program that reads two strings up to 100 characters, calls the function, and displays the return value.

```c
#include <stdio.h>
#include <string.h>

int str_cmp(const char *str1, const char *str2);

int main()
{
  int ret;
  char buf1[100], buf2[100];

  printf("Enter first string: ");
  gets(buf1);
  printf("Enter second string: ");
  gets(buf2);

  ret = str_cmp(buf1, buf2);
```

```
  if(ret == 0)
    printf("%s = %s\n", buf1, buf2);
  else if(ret < 0)
    printf("%s < %s\n", buf1, buf2);
  else
    printf("%s > %s\n", buf1, buf2);
  return 0;
}

int str_cmp(const char *str1, const char *str2)
{
  while(*str1 == *str2)
  {
    if(*str1 == '\0')
      return 0;
    str1++;
    str2++;
  }
  return *str1 - *str2;
}
```

Comments: If two different characters are found, the `while` loop terminates and the function returns their ASCII difference.

The implementation of the `str_cmp()` function is similar to the implementation of the `strcmp()` library function.

11.28 What is the output of the following program?

```
#include <stdio.h>

int *test(int *ptr1, int *ptr2);

int main()
{
  int *ptr, i = 1, arr[] = {10, 20, 30, 40, 50, 60, 70};

  ptr = test(arr+2, &i);
  printf("%d %d\n", arr[4], *ptr);
  return 0;
}

int *test(int p[], int *ptr2)
{
  p[2] = 200;
  return p+*ptr2;
}
```

Answer: When `test()` is called, we have `p = arr+2` and `ptr2 = &i`. Therefore, `p[0]` is equal to `arr[2]`, `p[1]` equals `arr[3]`, and `p[2]` equals `arr[4]`. With the statement `p[2] = 200;` the value of `arr[4]` becomes 200.

Since `ptr2` points to the address of `i`, the expression `p+*ptr2` is equivalent to `p+i = p+1`. Since `p` points to the third element of the array `arr`, `test()` returns a pointer to its fourth element.

Therefore, the program displays 200 40.

11.29 Write a function that takes as parameters a character, an integer, and a string and uses it as a pointer to check whether the character exists in the string or not. If not, it should return NULL. Otherwise, if the integer is 0, it should return a pointer to its first appearance, otherwise to the last one. Write a program that reads a string (up to 100 characters), a character, and an integer, calls the function and displays the part of the string after the appearance of the character. For example, if the user enters "bootstrap", 't' and 0, the program should display tstrap. If it is "bootstrap", 't' and 3, the program should display trap.

```c
#include <stdio.h>

char *str_chr(char str[], char ch, int f);
int main()
{
    char *ptr, ch, str[100];
    int flag;

    printf("Enter text: ");
    gets(str);

    printf("Enter character to search: ");
    scanf("%c", &ch);
    printf("Enter choice (0-first, other-last): ");
    scanf("%d", &flag);

    ptr = str_chr(str, ch, flag);
    if(ptr == NULL)
        printf(" '%c' is not included in the text\n", ch);
    else
        printf("The rest string is:%s\n", ptr);
    return 0;
}

char *str_chr(char str[], char ch, int f)
{
    char *tmp = NULL; /* If the character is not found, the function
        returns NULL. */
    while(*str != '\0')
    {
        if(*str == ch)
        {
            tmp = str;
            if(f == 0)/* If the character is found and the choice is
                0, the loop terminates and the function returns the pointer.
                If it isn't 0, tmp points to the place of its last
                appearance. */
            break;
        }
        str++;
    }
    return tmp;
}
```

11.30 The following program modifies the content of a string. Is there any programming bug?

```c
#include <stdio.h>
#include <string.h>

char *test();
int main()
{
  char ptr[100] = "sample";

  strcpy(ptr, test());
  printf("%s\n", ptr);
  return 0;
}

char *test()
{
  char str[] = "This is the text";
  return str;
}
```

Answer: When `test()` is called, the compiler allocates memory for the `str` array and stores the string into it. This memory location is returned.

Remember that the memory of a local variable is deallocated when the function terminates. Therefore, its content may be lost and the program might not display `"This is the text"`.

Just remember, don't return the address of a local variable unless it is declared as **static**.

If you want to change the contents of an array, the simplest way is to pass the array as an argument. For example, `test()` is modified like this:

```c
void test(char str[])
{
  strcpy(str, "This is the text");
}
```

11.31 Write a function that takes as parameters an array that contains the prices of some products in a shop and its size and returns the minimum, the maximum, and the average of the prices. Write a program that reads the prices of up to 100 products and stores them in an array. If the user enters –1, the insertion of prices should terminate. The program should use the function to display the minimum, the maximum, and the average of the prices.

```c
#include <stdio.h>

void stat_arr(float arr[], int size, float *min, float *max,
  float *avg);

int main()
{
  int i;
  float min, max, avg, arr[100];

  for(i = 0; i < 100; i++)
  {
    printf("Enter price: ");
    scanf("%f", &arr[i]);
```

```
    if (arr[i] == -1)
      break;
  }
  if (i == 0)
    return 0;
  /* The variable i indicates the number of the elements stored
     into the array. For example, if the user doesn't enter the value
     -1, i would be equal to 100. */
  stat_arr(arr, i, &min, &max, &avg);
  printf("Max=%.2f Min=%.2f Avg=%.2f\n", max, min, avg);
  return 0;
}
void stat_arr(float arr[], int size, float *min, float *max,
  float *avg)
{
  int i;
  float sum;
  sum = *min = *max = arr[0];
  for (i = 1; i < size; i++)
  {
    if (arr[i] > *max)
      *max = arr[i];
    if (arr[i] < *min)
      *min = arr[i];
    sum += arr[i];
  }
  *avg = sum/size;
}
```

Comments: Since the **return** statement returns one value at most, we have to pass pointers as additional arguments.

11.32 What is the output of the following program?

```
#include <stdio.h>

void test(int *arg);

int var = 100;

int main()
{
  int *ptr, i = 30;

  ptr = &i;
  test(ptr);
  printf("Val = %d\n", *ptr);
  return 0;
}

void test(int *arg)
{
  arg = &var;
}
```

Answer: That's a tricky one. Since the value of ptr and not its address is passed to test(), any changes in the value of arg don't affect the value of ptr.
  Therefore, the program displays Val = 30.

11.33 What is the output of the following program?

```
#include <stdio.h>

float *test(float *ptr1, float *ptr2);

int main()
{
  float a = 1.2, b = 3.4;

  *test(&a, &b) = 5.6;
  printf("val1 = %.1f val2 = %f.1\n", a, b);

  *test(&a, &b) = 7.8;
  printf("val1 = %.1f val2 = %f.1\n", a, b);
  return 0;
}

float *test(float *ptr1, float *ptr2)
{
  if(*ptr1 < *ptr2)
    return ptr1;
  else
    return ptr2;
}
```

Answer: test() returns a pointer to the parameter with the smaller value. In the first call, we have ptr1 = &a, so *ptr1 = a = 1.2. Similarly, we have ptr2 = &b, so *ptr2 = b = 3.4. Therefore, test() returns the pointer ptr1.

Since test() returns a pointer to the address of a, the statement *test(&a, &b) = 5.6; makes a equal to 5.6.

Therefore, the program displays val1 = 5.6 val2 = 3.4.

The second call of test() returns the pointer ptr2 because *ptr1 = a = 5.6 and *ptr2 = b = 3.4.

Like before, since test() returns a pointer to the address of b, the value of b becomes 7.8 and the program displays val1 = 5.6 val2 = 7.8.

11.34 Write a function that calculates the maximum common divisor (MCD) of two positive integers, according to the following Euclid's algorithm: suppose we have the integers a and b, with a > b. If b divides a precisely, then this is the MCD. If the remainder r of the division a/b isn't 0, then we divide b with r. If the new remainder of the division is 0, then the MCD is r, otherwise this procedure is repeated. Write a program that reads two positive integers and uses the function to calculate their MCD.

```
#include <stdio.h>

int mkd(int a, int b);

int main()
{
  int num1, num2;

  do
  {
    printf("Enter the first number: ");
    scanf("%d", &num1);
```

```
      printf("Enter the second number (equal or less than the
        first one): ");
      scanf("%d", &num2);
    } while((num2 > num1) || (num1 <= 0) || (num2 <= 0));

    printf("MKD of %d and %d is %d\n", num1, num2, mkd(num1, num2));
    return 0;
}

int mkd(int a, int b)
{
    int r;

    while(1)
    {
      r = a%b;
      if(r == 0)
        return b;
      else/* According to the algorithm we should divide b with r, so
        we change the values of a and b, respectively. */
      {
        a = b;
        b = r;
      }
    }
}
```

11.35 What is the output of the following program?

```
#include <stdio.h>

int *test(int *ptr1, int *ptr2);

int main()
{
    int arr[] = {1, 2, 3, 4};

    *test(arr, arr+3) = 30;
    printf("%d %d %d %d\n", arr[0], arr[1], arr[2], arr[3]);
    return 0;
}

int *test(int *ptr1, int *ptr2)
{
    *(ptr1+1) = 10;
    *(ptr2-1) = 20;
    return ptr1+3;
}
```

Answer: When test() is called, we have ptr1 = arr, so the pointer ptr1+1 points to arr[1]. Therefore, the statement *(ptr1+1) = 10; makes the value of arr[1] equal to 10.

Similarly, we have ptr2 = arr+3, so ptr2 points to arr[3]. Therefore, the statement *(ptr2-1) = 20; makes the value of arr[2] equal to 20.

Since ptr1 points to arr, the expression ptr1+3 returns a pointer to arr[3]. Therefore, the value of arr[3] becomes 30.

As a result, the program displays 1 10 20 30.

11.36   Write a function that takes as parameter an integer (i.e., N) and calculates the Nth term of the Fibonacci sequence, according to the formula F(N) = F(N-1)+F(N-2), where F(0) = 0 and F(1) = 1. Write a program that reads an integer N between 2 and 40 and uses the function to display the Nth term.

```c
#include <stdio.h>

unsigned int fib(int num);

int main()
{
  int num;

  do
  {
    printf("Enter a number between 2 and 40: ");
    scanf("%d", &num);
  } while(num < 2 || num > 40);

  printf("F(%d) = %u\n", num, fib(num));
  return 0;
}

unsigned int fib(int num)
{
  unsigned int term1, term2, sum;

  term1 = 1;
  term2 = 0;
  while(num > 1)
  {
    sum = term1 + term2;

    term2 = term1;
    term1 = sum;

    num-- ;
  }
  return sum;
}
```

Comments: The first terms of the Fibonacci sequence are 0, 1, 1, 2, 3, 5, 8, 13, 21, 34, 55, 89, 144,... For example, F(7) = 13, which is the sum of F(6) and F(5).

To calculate the Fibonacci term, we use the variable term1 to store the last sum, while the previous one is stored into term2.

11.37   What is the output of the following program?

```c
#include <stdio.h>

double *f(double ptr[]);

int main()
{
  int i;
  double a[8] = {0.1, 0.2, 0.3, 0.4, 0.5, 0.6, 0.7, 0.8};

  printf("Val = %.1f\n", *f(f(f(a))));
```

```
    for (i = 0; i < 8; i++)
      printf ("a[%d] = %.1f\n", i, a[i]);
    return 0;
}

double *f(double ptr[])
{
    (*ptr)++;
    return ptr+2;
}
```

Answer: The calls to f() are executed from the inner to the outer. When f() is first called, we have ptr = a = &a[0]. Therefore, the statement (*ptr)++; is equivalent to a[0]++, and the value of a[0] becomes 1.1.

The return value a+2 is used as an argument in the second call, meaning that the expression f(f(a)) is equivalent to f(a+2). Therefore, the value of a[2] becomes 1.3. Similarly, the return value a+4 is used as an argument in the third call and the value of a[2] becomes 1.5. Since the last call to f() returns the address of a[6], we use the * operator to display its value Val = 0.7.

Then, the program displays the elements of a:

```
    1.1 0.2 1.3 0.4 1.5 0.6 0.7 0.8
```

11.38 What is the output of the following program?

```
#include <stdio.h>

void test (int **arg);

int var = 100;

int main ()
{
    int *ptr, i = 30;

    ptr = &i;
    test (&ptr); /* The value of &ptr is the memory address of ptr,
       which points to the address of i. So, the type of this argument
       is a pointer to a pointer to an integer variable and complies to
       the function declaration. */
    printf ("Val = %d\n", *ptr);
    return 0;
}

void test (int** arg)
{
    *arg = &var;
}
```

Answer: Since the memory address of ptr is passed to test(), test() may change its value.

When test() is called, we have arg = &ptr and equivalently *arg = ptr. Therefore, the statement *arg = &var; is equivalent to ptr = &var; which means that the value of ptr is changed and it points to the address of var. Therefore, the program displays Val = 100.

11.39 An application field of the C language is the area of "Communication Protocols."
For example, we are going to show you a method to create network data frames.

Two computers, A and B, that reside in the same network may communicate only
if both know the physical address of each other. The physical address is called
MAC. For example, when A needs to communicate with B, it must know the MAC
address of B. In order to learn it, A should transmit a MAC frame to B, which contains a special message, called `ARP_Request`.

When B gets this MAC frame, it replies to A with another MAC frame, which
contains a special message, called `ARP_Reply`. The `ARP_Reply` contains the
MAC address of B. In this way, A learns the MAC address of B, so they can
communicate.

Write a program that reads the MAC address of A, the IP addresses of A and B
and creates a MAC frame which contains the ARP_Request message. The program
should display the content of the MAC frame in lines, where each line should
contain 16 bytes in hex format.

The MAC address consists of six bytes and it should be entered in the `x.x.x.x.x.x`
form (each x is an integer in [0, 255]), while the IP address consists of four bytes and
it should be entered in the `x.x.x.x` form.

Figure 11.2 shows the format of the MAC frame.

(a) Fill the first seven bytes of the `Preamble` field with `85` and the eighth byte
with `171`.

(b) The six bytes of the MAC destination address are equal to `255`.

(c) The first byte of the `Type` field is `8` and the second one is equal to `6`.

(d) Fill the four bytes of the `CRC` field with `1`s.

Figure 11.3 depicts the format of the `ARP_Request` message:

(a) The length of the `Hardware Type` field is two bytes. The first byte is `0` and
the second one is `1`.

(b) The length of the `Protocol Type` field is two bytes. The first byte is `8` and
the second one is `6`.

(c) The length of the `Hardware Length` field is one byte with value `6`.

(d) The length of the `Protocol Length` field is one byte with value `4`.

(e) The length of the `Operation` field is two bytes.

(f) Fill the six bytes of the `Target hardware address` with `0`s.

(g) Fill the four bytes of the `Target protocol address` with the IP address of B.

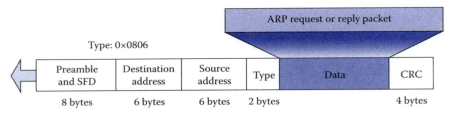

FIGURE 11.2
MAC frame format.

Hardware type		Protocol type
Hardware length	Protocol length	Operation Request 1, Reply 2
Sender hardware address (for example: 6 bytes for Ethernet)		
Sender protocol address (for example: 4 bytes for IP)		
Target hardware address (for example: 6 bytes for Ethernet) (it is not filled in a request)		
Target protocol address (for example: 4 bytes for IP)		

FIGURE 11.3
ARP_Request message format.

```c
#include <stdio.h>
#include <stdlib.h>

void Build_Frm(int MAC_src[], int IP_src[], int IP_dst[]);
void Show_Frm(unsigned char pkt[], int len);

int main()
{
  int MAC_src[6], IP_src[4], IP_dst[4];

  printf("Enter src MAC (x.x.x.x.x.x): ");
  scanf("%d.%d.%d.%d.%d.%d", &MAC_src[0], &MAC_src[1], &MAC_src[2],
    &MAC_src[3], &MAC_src[4], &MAC_src[5]);

  printf("Enter src IP (x.x.x.x): ");
  scanf("%d.%d.%d.%d", &IP_src[0], &IP_src[1], &IP_src[2], &IP_
    src[3]);

  printf("Enter dst IP (x.x.x.x): ");
  scanf("%d.%d.%d.%d", &IP_dst[0], &IP_dst[1], &IP_dst[2], &IP_
    dst[3]);

  Build_Frm(MAC_src, IP_src, IP_dst);
  return 0;
}
void Build_Frm(int MAC_src[], int IP_src[], int IP_dst[])
{
  unsigned char pkt[54] = {85, 85, 85, 85, 85, 85, 85, 171}; /*
    Initialize the first eight octets of the frame and zero the
    rest. */
  int i, j;

  for(i = 8; i < 14; i++)
    pkt[i] = 255; /* Broadcast MAC address. */
  for(i = 14, j = 0; i < 20; i++, j++)
    pkt[i] = MAC_src[j]; /* MAC source. */
```

```
pkt[20] = 8; /* Type. */
pkt[21] = 6;

pkt[22] = 0; /* Hardware Type. */
pkt[23] = 1;

pkt[24] = 8; /* Protocol Type. */
pkt[25] = 6;

pkt[26] = 6; /* Hardware Length. */
pkt[27] = 4; /* Protocol Length. */

pkt[28] = 0; /* Operation (ARP_Request). */
pkt[29] = 1;

for(i = 30, j = 0; i < 36; i++, j++)
  pkt[i] = MAC_src[j]; /* MAC source. */
for(i = 36, j = 0; i < 40; i++, j++)
  pkt[i] = IP_src[j]; /* IP source. */
/* The MAC destination in places [40-45] is initialized to 0. */
for(i = 46, j = 0; i < 50; i++, j++)
  pkt[i] = IP_dst[j]; /* IP destination. */
for(i = 50; i < 54; i++)
  pkt[i] = 1; /* CRC. */

Show_Frm(pkt, i);
}

void Show_Frm(unsigned char pkt[], int len)
{
  int i;
  for(i = 0; i < len; i++)
  {
    if((i > 0) && (i%16 == 0))
      printf("\n");
    printf("%02X ", pkt[i]);
  }
}
```

Comments: In a real network application, the network card of the system transmits this MAC frame to the Ethernet network.

Function Call with Parameter Two-Dimensional Array

The most common way to declare a function that takes as parameter a two-dimensional array is to write the name of the array followed by its dimensions. For example, test() takes as parameter a two-dimensional integer array with 5 rows and 10 columns.

```
void test(int arr[5][10]);
```

Alternatively, we can omit the first dimension and write

```
void test(int arr[][10]);
```

Exercises

11.40 Write a function that takes as parameters three two-dimensional 2 × 4 integer arrays, calculates the sum of the first two, and stores it in the third one. Write a program that reads 16 integers, stores them in two 2 × 4 arrays, and uses the function to calculate the sum of the first two.

```c
#include <stdio.h>

#define ROWS 2
#define COLS 4

void add_arrays(int arr1[][COLS], int arr2[][COLS], int arr3[]
  [COLS]);

int main()
{
int i, j, arr1[ROWS][COLS], arr2[ROWS][COLS], arr3[ROWS][COLS];

  printf("***** First array *****\n");
  for(i = 0; i < ROWS; i++)
    for(j = 0; j < COLS; j++)
    {
      printf("arr1[%d][%d] = ", i, j);
      scanf("%d", &arr1[i][j]);
    }

  printf("***** Second array *****\n");
  for(i = 0; i < ROWS; i++)
    for(j = 0; j < COLS; j++)
    {
      printf("arr2[%d][%d] = ", i, j);
      scanf("%d", &arr2[i][j]);
    }

  add_arrays(arr1, arr2, arr3);

  printf("***** Sum array *****\n");
  for(i = 0; i < ROWS; i++)
    for(j = 0; j < COLS; j++)
      printf("sum[%d][%d] = %d\n", i, j, arr3[i][j]);
  return 0;
}

void add_arrays(int arr1[][COLS], int arr2[][COLS], int arr3[]
  [COLS])
{
  int i, j;
  for(i = 0; i < ROWS; i++)
    for(j = 0; j < COLS; j++)
      arr3[i][j] = arr1[i][j] + arr2[i][j];
}
```

11.41 Write a function that takes as parameters an array of names and another name. The function should check if that name is contained in the array. If it does, the function should return a pointer to the position of that name in the array, otherwise NULL.

Write a program that reads the names of 20 students (up to 100 characters) and stores them in an array. Then, it reads another name and uses the function to check if that name is contained in the array.

```c
#include <stdio.h>
#include <string.h>

#define NUM 20
#define SIZE 100

char *find_name(char name[][SIZE], char str[]);

int main()
{
  char *ptr, str[SIZE], name[NUM][SIZE]; /* Declare an array of NUM
    rows and SIZE columns. The names of the students are stored in
    the array's rows. */
  int i;

  for(i = 0; i < NUM; i++)
  {
    printf("Enter name: ");
    gets(name[i]); /* We use the name[i] as a pointer to the
      respective i row of SIZE characters. */
  }
  printf("Enter name to search: ");
  gets(str);

  ptr = find_name(name, str);
  if(ptr == NULL)
    printf("%s is not contained\n", str);
  else
    printf("%s is contained\n", ptr);
  return 0;
}

char *find_name(char name[][SIZE], char str[])
{
  int i;
  for(i = 0; i < NUM; i++)
    if(strcmp(name[i], str) == 0)
      return name[i];
  return NULL; /* If this point is reached, the name isn't found in
    the array. */
}
```

11.42 A popular card game among children is a memory game. The game starts with a deck of identical pairs of cards face down on a table. The player selects two cards and turns them over. If they match, they remain face up. If not, they are flipped face down. The game ends when all cards are face up.

To simulate that game, write a program that uses the elements of a two-dimensional array as the cards. To test your program, use a 4 × 4 array and assign the values 1–8 to its elements (cards). Each number must appear twice. Set the values in random positions. An example of the array might be

$$\begin{bmatrix} 5 & 3 & 4 & 8 \\ 4 & 2 & 6 & 1 \\ 3 & 8 & 7 & 6 \\ 2 & 5 & 1 & 7 \end{bmatrix}$$

The program should prompt the user to select the positions of two cards and display a message to indicate if they match or not. The program ends when all cards are matched.

```c
#include <stdio.h>
#include <stdlib.h>
#include <time.h>

#define ROWS 4
#define COLS 4
void show_board(int c[][COLS], int s[][COLS]);
void sel_card(int c[][COLS], int s[][COLS], int *row, int *col);
int main()
{
  int i, j, m, r, c, r2, c2, cnt, cards[ROWS][COLS], status[ROWS]
    [COLS] = {0}; /* The status array indicates if a card faces up
    or down (0 is for down). */
  cnt = 0; /* This variable counts the number of the faced up
    cards. */
  for(i = r = 0; i < ROWS; i++)/* Assign the values 1 to 8,
    sequentially. */
  {
    for(j = 0; j < COLS; j += 2)
    {
      cards[i][j] = cards[i][j+1] = r+1;
      r++;
    }
  }
  /* Now, shuffle the cards. */
  srand((unsigned)time(NULL));
  for(i = 0; i < ROWS; i++)
  {
    for(j = 0; j < COLS; j++)
    {
      c = cards[i][j];
      m = rand()%ROWS;
      r = rand()%COLS;
      cards[i][j] = cards[m][r];
      cards[m][r] = c;
    }
  }
  show_board(cards, status);
  m = 0;
  while(cnt != ROWS*COLS)/* The game ends when all cards are faced
    up. */
  {
    sel_card(cards, status, &r, &c);
    printf("Card_1 = %d\n", cards[r][c]);
```

```c
    sel_card(cards, status, &r2, &c2);
    printf("Card_2 = %d\n", cards[r2][c2]);

    for(i = 0; i < 18; i++)/* Blank lines to delete history and make
      harder for the player to remember the card positions. */
      printf("\n");

    if(cards[r][c] == cards[r2][c2])
    {
      printf("Cards matched !!!\n");
      cnt += 2;
    }
    else
    {
      printf("Sorry. No match !!!\n");
      status[r][c] = status[r2][c2] = 0; /* Reset the cards to face
        down condition. */
    }
    m++;
    show_board(cards, status);
  }
  printf("Congrats: You did it in %d tries\n", m);
  return 0;
}

void show_board(int c[][COLS], int s[][COLS])
{
  int i, j;
  for(i = 0; i < ROWS; i++)
  {
    for(j = 0; j < COLS; j++)
    {
      if(s[i][j] == 1)
        printf("%d ", c[i][j]);
      else
        printf("* ");
    }
    printf("\n");
  }
}

void sel_card(int c[][COLS], int s[][COLS], int *row, int *col)
{
  while(1)
  {
    printf("Enter row and column: ");
    scanf("%d %d", row, col);
    (*row)-- ; /* Subtract 1, because the user doesn't start
      counting from 0. */
    (*col)-- ;
    if(*row >= ROWS || *row < 0 || *col >= COLS || *col < 0)
    {
      printf("Out of bound dimensions\n");
      continue;
    }
```

```
      if(s[*row][*col] == 1)
      {
        printf("Error: This card is already flipped\n");
        continue;
      }
      s[*row][*col] = 1; /* Change the card position to face up. */
      return;
    }
  }
```

Passing Data in `main()` Function

When we run a program from the command line, we can pass data to it. For example, suppose that the executable file `hello.exe` is stored in C disk. The command line

```
C:\>hello 100 200
```

executes the program `hello` and passes the values `100` and `200` to `main()`. To pass data to `main()`, we should define it as a function with parameters:

```
int main(int argc, char *argv[])
```

Though you can use any names, `argc` and `argv` are by convention the typical choice.

The value of `argc` is equal to the number of the command line arguments, including the name of the program itself. For example, in the aforementioned command line, the value of `argc` is 3.

`argv` is defined as an array of pointers to the command line arguments, which are stored in a string form. The `argv[0]` pointer points to the name of the program, while the pointers `argv[1]` to `argv[argc-1]` point to the rest arguments. The last `argv` element is the `argv[argc]`, whose value is NULL.

For example, using the same command line, the arguments `hello`, `100`, and `200` are passed to `main()` as strings. Therefore, `argv[0]` points to the string `"hello"`, `argv[1]` points to `"100"`, and `argv[2]` points to `"200"`, respectively. The value of `argv[3]` is NULL.

For example, the following program checks if the user entered the correct user name and password. Suppose that these are `"user"` and `"pswd"`, respectively.

```
#include <stdio.h>
#include <string.h>

int main(int argc, char *argv[])
{
  if(argc == 1)
    printf("Error: missing user name and password\n");
  else if(argc == 2)
    printf("Error: missing password\n");
  else if(argc == 3)
  {
    if(strcmp(argv[1], "user") == 0 &&
```

```
      strcmp(argv[2], "pswd") == 0)
      printf("Valid user. The program " "%s" " will be executed...\n",
        argv[0]);
    else
      printf("Wrong input\n");
  }
  else
    printf("Error: too many parameters\n");
  return 0;
}
```

The **if** statements check the value of argc. If it is 3, the program checks the validity of the user name and password. If not, the program displays an informative message.

Exercises

11.43   What is the output of the following program?

```
#include <stdio.h>
int main(int argc, char *argv[])
{
  while(--argc)
    printf((argc > 1) ? "%s " : "%s\n", *++argv);
  return 0;
}
```

Answer: argc equals the number of the command line arguments. If the only argument is the name of the program, argc would be 1 and the **while** loop will not be executed because its value becomes 0.

If there are more arguments, the program displays them, separated by a space, and it adds a new line character after printing the last one.

For example, suppose that the program accepts the command line arguments one and two. In that case, argc would be 3 and the first execution of the **while** loop makes it 2. Since the expression (argc > 1) is true, the "%s   " will be replaced by the *++argv and a space.

Since argv points to argv[0], the expression ++argv makes it point to argv[1]. Therefore, *++argv is equivalent to argv[1], and the program displays one. In the next iteration of the **while** loop, argc becomes 1, so the "%s\n" will be replaced by the argv[2]. Therefore, the program prints two and a new line character.

Note that if we write *argv++ instead of *++argv, the program would display the command line arguments, but the last one.

11.44   The popular ping command is used to test the communication between two systems in an IP network. An example of its use is ping www.ntua.gr.

Write a program that reads the command line argument and checks if it is a valid hostname, meaning that it should begin with "www.", and the part after the second dot should be two or three characters long.

```
#include <stdio.h>
#include <string.h>
```

```c
int main(int argc, char *argv[])
{
  int i, len;

  if(argc != 2)
  {
    printf("Wrong number of arguments\n");
    return 0;
  }
  if(strncmp(argv[1], "www.", 4) != 0)
  {
    printf("Name must begin with www.\n");
    return 0;
  }
  len = strlen(argv[1]);
  for(i = 4; i < len; i++)
    if(argv[1][i] == '.')
      break;

  if(i == len)
  {
    printf("Second . is missing\n");
    return 0;
  }
  if((len-i-1) != 2 && (len-i-1) != 3)
  {
    printf("The last part should be two or three characters
      long\n");
    return 0;
  }
  printf("The hostname %s is valid\n", argv[1]);
  return 0;
}
```

Functions with Variable Number of Parameters

A function may accept a variable number of parameters. To declare such a function, we put first the fixed parameters and then the ... symbol, called *ellipsis*. For example, the declaration

```c
void test(int num, char *str,...);
```

indicates that test() takes two fixed parameters, an integer and a pointer to a character, which may be followed by a variable number of additional parameters.

In practice, you'll rarely need to write functions with a variable number of parameters. However, two of the most used functions, printf() and scanf(), accept a variable number of parameters.

A function with a variable number of parameters must have one fixed parameter at least.

To call such a function, first we write the values of the fixed parameters and then the values of the optional parameters. For example, we could call test() like this:

```
test(3, "example", 5, 8.9, "sample");
```

The data types of the optional parameters are **int**, **float**, and **char*** with values 5, 8.9, and "sample", respectively.

To handle the optional parameter list, we use the following macros:

1. va_list. The va_list type is defined in stdarg.h as a pointer the optional parameter list.

2. va_start. The va_start macro is defined in stdarg.h and it takes two parameters. The first one is a pointer of type va_list, while the *name* of the second one should be the same with the name of the last fixed parameter. For example, the name of the last fixed parameter in test() is str.

   After calling va_start, va_list points to the first optional parameter.

3. va_arg. The va_arg macro is defined in stdarg.h and it takes two parameters. The first one is a pointer of type va_list and the second one is the type of an optional parameter. For example, the type of the first optional parameter in test() is **int**.

   va_arg() returns the value of the optional parameter and advances va_list to point to the next optional parameter.

   In order to get the values of all optional parameters, we should call va_arg() as many times as the number of the optional parameters.

4. va_end. The va_end() macro is defined in stdarg.h and it takes as parameter a pointer of type va_list. It should be called to end the processing of the optional parameter list.

In the following program, test() takes a variable number of parameters of type **char***. The fixed parameter num indicates their number.

```
#include <stdio.h>
#include <stdarg.h>

void test(int num,…);

int main()
{
  test(3, "text_1", "text_2", "text_3");
  return 0;
}

void test(int num,…)
{
  char *str;
  int i;
  va_list arg_ptr;

  va_start(arg_ptr, num); /* arg_ptr points to the first optional
    parameter. Notice that the name of the second parameter should be the
    same with the name of the last fixed parameter in the declaration of
    test(). */
```

```
for (i = 0; i < num; i++)
{
  str = va_arg(arg_ptr, char*); /* Each call of va_arg() returns the
    value of the respective optional parameter of type char*, and arg_
    ptr advances to point to the next optional parameter. The second
    parameter is char*, because it's said that the type of all optional
    parameters is char*. */
  printf("%s ", str);
}
va_end(arg_ptr); /* The value of arg_ptr pointer is reset to NULL. */
}
```

The **for** loop gets the values of all optional parameters and prints them. Therefore, the program displays text_1 text_2 text_3.

The main difficulty in handling a function with a variable parameter list is that there is no easy way to determine the number of its optional parameters. A simple solution is to add a fixed parameter, which declares their number. In the previous example, we used the parameter num.

Recursive Functions

A function that calls itself is called recursive. For example,

```
#include <stdio.h>

void show(int num);

int main()
{
  int i;

  printf("Enter number: ");
  scanf("%d", &i);

  show(i);
  return 0;
}

void show(int num)
{
  if (num > 1)
    show(num-1);

  printf("val = %d\n", num);
}
```

To see how recursion works, assume that the user enters a number greater than 1, for example, 3.

(a) In the first call of show(), since num = 3 > 1, show() calls itself with argument num-1 = 3-1 = 2. printf() isn't executed. Since show() isn't terminated, the allocated memory for the variable num with value 3 is not deallocated.

(b) In the second call of `show()`, since num = 2 > 1, `show()` calls again itself with argument num−1 = 2−1 = 1. Like before, `printf()` isn't executed and the allocated memory for the new variable num with value 2 is not deallocated.

(c) In the third call of `show()`, `show()` isn't called again because num isn't greater than 1. Therefore, this `printf()` displays val = 1.

Then, all unexecuted `printf()` will be executed one by one, starting from the last one. In each termination of `show()`, the memory of the respective num variable is deallocated. Therefore, the program displays

```
val = 1
val = 2
val = 3
```

A recursive function should contain a termination statement in order to prevent infinite recursion.

In the previous example, this statement was the condition **if**(num > 1).

When a recursive function is called, new memory is allocated from the *stack* to store its nonstatic variables. The information for which part of the code is left unexecuted is also stored in the *stack*. That code will be executed when the function won't call itself again.

However, the size of the stack isn't particularly large, meaning that it can't store a large number of variables. For example, if the user enters a large value in the previous program, for example, 50000, it is very likely that the program won't execute and the message "stack overflow" appears. This message indicates that there is no available memory in the stack to store the new copies of num and the information for the unexecuted part of the code.

Be careful when using a recursive function, because if it calls itself many times, the execution time may be significantly high and the available memory in the stack may be exhausted.

In practice, recursion is often needed in the development of math algorithms. However, if you can use an iteration loop instead, it'd be better.

Exercises

11.45 What is the output of the following program?

```c
#include <stdio.h>

int a = 4; /* Global variable. */

int main()
{
  if(a == 0)
    return 0;
```

```
    else
    {
      printf("%d ", a-- );
      main();
    }
    return 0;
}
```

Answer: Notice that main() can be also called recursively. In each call, the value of a is decremented by 1. The program stops calling main() once its value becomes 0. Therefore, the program displays 4 3 2 1.

11.46 What is the output of the following program?

```
#include <stdio.h>

int unknown(int num1, int num2);

int main()
{
  int num1, num2;

  printf("Enter first number: ");
  scanf("%d", &num1);

  do
  {
    printf("Enter second number (greater than 0): ");
    scanf("%d", &num2);
  } while(num2 <= 0);

  printf("%d\n", unknown(num1, num2));
  return 0;
}

int unknown(int num1, int num2)
{
  if(num2 == 1)
    return num1;
  else
    return num1 + unknown(num1, num2 - 1);

}
```

Answer: This program outputs the product of num1*num2.
For example, if the user enters the numbers num1 = 10 and num2 = 4, then the call to unknown(num1,num2) returns

```
num1 + unknown(num1, num2-1 = 3) =
num1 + num1 + unknown(num1, num2-1 = 2) =
num1 + num1 + num1 + unknown(num1,num2-1 = 1)
```

The last call of unknown(num1,1) returns num1, because num2 = 1.
Therefore, the return value is num1+num1+num1+num1 = 4*num1 = num2*num1.

11.47 What is the output of the following program?

```
#include <stdio.h>

int unknown(int arr[], int num);
```

```c
int main()
{
  int arr[] = {10, 20, 30, 40};

  printf("%d\n", unknown(arr, 4));
  return 0;
}
int unknown(int arr[], int num)
{
  if(num == 1)
    return arr[0];
  else
    return arr[num-1] + unknown(arr, num-1);
}
```

Answer: When unknown() is called, it returns

```
arr[4-1 = 3]  + unknown(arr,4-1 = 3) =
arr[3]  + (arr[3-1] + unknown(arr,3-1)) =
arr[3]  + arr[2]  + (arr[2-1] + unknown(arr,2-1)) =
arr[3]  + arr[2]  + arr[1]  + unknown(arr,1)
```

The last call of unknown(arr,1) returns arr[0] because num = 1. Therefore, the return value is arr[3]+arr[2]+arr[1]+arr[0] and the program displays 100.

11.48 Write a recursive function that takes as parameter an integer value n and returns its factorial n! using the formula n! = n*(n–1)! Write a program that reads a positive integer less than 170 and uses the function to display its factorial.

```c
#include <stdio.h>

double fact(int num);

int main()
{
  int num;

  do
  {
    printf("Enter a positive integer less than 170: ");
    scanf("%d", &num);
  } while(num < 0 || num > 170);

  printf("Factorial of %d is %e\n", num, fact(num));
  return 0;
}

double fact(int num)
{
  double val;

  if((num == 0) || (num == 1))
    val = 1;
  else
    val = num * fact(num - 1);
  return val;
}
```

Comments: Notice that for large values of num the calls to factorial() increase, therefore the time to calculate its factorial also increases. In that case, the alternative solution with the **for** loop in 6.14 (Exercise) calculates the factorial's number faster.

11.49 Write a recursive function that takes as parameter an integer (i.e., N) and returns the Nth term of the Fibonacci sequence using the formula $F(N) = F(N-1)+F(N-2)$, where $F(0) = 0$ and $F(1) = 1$. Write a program that reads an integer N between 2 and 40 and uses the function to display the Nth term.

```c
#include <stdio.h>

unsigned int fib(int num);

int main()
{
  int num;

  do
  {
    printf("Enter a number between 2 and 40: ");
    scanf("%d", &num);
  } while(num < 2 || num > 40);

  printf("F(%d) = %d\n", num, fib(num));
  return 0;
}

unsigned int fib(int num)
{
  if(num == 0)
    return 0;
  else if(num == 1)
    return 1;
  else
    return fib(num-1) + fib(num-2);
}
```

Comments: Notice that for large values of num the execution time of fib() is significantly increased. In that case, the code described earlier in 11.36 (Exercise) is executed faster.

11.50 Sometimes in math it is very difficult to prove some problems that seem quite simple like the one of the German mathematician Lothar Collatz, who first proposed it in 1937.

Think of a positive integer n and execute the following algorithm:

(a) If it is even, divide it by two (n/2).

(b) If it is odd, triple it and add one (3n+1).

Repeat the process for each new number and you'll come to a surprising result: for any integer you choose, you'll always end up with ...1!!!

For example, if you choose the number 53, the produced numbers are 53 -> 160 -> 80 -> 40 -> 20 -> 10 -> 5 -> 16 -> 8 -> 4 -> 2 -> 1.

This math problem, well known as "Collatz conjecture," remains unsolved, although the use of computing machines confirms that for any initial value up to 2^{60} we'll eventually reach 1.

What we are asking is to write a recursive function that takes as parameter a positive integer and displays the produced sequence of numbers to confirm the "Collatz conjecture."

```c
#include <stdio.h>

int collatz(int n);

int main()
{
  int a;

  do
  {
    printf("Enter a positive integer: ");
    scanf("%d", &a);
  } while(a <= 0);

  printf("The result is %d indeed!!!\n", collatz(a));
  return 0;
}
int collatz(int n)
{
  printf("%d\n", n);

  if(n == 1)
    return 1;
  else if(n & 1)/* If n is odd. */
    return collatz(3*n+1);
  else/* If n is even. */
    return collatz(n/2);
}
```

Comments: Execute the program for several positive integers. The result is amazing, indeed. You'll always reach 1.

Unsolved Exercises

11.1   Write a function that takes as parameters three integers and checks if the sum of the first two numbers is equal to the third one. If it is, the function should return the greater of the first two numbers, otherwise the less of the second and the third one. Write a program that reads three integers, calls the function, and displays the return value.

11.2   Write the functions f() and g(), as follows:

$$f(x)=\begin{cases} x+2, & x>0 \\ -3x+7, & x\leq0 \end{cases} \quad g(x)=\begin{cases} x^2+2, & x>0 \\ 7x-5, & x\leq0 \end{cases}$$

Write a program that reads an integer (i.e., x) and uses f() and g() to display the result of f(g(x)), with the restriction that g() must be called from inside f().

11.3 Write a function that takes as parameters two integers (i.e., a and b), reads 100 integers, and displays the minimum of those within [a, b]. Write a program that reads two integers and calls the function. (Note: the first argument should be less than the second.)

11.4 Write a **void** function that takes as parameters two arrays and the number of the elements to compare. If they are the same, the function should return 1, 0 otherwise. Write a program that reads 200 **double** numbers and splits them in two arrays. Then, the program should read the number of the elements to be compared and use the function to compare them. (*Hint:* since the return type is **void** add a pointer argument to return that value.)

11.5 Write a **void** function that takes as a parameter a string and returns the number of its lowercase, uppercase letters and digits. Write a program that reads a string up to 100 characters, and if it begins with 'a' and ends with 'q', the program should call the function and display the return values.

11.6 Write a function that takes as parameters an array and checks if there are duplicated values. If so, the function should return a pointer to the element which appears the most times, otherwise NULL. Write a program that reads 100 **double** numbers, stores them in an array and uses the function to find the element with the most appearances. *Note:* if more than one element appear the same most times, the function should return the first found.

11.7 Write a function that takes as parameters two strings and returns a pointer to the longest part in the first string that doesn't contain any character of the second string. If none part is found, the function should return NULL. Write a program that reads two strings up to 100 characters, calls the function, and displays that part.

11.8 Write a function that takes as parameters a two-dimensional array and a **double** number and returns a pointer to the row in which that number appears the most times. If the number isn't found, the function should return NULL. Write a program that assigns random values to a 5x5 array of doubles, then it reads a double number and uses the function to find the row with the most appearances.

11.9 Write a program that accepts three command line arguments and displays them in alphabetical ascending order.

11.10 Write a program that displays the characters of its command line arguments in reverse order. For example, if the arguments are one two, the program should display owt eno.

11.11 Write a program that accepts as command line arguments the sign of a math operation, two one-digit numbers, and displays the result of the operation. For example, if the arguments are +, 5, –3, the program should display 2.

11.12 Write a function that accepts a variable number of pointers to integer arguments and returns the pointer to the greatest number. Write a program that reads three integers and uses the function to display the greatest.

11.13 Write a function (i.e., f()) that takes as parameters an integer (i.e., a) and a pointer to another function (i.e., g()), which accepts an integer argument, and if that argument is positive it returns the corresponding negative, otherwise it returns the argument as is. If a is even, f() should use the pointer to call g() and display the return value. If it is odd, it should make it even and then call g(). Write a program that reads an integer and uses a pointer variable to call f(). (*Note*: re-visit the section "Pointer to Function" in Chapter 8.)

11.14 In math, a triangular number counts the objects that can form an equilateral triangle, as shown next.

$$T_0 = 0 \qquad T_1 = 1 \qquad T_2 = 3 \qquad T_3 = 6$$

The $T(n)$ triangular number is the number of the objects composing the equilateral triangle and it is equal to the sum of the n numbers from 1 to n. Therefore, the $T(n)$ triangular number is expressed as: $T(n)=1+2+3+\cdots+(n-1)+n$ and in a recursive form as: $T(n)=\begin{cases} n, & \text{for } n=0 \text{ or } n=1 \\ n+T(n-1), & \text{for } n>1 \end{cases}$

Write a program that reads a positive integer (n) up to 20 and uses a recursive function to display the $T(n)$ triangular number.

11.15 Modify the power() function of 11.14 (Exercise) and use the formula $m^n = m*m^{n-1}$ to compute the result of m^n recursively.

11.16 Image editing programs often use the "flood fill" algorithm to fill similarly colored connected areas with a new color. Suppose that the two-dimensional 8×8 array of Figure 11.4a represents the pixels of an image, where 0 represents the black color, 1: white, 2: red, 3: green, and 4: blue, while a pixel is similarly colored connected with another, if they have the same color and it is adjacent to it. The similarly colored areas are depicted in Figure 11.4b.

To implement the "flood fill" algorithm, write a recursive floodfill() function which change the color (i.e., c) of a pixel at a location (i.e., i, j) to a new color (i.e., nc) and then it changes the color of its neighbouring pixels (i.e., the pixels to the left, right, above, and below the pixel at (i, j)) whose color is also c. This process continues recursively on the neighbours of the changed pixels until there are no more pixel locations to consider.

For example, if we choose to change the color of the pixel in the position 0, 0 from black (i.e., 0) to green (i.e., 3), the color of the top-left area of four pixels changes to green, as shown in Figure 11.4c.

Write a program that creates a two-dimensional 8×8 array of integers and assigns to its elements random values in [0, 4]. Then, the program should read the location of a pixel and a new color and use floodfill() to change the existing color of its similarly colored area with the new one.

```
0 0 0 1 1 1 2 2      0 0 0 1 1 1 2 2      3 3 3 1 1 1 2 2
0 4 4 4 1 1 1 2      0 4 4 4 1 1 1 2      3 4 4 4 1 1 1 2
4 4 4 4 4 2 2 2      4 4 4 4 4 2 2 2      4 4 4 4 4 2 2 2
0 0 3 3 3 3 3 3      0 0 3 3 3 3 3 3      0 0 3 3 3 3 3 3
0 0 0 0 3 3 1 1      0 0 0 0 3 3 1 1      0 0 0 0 3 3 1 1
0 0 3 3 3 1 1 1      0 0 3 3 3 1 1 1      0 0 3 3 3 1 1 1
0 1 1 1 1 1 1 1      0 1 1 1 1 1 1 2      0 1 1 1 1 1 1 2
1 2 2 2 2 2 2 2      1 2 2 2 2 2 2 2      1 2 2 2 2 2 2 2
(a)                  (b)                  (c)
```

FIGURE 11.4
"Flood fill" algorithm description.

12

Searching and Sorting Arrays

This chapter describes the most common and simplest algorithms used for searching a value in an array and for sorting the elements of an array in ascending or descending order.

Searching Arrays

To check if a value is stored into an array, we are going to describe the linear and binary search algorithms.

Linear Search

The linear search algorithm (also called sequential search) is the simplest algorithm to search for a value in a *nonsorted* array. The searched value is compared against the value of each element until a match is found.

In an array of n elements, the maximum number of searches is n. This may occur if the searched value is not found or is equal to the last element.

In the following program, the linear _ search() function implements the linear search algorithm.

Exercises

12.1 Write a function that searches for a number in an array of doubles. If the number is stored, the function should return the number of its occurrences and the position of its first occurrence, otherwise −1. Write a program that reads up to 100 doubles and stores them in an array. If the user enters −1, the insertion of numbers should terminate. Then, the program should read a double and use the function to display the number of its occurrences and the position of its first occurrence.

```c
#include <stdio.h>

int linear_search(double arr[], int size, double num, int *t);

int main()
{
  int i, times, pos;
  double num, arr[100];
```

```c
  for(i = 0; i < 100; i++)
  {
    printf("Enter number: ");
    scanf("%lf", &num);
    if(num == -1)
      break;
    arr[i] = num;
  }
  printf("Enter number to search: ");
  scanf("%lf", &num);

  pos = linear_search(arr, i, num, &times); /* The variable i
    indicates the number of the array's elements. */
  if(pos == -1)
    printf("%f isn't found\n", num);
  else
    printf("%f appears %d times (first pos = %d)\n", num, times, pos);
  return 0;
}

int linear_search(double arr[], int size, double num, int *t)
{
  int i, pos;

  pos = -1;
  *t = 0;
  for(i = 0; i < size; i++)
  {
    if(arr[i] == num)
    {
      (*t)++;
      if(pos == -1)/* Store the position of the first occurrence. */
        pos = i;
    }
  }
  return pos;
}
```

12.2 Write a function that takes as parameter an integer array and returns the maximum number of the same occurrences. For example, if the array is {1, 9, –3, 5, –3, 8}, the function should return 2 because –3 appears the most times, that is, 2. Write a program that reads 10 integers, stores them in an array, and uses the function to display the maximum number of the same occurrences.

```c
#include <stdio.h>

#define SIZE 10

int num_occurs(int arr[]);

int main()
{
  int i, arr[SIZE];

  for(i = 0; i < SIZE; i++)
  {
    printf("Enter number: ");
```

```
      scanf("%d", &arr[i]);
    }
    printf("\nMax occurrences is %d\n", num_occurs(arr));
    return 0;
  }

  int num_occurs(int arr[])
  {
    int i, j, k, max_times;

    max_times = 0;
    for(i = 0; i < SIZE; i++)
    {
      k = 0;
      for(j = i; j < SIZE; j++) /* Compare arr[i] against the rest
        elements. */
      {
        if(arr[i] == arr[j])
          k++; /* If another element has the same value, increase the
            counter. */
      }
      if(k > max_times)
        max_times = k;
    }
    return max_times;
  }
```

Binary Search

The binary search algorithm is used for searching for a value in a *sorted* array (either in ascending or descending order). To see how this algorithm works, assume that we are searching for a value in an array sorted in ascending order:

Step 1: We use two variables, that is, start and end, which indicate the start and the end of the part of the array, in which we are searching for the value. We use the variable middle to calculate the middle position of that part: middle = (start+end)/2. For example, if we have a sorted array of 100 integers, start should be initialized to 0, end to 99, so middle becomes 49.

Step 2: We compare the value we are searching for against the middle element:

(a) If they are equal, then the searched value is found and the algorithm terminates.

(b) If it is greater, the algorithm is repeated at the part of the array starting from the middle position and up to the end. Therefore, start becomes start = middle+1 and the algorithm goes back to Step 1.

(c) If it is less, the algorithm is repeated at the part of the array starting from the start position and up to the middle. Therefore, end becomes end = middle−1 and the algorithm goes back to Step 1.

In short, the binary search algorithm divides the array into two parts. Then, the searched value is compared with the middle element and the same process is repeated in the respective part.

The algorithm terminates if either the searched value is found or `start` becomes greater than end.

In the following program, the `binary _ search()` function implements the binary search algorithm.

Exercises

12.3 Write a function that searches for a number in an array of integers. If the number is stored, the function should return the position of its first occurrence, otherwise −1. Write a program that declares an array of integers sorted in ascending order. The program should read an integer and use the function to display its array position.

```c
#include <stdio.h>

int binary_search(int arr[], int size, int num);

int main()
{
  int num, pos, arr[] = {10, 20, 30, 40, 50, 60, 70};

  printf("Enter number to search: ");
  scanf("%d", &num);

  pos = binary_search(arr, sizeof(arr)/sizeof(int), num);
  if(pos == -1)
    printf("%d isn't found\n", num);
  else
    printf("%d is found in position %d\n", num, pos);
  return 0;
}

int binary_search(int arr[], int size, int num)
{
  int start, end, middle;

  start = 0;
  end = size - 1;

  while(start <= end)
  {
    middle = (start + end)/2;

    if(num < arr[middle])
      end = middle - 1;
    else if(num > arr[middle])
      start = middle + 1;
    else
      return middle;
  }
  return -1; /* If the execution reaches this point means that the
    number was not found. */
}
```

Comments: When binary _ search() is called, the expression **sizeof**(arr)/ **sizeof**(**int**) calculates the number of the array elements. Since the size of arr is 28 bytes, the number of its elements is 28/4 = 7.

Let's see how the algorithm works by assuming that the user enters the number 45.

First iteration. The initial value of start is 0 and end is 6. middle becomes(start+end)/2 = 6/2 = 3. Since arr[middle] is 40, less than 45, the next statement will be start = middle+1 = 3+1 = 4.

Second iteration. middle becomes (start+end)/2 = (4+6)/2 = 5. Since arr[middle] is 60, greater than 45, the next statement will be end = middle-1 = 5-1 = 4.

Third iteration. middle becomes (start+end)/2 = (4+4)/2 = 4. Since arr[middle] is 50, greater than 45, the next statement will be end = middle-1 = 4-1 = 3.

Since start is greater than the end, the loop terminates and the function returns -1.

12.4 Write a program that reads an integer within [0, 1000] and uses the binary search algorithm to "guess" that number. The program should make questions to determine if the number we are searching for is less or more than the middle of the examined interval. The answers must be given in the form of 0 (no) or 1 (yes). The program should display how many tries were needed to find the number.

```c
#include <stdio.h>
int main()
{
  int x, ans, low, high, middle, times;

  do
  {
    printf("Enter number in [0, 1000]: ");
    scanf("%d", &x);
  } while(x < 0 || x > 1000);
  times = 1;
  low = 0;
  high = 1000;
  middle = (high+low)/2;

  while(high >= low)
  {
    printf("Is %d the hidden number (0 = No, 1 = Yes) ? ", middle);
    scanf("%d", &ans);
    if(ans == 1)
    {
      printf("Num = %d is found in %d tries\n", x, times);
      return 0;
    }
    times++;
    printf("Is the hidden number < %d (0 = No, 1 = Yes) ? ", middle);
    scanf("%d", &ans);
    if(ans == 1)
```

```
    {
      high = middle - 1;
      middle = (high + low)/2;
    }
    else
    {
      low = middle + 1;
      middle = (high + low)/2;
    }
  }
  printf("Num = %d isn't found. You probably gave a wrong answer\n",
  x);
  return 0;
}
```

Sorting Arrays

There are several algorithms to sort an array. We are going to describe the selection sort, the insertion sort, and the bubble sort algorithms.

Selection Sort

To describe the algorithm, we'll show you how to sort an array in ascending order.

At first, we find the element with the minimum value and we swap it with the first element of the array. Therefore, the minimum value is stored in the first position.

Then, we find the minimum value among the rest elements, except the first one. Like before, we swap that element with the second element of the array. Therefore, the second minimum value is stored in the second position.

This procedure is repeated with the rest elements and the algorithm terminates once the last two elements are compared.

To sort the array in descending order, we find the maximum value instead of the minimum.

In the following program, the sel_sort() function implements the selection sort algorithm to sort an array in ascending order.

Exercises

12.5 Write a function that takes as parameters an array of doubles and uses the selection sort algorithm to sort it in ascending order. Write a program that reads 10 doubles, stores them in an array, and uses the function to sort it.

```
#include <stdio.h>

#define SIZE 10

void sel_sort(double arr[], int size);
```

```c
int main()
{
  int i;
  double a[SIZE];

  for(i = 0; i < SIZE; i++)
  {
    printf("Enter number: ");
    scanf("%lf", &a[i]);
  }
  sel_sort(a, SIZE);

  printf("\n***** Sorted array *****\n");
  for(i = 0; i < SIZE; i++)
    printf("%f\n", a[i]);
  return 0;
}

void sel_sort(double arr[], int size)
{
  int i, j;
  double temp;

  for(i = 0; i < size; i++)
  {
    for(j = i+1; j < size; j++)
    {
      if(arr[i] > arr[j])
      {
        /* Swap values. */
        temp = arr[i];
        arr[i] = arr[j];
        arr[j] = temp;
      }
    }
  }
}
```

Comments: In each iteration of the inner **for** loop, arr[i] is compared against the elements from i+1 up to SIZE-1. If an element is less than arr[i], their values are swapped. Therefore, in each iteration of the outer **for** loop, the minimum value of the elements from i up to SIZE-1 is stored at arr[i].

To sort the array in descending order, change the **if** statement to:

```c
if(arr[i] < arr[j])
```

12.6 Write a program that reads 10 names up to 100 characters each and displays them in ascending alphabetical order.

```c
#include <stdio.h>
#include <string.h>

#define NUM 10

int main()
```

```
{
  char temp[100], name[NUM][100]; /* Array of NUM rows (each row
    contains up to 100 characters). For example, the first name is
    stored at name[0] and its first character at name[0][0]. Each
    name[i] can be used as a pointer to the corresponding row. */
  int i, j;

  for(i = 0; i < NUM; i++)
  {
    printf("Enter name: ");
    gets(name[i]);
  }
  for(i = 0; i < NUM; i++)
  {
    for(j = i+1; j < NUM; j++)
    {
      if(strcmp(name[i], name[j]) > 0)/* Swap the names. */
      {
        strcpy(temp, name[j]);
        strcpy(name[j], name[i]);
        strcpy(name[i], temp);
      }
    }
  }
  printf("\n***** Names in increase order *****\n");
  for(i = 0; i < NUM; i++)
    printf("%s\n", name[i]);
  return 0;
}
```

12.7 Write the add _ sort() function that inserts a number into a sorted array, so that the array remains sorted. Write a program that reads 9 integers, stores them in an array of 10 integers, and uses the sel _ sort() of the previous exercise to sort the array in ascending order. Then, the program should read the 10th integer and use the add _ sort() to insert it in the array.

```
#include <stdio.h>
#define SIZE 10

void sel_sort(int arr[], int size);
void add_sort(int arr[], int size, int num);

int main()
{
  int i, num, p[SIZE];

  for(i = 0; i < SIZE-1; i++)/* Read 9 integers and store them in
    the array. */
  {
    printf("Enter number: ");
    scanf("%d", &p[i]);
  }
  sel_sort(p, SIZE-1); /* Sorting the 9 elements of the array. */

  printf("Insert number in sorted array: ");
  scanf("%d", &num);
```

```
  add_sort(p, SIZE-1, num); /* Insert the 10th integer in the array. */
  for(i = 0; i < SIZE; i++)
    printf("%d\n", p[i]);
  return 0;
}

void add_sort(int arr[], int size, int num)
{
  int i, pos;

  if(num <= arr[0])
    pos = 0;
  else if(num >= arr[size-1]) /* If it greater than the last one,
    store it in the last position and return. */
  {
    arr[size] = num;
    return;
  }
  else
  {
    for(i = 0; i < size-1; i++)
    {
    /* Check all adjacent pairs up to the last one at positions
      SIZE-3 and SIZE-2 to find the position in which num should be
      inserted. */
      if(num >= arr[i] && num <= arr[i+1])
        break;
    }
    pos = i+1;
  }
  for(i = size; i > pos; i--)
    arr[i] = arr[i-1]; /* The elements are shifted one position to
      the right, starting from the last position of the array, that
      is [SIZE-1], up to the position in which the new element will
      be inserted. For example, in the last iteration, i = pos+1, so,
      arr[pos+1] = arr[pos]. */
  arr[pos] = num; /* Store the inserted number. */
}
void sel_sort(int arr[], int size)
{
  int i, j, temp;
  for(i = 0; i < size; i++)
  {
    for(j = i+1; j < size; j++)
    {
      if(arr[i] > arr[j])
      {
        /* Swap values. */
        temp = arr[i];
        arr[i] = arr[j];
        arr[j] = temp;
      }
    }
  }
}
```

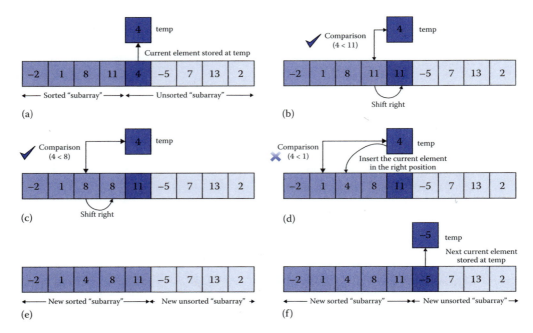

FIGURE 12.1
Example of insertion sort algorithm.

Insertion Sort

To describe the algorithm, we'll sort an array in ascending order.

The algorithm is based on sequential comparisons between each element (starting from the second and up to the last element) and the elements on its left, which form the "sorted subarray." The elements on its right form the "unsorted subarray."

In particular, the most left element of the "unsorted subarray" (i.e., the examined element) is compared against the elements of the "sorted subarray" from right to left, according to that:

Step 1: At first, it is stored in a temporary variable (Figure 12.1a).

Step 2a: If it is greater than the most right element of the "sorted subarray", its position doesn't change and the algorithm continues with the testing of the next most left element.

Step 2b: If it is less than the most right element of the "sorted subarray," the latter is shifted one position to the right while the examined element is compared against the most right but one element (Figure 12.1b and c). If it is greater, the examined element is stored at the position of the last shifted element, otherwise the same procedure is repeated (the most right but one element is shifted one position to the right and the examined element is compared against the most right but two elements) (Figure 12.1c and d). Step 2b terminates either if the examined element is greater than an element of the "sorted subarray" (Figure 12.1d and e) or the first element of the "sorted subarray" is reached. Then, Step 1 is repeated for the new most left element (Figure 12.1e and f).

The algorithm terminates when the most right element of the "unsorted subarray" is tested.

The algorithm resembles the way that a card player would sort a card game hand, assuming that he starts with an empty left hand and all cards face down on the table. The player picks up a card with his right hand and inserts it in the correct position in the left hand. To find the correct position, that card is compared against the cards in his left hand, from right to left.

To sort the array in descending order, an element of the "sorted subarray" is shifted one position to the right if it is less, and not if it is greater, than the examined element.

In the following program, the `insert _ sort()` function implements the insertion sort algorithm to sort an array in ascending order.

Exercise

12.8 Write a function that takes as parameter an array of integers and uses the insertion sort algorithm to sort it in ascending order. Write a program that reads 5 integers, stores them in an array, and uses the function to sort it.

```c
#include <stdio.h>

#define SIZE 5

void insert_sort(int arr[], int size);

int main()
{
  int i, a[SIZE];

  for(i = 0; i < SIZE; i++)
  {
    printf("Enter number: ");
    scanf("%d", &a[i]);
  }
  insert_sort(a, SIZE);
  printf("\n***** Sorted array *****\n");
  for(i = 0; i < SIZE; i++)
    printf("%d\n", a[i]);
  return 0;
}

void insert_sort(int arr[], int size)
{
  int i, j, temp;

  for(i = 1; i < size; i++)
  {
    temp = arr[i];
    j = i;
    while((j > 0) && (arr[j-1] > temp))
    {
      arr[j] = arr[j-1]; /* Shift this element one position to the
        right. */
```

```
    j-- ;
  }
  arr[j] = temp;
 }
}
```

Comments: The **for** loop compares the elements starting from the second one. In each iteration, temp holds the examined element. The **while** loop shifts one position to the right the elements that are on the left of the examined element and they are also greater than it.

Let's assume that the array elements are 7, 3, 1, 9, 4.

*First **for** loop iteration* (i = 1 *and* temp = arr[1] = 3)

(j = 1) first **while** loop iteration: 7--> 3 1 9 4

So, the array is transformed to **3 7** 1 9 4.

*Second **for** loop iteration* (i = 2 *and* temp = arr[2] = 1)

(j = 2) first **while** loop iteration: 3 7--> 1 9 4
(j = 1) second **while** loop iteration: 3--> 7 1 9 4

So, the array is transformed to **1 3 7** 9 4.

*Third **for** loop iteration* (i = 3 *and* temp = arr[3] = 9)

(j = 3) No shifting takes place because the fourth array element (i.e., arr[3] = 9) is greater than the "most right element" of its left elements (i.e., arr[2] = 7).

So, the array remains the same: **1 3 7 9** 4.

*Fourth **for** loop iteration* (i = 4 *and* temp = arr[4] = 4)

(j = 4) first **while** loop iteration: 1 3 7 9--> 4
(j = 3) second **while** loop iteration: 1 3 7--> 9 4

The sorting is completed and the array is transformed to **1 3 4 7 9**.
To sort the array in descending order, change the **while** statement to

```
while((j > 0) && (arr[j-1] < temp))
```

in order to shift one position to the right the elements of the "sorted subarray," which are less than (and not greater than) the examined element.

Bubble Sort

The bubble sort algorithm is based on sequential comparisons between adjacent array elements. Each element "bubbles" up and is stored in the proper position. For example, suppose that we want to sort an array in ascending order.

At first, the last element is compared against the last by one. If it less, the elements are swapped, so the smaller value "bubbles" up. Then, the last by one element is compared against the last by two. Like before, if it is less, the elements are swapped, so the smaller value keeps "bubbling" up. The comparisons continue up to the beginning of the array, and eventually the smallest value "bubbles" to the top of the array and it is stored at its first position.

This procedure is repeated from the second element and up to the last one, so the second smallest value of the array "bubbles" to the top and stored in the second position. The same is repeated for the part of the array from its third element and up to the last one, and so forth.

The algorithm terminates when none element "bubbles" to the top.

To sort the array in descending order, the "bubbling" value is the greatest and not the smallest one.

In the following program, the bubble _ sort() function implements the bubble sort algorithm to sort an array in ascending order.

Exercise

12.9 Write a function that takes as parameter an array of integers and uses the bubble sort algorithm to sort it in ascending order. Write a program that reads 5 integers, stores them in an array, and uses the function to sort it.

```c
#include <stdio.h>

#define SIZE 5

void bubble_sort(int arr[]);

int main()
{
  int i, p[SIZE];

  for(i = 0; i < SIZE; i++)
  {
    printf("Enter number: ");
    scanf("%d", &p[i]);
  }
  bubble_sort(p);

  printf("\n***** Sorted array *****\n");
  for(i = 0; i < SIZE; i++)
    printf("%d\n", p[i]);
  return 0;
}
void bubble_sort(int arr[])
{
  int i, j, temp, reorder;

  for(i = 1; i < SIZE; i++)
  {
    reorder = 0;

    for(j = SIZE-1; j >= i; j-- )
    {
      if(arr[j] < arr[j-1])
      {
        /* Swap values. */
        temp = arr[j];
```

```
      arr[j] = arr[j-1];
      arr[j-1] = temp;
      reorder = 1;
    }
  }
  if(reorder == 0)
    return;
  }
}
```

Comments: The reorder variable checks if the sorting is completed in order to avoid unnecessary iterations. If two elements are swapped, it is set to 1. Otherwise, the value 0 means that the array is sorted and the function terminates.

Let's assume that the array elements are 10, 9, 4, 7, 6.

First iteration of the outer **for** *loop* (i = 1)

First iteration of the inner **for** loop (j = 4): 10 9 4 6 <-> 7
Second iteration of the inner **for** loop (j = 3): 10 9 4 6 7
Third iteration of the inner **for** loop (j = 2): 10 4 <-> 9 6 7
Fourth iteration of the inner **for** loop (j = 1): 4 <-> 10 9 6 7

So, the array is transformed to **4** 10 9 6 7.

Second iteration of the outer **for** *loop* (i = 2)

First iteration of the inner **for** loop (j = 4): 4 10 9 6 7
Second iteration of the inner **for** loop (j = 3): 4 10 6 <-> 9 7
Third iteration of the inner **for** loop (j = 2): 4 6<-> 10 9 7

So, the array is transformed to **4 6** 10 9 7.

Third iteration of the outer **for** *loop* (i = 3)

First iteration of the inner **for** loop (j = 4): 4 6 10 7 <-> 9
Second iteration of the inner **for** loop (j = 3): 4 6 7 <-> 10 9

So, the array is transformed to **4 6 7** 10 9.

Fourth iteration of the outer **for** *loop* (i = 4)

First iteration of the inner **for** loop (j = 4): 4 6 7 10 <-> 9

The sorting is completed and the array is transformed to **4 6 7 9 10**.
To sort the array in descending order, change the **if** statement to

```
if(arr[j] > arr[j-1])
```

bsearch() and qsort() Library Functions

The bsearch() library function uses the binary search algorithm to search for a value in a sorted array. The qsort() library function uses another algorithm, called quick sort, to sort an array. A short description for both of them is provided in Appendix C.

Exercise

12.10 Write a program that reads 10 integers and stores them in an array. The program should use the qsort() to sort the array in ascending order. Then, the program should read an integer and use the bsearch() to check if it exists in the array.

```c
#include <stdio.h>
#include <stdlib.h>

#define NUM 10

int compare(const void *elem1, const void *elem2);

int main()
{
  int *pos, i, arr[NUM];

  for(i = 0; i < NUM; i++)
  {
    printf("Enter number: ");
    scanf("%d", &arr[i]);
  }
  qsort(arr, NUM, sizeof(int), compare);

  printf("\nSorted array: ");
  for(i = 0; i < NUM; i++)
    printf("%d ", arr[i]);

  printf("\n\nEnter number to search: ");
  scanf("%d", &i);

  pos = (int*)bsearch(&i, arr, NUM, sizeof(int), compare);

  if(pos == NULL)
    printf("\n%d isn't found\n", i);
  else
    printf("\n%d is in %d position\n", i, pos-arr+1);
  return 0;
}

int compare(const void *elem1, const void *elem2)
{
  if(*(int*)elem1 < *(int*)elem2)
    return -1;
  else if(*(int*)elem1 > *(int*)elem2)
    return 1;
  else
    return 0;
}
```

Comments: Appendix C describes the parameters for both qsort() and bsearch().

Note that the name of the compare() function is used as a pointer to that function. According to the declarations of qsort() and bsearch(), the parameters of compare() must be declared as **const void*** pointers. Since we compare integers, we typecast them to **int***. Then, the values that the two pointers point to are compared and compare() returns a value, according to the description of bsearch().

If the searched value is found, bsearch() returns a pointer to the respective element, otherwise it returns NULL. Like before, we typecast the return type from **void*** to **int*** and the program uses pointer arithmetic to display its position, if found.

Note that we could replace the **if-else** statement in compare() with that:

```
int compare(const void *elem1, const void *elem2)
{
  return *(int*)elem1 - *(int*)elem2; /* For ascending order. */
}
```

To sort the array in descending order, just reverse the comparison:

```
int compare(const void *elem1, const void *elem2)
{
  return *(int*)elem2 - *(int*)elem1; /* For descending order. */
}
```

13

Structures and Unions

This chapter introduces two new types: *structures* and *unions*. Like arrays, structures and unions aggregate a set of values into a single entity. However, their properties are quite different from an array's. Unlike arrays, the elements of a structure or union may have *different* types. Furthermore, to access a member of a structure or union, we specify its *name*, and not its position as an integer subscript. This chapter discusses how to define structures and unions types, declare variables, and perform operations on them.

Structures

When we want to group related data items within a single entity, a structure is a typical choice. For example, a structure may hold information for a company, such as its name, business core, tax number, number of employees, contact information, and other data.

Defining a Structure

To define a named structure type, we specify the name of the structure followed by its elements enclosed in braces, like this:

```
struct struct_name
{
  type_1 name_1;
  type_2 name_2;
  ...,
  type_n name_n;
};
```

The definition of a structure type begins with the **struct** word and must end with a semicolon. A structure may contain elements of different data types. For example, suppose we need to store information about a company, we could define a structure type named company, like this:

```
struct company
{
  char name[50];
  int start_year;
  int field;
  int tax_num;
  int num_empl;
  char addr[50];
  float balance;
};
```

The elements of a structure, also called *members* or *fields*, are used like the ordinary variables.

Typically, a structure type is defined with global scope. If it is defined inside a function, other parts of the programs would ignore its existence and won't be able to declare variables of that structure type.

Declaring Structure Variables

Once a named structure type is defined, we can use that name to declare structure variables. For example, to declare two structure variables of the type company we write

```
struct company comp_1, comp_2;
```

Alternatively, a structure variable can be declared together with the definition of its type. For example,

```
struct book
{
  char title[100];
  int year;
  float price;
} book_1, book_2;
```

The variables book _ 1 and book _ 2 are declared as structures variables of the type book.

When a structure variable is declared, the compiler allocates memory to store its fields in the order in which they appear. For example, the following program displays how many bytes the variable date _ 1 allocates:

```
#include <stdio.h>

struct date
{
  int day;
  int month;
  int year;
};

int main()
{
  struct date date_1;

  printf("%d\n", sizeof(date_1));
  return 0;
}
```

Since date _ 1 consists of three integer fields, the allocated memory is 3*4 = 12 bytes.

However, the allocated size may be larger than the sizes of the fields added together. For example, if we change the type of the day field from **int** to **char**, the program may display again 12, not 9. This may happen when the compiler requires that each field is stored in an address multiple of some number (typically four). This requirement may be set for a faster access of the structure fields. If we assume that the month field should be stored in an address multiple of four, the compiler would allocate three more bytes right after the day field. In that case, the program would display 12.

To calculate the memory size of a structure, always use the **sizeof** *operator; don't add the sizes of the fields.*

A structure type can be also defined with the **typedef** specifier. In that case, add the name of the structure after the right brace. For example:

```
typedef struct
{
  char title[100];
  int year;
  float price;
} book;
```

However, it is allowed to name a structure type and use the **typedef** specifier, as well. For example:

```
typedef struct book
{
  char title[100];
  int year;
  float price;
} book;
```

In fact, we'll use this method at Chapter 14, when declaring structures to be used in a linked list.

If a structure type is defined with the **typedef** specifier, don't add the word **struct** when declaring a structure variable. For example,

```
book book_1, book_2;
```

Typically, the **typedef** specifier is used to create a synonym of a basic data type. For example, the statement

```
typedef unsigned int size_t;
```

creates a new data type named size_t as a synonym of the **unsigned int** type. Therefore, the declarations **unsigned int** i; and size_t i; are equivalent.

In another example, the statement

```
typedef int arr[100];
```

creates a new data type named arr, which is an array of 100 integers. Therefore, the statement arr arr1; declares arr1 as an array of 100 integers.

Accessing the Structure Fields

To access a structure field, we write the name of the structure variable followed by the operator (.) and the name of the field. For example, the following program defines the book structure, assigns values in the fields of book _ 1, and displays them:

```c
#include <stdio.h>
#include <string.h>

struct book
{
  char title[100];
  int year;
  float price;
};
int main()
{
  struct book book_1;

  strcpy(book_1.title, "Literature");
  book_1.year = 2010;
  book_1.price = 10.85;
  printf("%s %d %.2f\n", book_1.title, book_1.year, book_1.price);
  return 0;
}
```

Like an array, the fields of a structure variable can be initialized when it is declared. To do that, add the = operator after the name of the structure variable and enclose the values of its fields in braces, separated with a comma. The list of values must appear in the same order as the fields of the structure and should match their data types. Consider the following declaration:

```c
struct book book_1 = {"Literature", 2010, 10.85};
```

The value of book _ 1.title becomes "Literature", book _ 1.year becomes 2010, and book _ 1.price becomes 10.85.

As with arrays, any unassigned fields are given the value 0. For example, with the declaration

```c
struct book book_1 = {"Literature"};
```

the values of the fields year and price are set to 0.

Similarly, with the declaration

```c
struct book book_1 = {0};
```

the values of the book _ 1 fields are set to 0.

Also, a structure variable may be declared and initialized when its type is defined, as shown here:

```c
struct book
{
  char title[100];
  int year;
```

```
  float price;
} book_1 = {"Literature", 2010, 10.85};
```

Besides the (.) operator, we'll see next how to use the -> operator to access the fields of a structure.

Pointer to a Structure Field

A pointer to a structure field is used like an ordinary pointer. For example, the following program uses pointer variables to display the values of the book _ 1 fields:

```
#include <stdio.h>
#include <string.h>

struct book

{
  char title[100];
  int year;
  float price;
};

int main()
{
  char *ptr1;
  int *ptr2;
  float *ptr3;
  struct book book_1;

  strcpy(book_1.title, "Literature");
  book_1.year = 2010;
  book_1.price = 10.8;

  ptr1 = book_1.title;
  ptr2 = &book_1.year;
  ptr3 = &book_1.price;

  printf("%s %d %.2f\n", ptr1, *ptr2, *ptr3);
  return 0;
}
```

To make a pointer variable to point to a field of a structure, it should be declared as a pointer of the same type. For example, ptr1 is declared as **char*** because the type of the title field is **char**.

Notice that in the statement ptr1 = book _ 1.title; we didn't add the & operator before the name of the structure because we use the name of the array as a pointer.

Structure Operations

Although we can't use the = operator to copy one array into another, we can use it to copy one structure into another of the same type. For example, the following program declares the variables s1 and s2 and copies s1 into s2:

```
#include <stdio.h>

struct student
```

```
{
  int code;
  float grd;
};
int main()
{
  struct student s1, s2;

  s1.code = 1234;
  s1.grd = 6.7;
  s2 = s1; /* Legal, since s1 and s2 are variables of the same structure
    type. */
  printf("C:%d G:%.2f\n", s2.code, s2.grd);
  return 0;
}
```

The statement s2 = s1; copies the values of s1 fields into the respective fields of s2 fields. Therefore, it is equivalent to

```
s2.code = s1.code;
s2.grd = s1.grd;
```

Besides assignment, no other operation can be performed on entire structures. For example, the operators == and != can't be used to check whether two structures are equal or not. Therefore, it isn't allowed to write

```
if(s1 == s2)
```

or

```
if(s1 != s2)
```

To test whether two structures are equal, you should compare their fields one by one. For example,

```
if((s1.code == s2.code) && (s1.grd == s2.grd))
```

Structures Containing Arrays

Since a structure may contain any type of data, it may contain one or more arrays. For example,

```
#include <stdio.h>
#include <string.h>

struct student
{
  char name[50];
  float grades[2];
};

int main()
{
  struct student s1;
```

```
    strcpy(s1.name, "somebody");
    s1.grades[0] = 8.5;
    s1.grades[1] = 7.5;
    printf("%s %c %c\n", s1.name, s1.name[0], *s1.name);
    return 0;
}
```

As you guess, an array field is treated like an ordinary array. For example, the statement strcpy(s1.name, "somebody"); copies the string "somebody" into the field name. The value of s1.name[0] becomes 's', s1.name[1] becomes 'o', and so on.

When using pointer arithmetic to handle the elements of an array field, the * operator must precede the name of the structure. For example, since the s1.name can be used as a pointer to its first character, the *s1.name is equal to s1.name[0], the *(s1.name+1) is equal to s1.name[1], and so on. Like ordinary arrays, the parentheses must be inserted for reasons of priority.

Therefore, the program displays somebody s s.

Structures Containing Pointers

A structure may contain one or more pointer fields. For example,

```
#include <stdio.h>

struct student

{
    char *name;
    float *avg_grd;
};

int main()
{
    float grd = 8.5;
    struct student s1;

    s1.name = "somebody";
    s1.avg_grd = &grd;
    printf("%s %.2f\n", s1.name+3, *s1.avg_grd);
    return 0;
}
```

With the statement s1.name = "somebody"; the compiler allocates memory to store the string "somebody" and then the name pointer points to that memory.

With the statement s1.avg _ grd = &grd; the avg _ grd points to the address of grd. To access the content of the memory pointed to by a pointer field, the * operator must precede the name of the structure.

Therefore, the program displays ebody 8.50.

Structures Containing Structures

A structure may contain one or more nested structures. A nested structure must be defined before the definition of the structure in which it is contained, otherwise the compiler will raise an error message.

For example, in the following program, prod _ 1 contains the nested structures
s _ date and e _ date. The s _ date keeps the production date of a product and the
e _ date its expiration date:

```c
#include <stdio.h>
#include <string.h>

struct date
{
  int day;
  int month;
  int year;
};

struct product/* Since the type date is defined, it can be used to
  declare nested structures. */
{
  char name[50];
  double price;
  struct date s_date;
  struct date e_date;
};

int main()
{
  struct product prod_1;

  strcpy(prod_1.name, "product");
  prod_1.s_date.day = 1;
  prod_1.s_date.month = 9;
  prod_1.s_date.year = 2012;

  prod_1.e_date.day = 1;
  prod_1.e_date.month = 9;
  prod_1.e_date.year = 2015;

  prod_1.price = 7.5;
  printf("The product's life is %d years\n", prod_1.e_date.year -
    prod_1.s_date.year);
  return 0;
}
```

Notice that for accessing a field of the nested structure the (.) operator must be used twice.
The program subtracts the respective fields and displays the product life.

Exercise

13.1 Define a structure type named computer with fields: manufacturer, model,
processor, and price. Write a program that uses a structure variable to read the
characteristics of a computer and display them.

```c
#include <stdio.h>

struct computer
```

```
    {
    /* Assume that 50 characters are enough to hold the computer's
       characteristics. */
       char comp[50];
       char model[50];
       char cpu[50];
       float prc;
    };
    int main()
    {
       struct computer pc;

       printf("Enter company: ");
       gets(pc.comp);

       printf("Enter model: ");
       gets(pc.model);

       printf("Enter cpu: ");
       gets(pc.cpu);

       printf("Enter price: ");
       scanf("%f", &pc.prc);

       printf("\nC:%s M:%s CPU:%s P:%.2f\n", pc.comp, pc.model, pc.cpu,
         pc.prc);
       return 0;
    }
```

Bit Fields

A structure may contain fields whose length is specified as a number of bits. A bit field is declared like this:

```
data_type field_name : bits_number;
```

The bit fields can be used just like any other structure field. Most compilers support the **int**, **char**, **short**, **long** data types for a bit field.

Here is an example of a structure with several bit fields:

```
struct person
{
  unsigned char sex : 1;
  unsigned char married : 1;
  unsigned char children : 4;
  char name[50];
};
```

Since the size of the sex and married fields is one bit, their values can be either 0 or 1. Since the size of the children field is four bits, it can take values between 0 and 15.

Since the type of the bit fields is **unsigned char**, the compiler allocates one byte. Actually, 1+1+4 = 6 bits will be used to store their values, while the other two won't be used. If we weren't using bit fields, the compiler would allocate three bytes. Therefore, we save two bytes of memory.

The main advantage of using bit fields is to save memory space. For example, if we use the `person` structure to store the data of 200000 people, we'd save 400000 bytes when using the bit fields.

When you assign a value to a bit field, be sure that this value fits in the bit field. For example, suppose that the value 2 is assigned in the `married` bit field.

```
struct person person_1;
person_1.married = 2; /* Wrong assignment. */
```

Since the value 2 is encoded in two bits (10), the value of `married` would be 0 (assume that bit fields are allocated from right to left), not 2.

Since the memory of a bit field is not allocated like the ordinary variables, it is not allowed to apply the & operator to a bit field. For example, this statement won't be compiled:

```
unsigned char *ptr = &person_1.married;
```

The value of a bit field can be either positive or negative since the compiler may treat the high-order bit as a sign bit. To avoid the case of negative values, declare its type to be **unsigned**. If the size of a bit field is one bit, then its type must be defined as **unsigned** since a single bit can't be signed.

Pointer to Structure

A pointer to a structure is used like a pointer to an ordinary variable. Consider the following program:

```
#include <stdio.h>
#include <string.h>

struct student
{
  char name[50];
  float grd;
};

int main()
{
  struct student *stud_ptr, stud;

  stud_ptr = &stud;

  strcpy((*stud_ptr).name, "somebody");
  (*stud_ptr).grd = 6.7;
  printf("N: %s G: = %.2f\n", stud.name, stud.grd);
  return 0;
}
```

The variable stud _ ptr is declared as a pointer to a structure variable of type student.

The statement `stud _ ptr = &stud;` makes it point to the address of the variable `stud`. In particular, it points to the address of its first field. Since `stud _ ptr` points to the address of `stud`, `*stud _ ptr` is equivalent to `stud`. Then, we are using the (.) operator to access its fields. The expression `(*stud _ ptr)` must be enclosed in parentheses because the (.) operator takes precedence over the * operator.

Alternatively, we can use the pointer and the `->` operator to access the fields of a structure. For example,

```c
#include <stdio.h>
#include <string.h>

struct student
{
  char name[50];
  float grd;
};

int main()
{
  struct student *stud_ptr, stud;

  strcpy(stud.name, "somebody");
  stud.grd = 6.7;

  stud_ptr = &stud;
  printf("N:%s G:%.2f\n", stud_ptr->name, stud_ptr->grd);
  return 0;
}
```

The expression `stud _ ptr->name` is equivalent to `(*stud _ ptr).name` and the expression `stud _ ptr->grd` is equivalent to `(*stud _ ptr).grd`.

Therefore, both programs display

```
N: somebody G: 6.70
```

When using a pointer variable to access the fields of a structure, we find it simpler to use the `->` operator.

The same rules concerning the arithmetic of an ordinary pointer variable apply to a structure pointer. For example, if a structure pointer is increased by one, its value will be increased by the size of the structure it points to.

Arrays of Structures

An array of structures is an array whose elements are structures. Typically, an array of structures is used when we need to store information about many items, like the data of a company's employees or students' data or the products of a warehouse. In fact, an array of structures may be used as a simple *database*. For example, with the statement

```c
struct student stud[100];
```

the variable `stud` is declared as an array of 100 structures of type `student`.

Like the ordinary arrays, an array of structures can be initialized when it is declared. For example, suppose that the type `student` is defined, like this:

```
struct student
{
  char name[50];
  int code;
  float grd;
};
```

An initialization example could be

```
struct student stud[] = {{"nick sterg", 1500, 7.3},
                         {"john theod", 1600, 5.8},
                         {"peter karast", 1700, 6.7}};
```

The value of stud[0].name field becomes "nick sterg", stud[0].code becomes 1500, and stud[0].grd becomes 7.3.

Note that the initialization of an array can be combined with the definition of the structure type. For example,

```
struct student
{
  char name[50];
  int code;
  float grd;
} stud[] = {{"nick sterg", 1500, 7.3},
            {"john theod", 1600, 5.8},
            {"peter karast", 1700, 6.7}};
```

The inner braces around each structure can be omitted; however, we prefer to use them to make clearer the initialization of each structure.

Like the ordinary arrays, the uninitialized fields are set to 0. For example, if we write

```
struct student stud[100] = {0};
```

the fields of all structures are initialized to 0.

The following program stores the data of 100 students to an array of structures of type student.

```
#include <stdio.h>

#define SIZE 100

struct student
{
  char name[50];
  int code;
  float grd;
};

int main()
{
  int i;
  struct student stud[SIZE];

  for(i = 0; i < SIZE; i++)
```

```
{
  printf("\nEnter name: ");
  gets(stud[i].name);

  printf("Enter code: ");
  scanf("%d", &stud[i].code);

  printf("Enter grade: ");
  scanf("%f", &stud[i].grd);

  printf("\nN: %s C: %d G: %.2f\n", stud[i].name, stud[i].code, stud[i].
    grd);
  getchar(); /* Read the '\n' character that is stored in stdin, after
    the grade is entered. */
}
  return 0;
}
```

Besides array subscripting, we can use pointer notation to access the elements. For example, the `*stud` is equivalent to `stud[0]`, `*(stud+1)` is equivalent to `stud[1]`, `*(stud+2)` is equivalent to `stud[2]`, and so on.

Therefore, to access the `code` field of the third student, we can write either `stud[2].code` or `(*(stud+2)).code`. The parentheses in the second case are required for reasons of priority.

As in the case of ordinary arrays, our preference is to use array subscripting in order to get a more readable code.

Exercises

13.2 Modify the aforementioned program and use a pointer to read and display the student data.

```
#include <stdio.h>

#define SIZE 100

struct student
{
  char name[50];
  int code;
  float grd;
};

int main()
{
  int i;
  struct student *stud_ptr, stud[SIZE];

  stud_ptr = stud; /* Equivalent to stud_ptr = &stud[0]; */
  for(i = 0; i < SIZE; i++)
```

```
    {
      printf("\nEnter name: ");
      gets(stud_ptr->name);

      printf("Enter code: ");
      scanf("%d", &stud_ptr->code);

      printf("Enter grade: ");
      scanf("%f", &stud_ptr->grd);

      printf("\nN:%s C:%d G:%.2f\n", stud_ptr->name, stud_ptr->code,
        stud_ptr->grd);
      getchar();

      stud_ptr++; /* The pointer points to the next element. */
    }
    return 0;
}
```

13.3 Define the structure type `stock` with fields: name, code, and price. Write a program that uses this type to read the data of `100` stocks. Next, the program should read a price and display the stocks that cost less.

```
#include <stdio.h>

#define SIZE 100

struct stock
{
  char name[50];
  int code;
  double prc;
};

int main()
{
  int i;
  double prc;
  struct stock s[SIZE];

  for(i = 0; i < SIZE; i++)
  {
    printf("\nEnter name: ");
    gets(s[i].name);

    printf("Enter code: ");
    scanf("%d", &s[i].code);

    printf("Enter price: ");
    scanf("%lf", &s[i].prc);

    getchar();
  }
  printf("\nEnter price to check: ");
  scanf("%lf", &prc);

  for(i = 0; i < SIZE; i++)
```

```
    {
      if(s[i].prc <= prc)
        printf("\nN: %s C: %d P: %f\n", s[i].name, s[i].code, s[i].prc);
    }
    return 0;
  }
```

Structures as Function Arguments

A structure variable can be passed to a function just like any other variable, either by value—meaning passing the structure itself—or by reference—meaning passing the memory address of the structure. Recall that when passing a variable by value the function can't modify its value. For example, consider the following program:

```
#include <stdio.h>
#include <string.h>

void test(struct student stud_1);

struct student
{
  char name[50];
  int code;
  float grd;
};

int main()
{
  struct student stud = {"somebody", 20, 5};

  test(stud);
  printf("N: %s C: %d G: %.2f\n", stud.name, stud.code, stud.grd);
  return 0;
}

void test(struct student stud_1)
{
  strcpy(stud_1.name, "new_name");
  stud_1.code = 30;
  stud_1.grd = 7;
}
```

When test() is called, the values of stud fields are copied to the respective fields of stud _ 1. Since the stud _ 1 and stud reside in different memory locations, any changes in the fields of stud _ 1 don't affect the fields of stud. Therefore, the program displays N: somebody C: 20 G: 5.00.

As you already know by now, even if we were using the name stud instead of stud _ 1, the result would be the same because they are still different variables.

On the other hand, when a variable is passed by reference to a function, the function may change its value because it has access to its memory address. For example, let's change test() in order to modify the fields of stud.

```
#include <stdio.h>
#include <string.h>
```

```
void test(struct student *stud_1);

struct student
{
  char name[50];
  int code;
  float grd;
};

int main()
{
  struct student stud = {"somebody", 20, 5};

  test(&stud);
  printf("N: %s C: %d G: %.2f\n", stud.name, stud.code, stud.grd);
  return 0;
}
void test(struct student *stud_ptr)
{
  strcpy(stud_ptr->name, "new_name");
  stud_ptr->code = 30;
  stud_ptr->grd = 7;
}
```

When `test()` is called, we have `stud_ptr = &stud`. Therefore, since `stud_ptr` points to the memory address of `stud`, the function may change the values of its fields.

As a result, the program displays N: `new_name` C: 30 G: 7.00

When a structure is passed to a function, its fields are copied to the fields of the respective parameter. This copy operation may impose a time overhead in the program, especially if the structure is large or the function is called many times. On the other hand, when the address of the structure is passed to the function, the fields are not copied.

> *For better performance, pass the address of the structure and not the structure itself, even if the function doesn't need to change the values of its fields.*

To prevent the function from changing the values of the structure fields, declare the pointer as **const**. For example,

```
void test(const struct student *stud_ptr);
```

Now, `test()` can't modify the fields of the structure pointed to by `stud_ptr`.

Exercises

13.4   What is the output of the following program?

```
#include <stdio.h>

struct student *test();

struct student
```

```
{
  char name[50];
  int code;
  float grd;
};

struct student stud_1 = {0}; /* Global structure variable, all its
fields are set to 0. */

int main()
{
  struct student stud = {"somebody", 1111, 7.5};
  struct student *ptr = &stud;

  *ptr = *test();
  printf("%s %d %d\n", ptr->name, ptr->code, ptr->grd);
  return 0;
}

struct student *test()
{
  return &stud_1;
}
```

Answer: Since test() returns a pointer to stud_1, the expression *test() is equivalent to stud_1. Also, since ptr points to the address of stud, *ptr is equivalent to stud.

Therefore, the expression *ptr = *test() is equivalent to stud = stud_1, and the fields of stud become 0.

As a result, the program displays zero values.

13.5 Define the structure type time with fields: hours, minutes, and seconds. Write a function that takes as parameter a pointer to an integer and converts that integer to hours, minutes, and seconds. These values should be stored into the fields of a structure of type time, and the function should return that structure. Write a program that reads an integer, calls the function, and displays the fields of the returned structure.

```
#include <stdio.h>

struct time mk_time(int *ptr);

struct time
{
  int hours;
  int mins;
  int secs;
};
int main()
{
  int secs;
  struct time t;

  printf("Enter seconds: ");
  scanf("%d", &secs);

  t = mk_time(&secs);
  printf("\nH:%d M:%d S:%d\n", t.hours, t.mins, t.secs);
```

```
    return 0;
}

struct time mk_time(int *ptr)
{
  struct time tmp;

  tmp.hours = *ptr/3600;
  tmp.mins = (*ptr%3600)/60;
  tmp.secs = *ptr%60;
  return tmp;
}
```

13.6 Define the structure type book with fields: title, code, and price. Write a function that takes as parameters two pointers to structures of type book and uses them to swap the structures. Write a program that reads and stores the data of two books into two structures of type book. Then, the program should call the function and display the fields of the two structures.

```
#include <stdio.h>

void swap(struct book *b1, struct book *b2);

struct book
{
  char title[50];
  int code;
  float prc;
};

int main()
{
  int i;
  struct book b[2];

  for(i = 0; i < 2; i++)
  {
    printf("\nEnter title: ");
    gets(b[i].title);

    printf("Enter code: ");
    scanf("%d", &b[i].code);

    printf("Enter price: ");
    scanf("%f", &b[i].prc);

    getchar();
  }
  swap(&b[0], &b[1]);
  for(i = 0; i < 2; i++)
    printf("\nN:%s C:%d P:%.2f\n", b[i].title, b[i].code, b[i].prc);
  return 0;
}

void swap(struct book *b1, struct book *b2)
{
  struct book temp;
```

```
   temp = *b1;
   *b1 = *b2;
   *b2 = temp;
}
```

13.7 Define the structure type `product` with fields: name, code, and price. Write a function that takes as parameters an array of such structures and an integer. The function should check if a product's code is equal to that number and, if it does, it should return a pointer to the respective structure, otherwise NULL. Write a program that uses this structure type to read the data of 100 products and store them in an array of such structures. Then, the program should read an integer, call the function, and if the function doesn't return NULL, the program should display the product's name and price.

```
#include <stdio.h>

#define SIZE 100

typedef struct
{
   char name[50];
   int code;
   float prc;
} product;

product *find_stud(product pro[], int code); /* Define a function
   that takes as parameters an array of structures and an integer and
   returns a pointer to a structure of type product. */
int main()
{
   int i, code;
   product *ptr, pro[SIZE];

   for(i = 0; i < SIZE; i++)
   {
      printf("\nName: ");
      gets(pro[i].name);

      printf("Code: ");
      scanf("%d", &pro[i].code);

      printf("Price: ");
      scanf("%f", &pro[i].prc);

      getchar();
   }
   printf("\nEnter code to search: ");
   scanf("%d", &code);

   ptr = find_stud(pro, code);
   if(ptr == NULL)
      printf("\nNo product with code = %d\n", code);
   else
      printf("\nN: %s C: %d P: %.2f\n", ptr->name, code, ptr->prc);
   return 0;
}

product *find_stud(product pro[], int code)
```

```
{
  int i;

  for(i = 0; i < SIZE; i++)
  {
    if(pro[i].code == code)
      return &pro[i]; /* If the code is found, the function
        terminates and returns the address of that structure. */
  }
  return NULL; /* If the code is not found, the function returns
    NULL. */
}
```

13.8   Define the structure type coord with fields the coordinates of a point, for example, x and y. Define a structure of type rectangle with fields two structures of type coord, for example, point_A and point_B. Write a function that takes as parameters the two endpoints of a rectangle's diagonal. Each endpoint should be a structure of type coord. The function should calculate and return the area of the rectangle. Write a program that uses the type rectangle to read the coordinates of a rectangle's diagonal and uses the function to display its area.

```
#include <stdio.h>

struct coord
{
  double x;
  double y;
};

struct rectangle
{
  struct coord point_A; /* First diagonal point. */
  struct coord point_B; /* Second diagonal point. */
};

double rect_area(struct coord *c1, struct coord *c2);

int main()
{
  struct rectangle rect;

  printf("Enter the x and y coords of the first point: ");
  scanf("%lf%lf", &rect.point_A.x, &rect.point_A.y);

  printf("Enter the x and y coords of the second point: ");
  scanf("%lf%lf", &rect.point_B.x, &rect.point_B.y);

  printf("Area:%f\n", rect_area(&rect.point_A, &rect.point_B));
  return 0;
}

double rect_area(struct coord *c1, struct coord *c2)
{
  double base, height;

  if(c1->x > c2->x)
    base = c1->x - c2->x;
  else
```

```
      base = c2->x - c1->x;

  if(c1->y > c2->y)
    height = c1->y - c2->y;
  else
    height = c2->y - c1->y;

  return base*height; /* Return the area. */
}
```

13.9 For a complex number z, we have z = a + bj, where j is the imaginary unit, a is the real part, and b is the imaginary part, respectively. Define the structure type `complex` with fields the float numbers `re` and `im`, which represent the real and imaginary parts of a complex number.

Write a function that takes as parameters two structures of type `complex` and a character, which represents the sign of a math operation. The function should perform the math operation and return the result as a structure of type `complex`. Write a program that uses the structure `complex` to read two complex numbers and a math sign (+, -, *,/) and uses the function to display the result of the math operation.

Remind that, if $z_1 = a + bj$ and $z_2 = c + dj$, we have

$$z = z_1 + z_2 = (a+c)+(b+d)j$$

$$z = z_1 - z_2 = (a-c)+(b-d)j$$

$$z = z_1 \times z_2 = (ac-bd)+(bc+ad)j$$

$$z = \left(\frac{z_1}{z_2}\right) = \left(\frac{ac+bd}{c^2+d^2}\right)+\left(\frac{bc-ad}{c^2+d^2}\right)j$$

```c
#include <stdio.h>

struct complex operation(struct complex a1, struct complex a2, char
sign);

struct complex
{
  double re; /* The real part of the complex number. */
  double im; /* The imaginary part of the complex number. */
};

int main()
{
  char sign;
  struct complex z1, z2, z;

  printf("Enter real and imaginary part of the first complex
number: ");
  scanf("%lf%lf", &z1.re, &z1.im);
  printf("z1 = %f%+fj\n", z1.re, z1.im);

  printf("Enter real and imaginary part of the second complex
    number: ");
```

```
    scanf("%lf%lf", &z2.re, &z2.im);
    printf("z2 = %f%+fj\n", z2.re, z2.im);

    printf("Enter sign (+, -, *,/): ");
    scanf(" %c", &sign); /* We add a space before %c, in order to
    bypass the new line character stored in stdin with the previous
    insertion. */
    if(sign == '+' || sign == '-' || sign == '*' || sign == '/')
    {
      if(sign == '/' && z2.re == 0 && z2.im == 0)
        printf("Division with zero is not allowed\n");
      else
      {
        z = operation(z1, z2, sign);
        printf("z = z1 %c z2 = %f%+fj\n", sign, z.re, z.im);
      }
    }
    else
      printf("Wrong sign\n");
    return 0;
}

struct complex operation(struct complex a1, struct complex a2, char
  sign)
{
  struct complex a;
  switch(sign)
  {
    case '+':
      a.re = a1.re + a2.re;
      a.im = a1.im + a2.im;
    break;

    case '-':
      a.re = a1.re - a2.re;
      a.im = a1.im - a2.im;
    break;

    case '*':
      a.re = (a1.re*a2.re) - (a1.im*a2.im);
      a.im = (a1.im*a2.re) + (a1.re*a2.im);
    break;

    case '/':
      a.re = ((a1.re*a2.re) + (a1.im*a2.im))/((a2.re*a2.re)+(a2.
        im*a2.im));
      a.im = ((a1.im*a2.re) - (a1.re*a2.im))/((a2.re*a2.re)+(a2.
        im*a2.im));
    break;
  }
  return a;
}
```

Comments: Note that we use the flag + in printf() to display the imaginary part of the complex number. Recall from Chapter 2 that when using the flag + the positive values are prefixed with +.

The reason we passed to `operation()` the structures and not their addresses is to show you an example of a function with parameter structures.

13.10 Define the structure type `student` with fields: name, code, and grade. Write a program that uses this type to read the data of 100 students and store them in an array of such structures. If the user enters the grade –1, the insertion of student data should terminate. Write a function to sort the structures in grade ascending order and another function to display the data of the students who got a higher grade than the average grade of all students.

```c
#include <stdio.h>

#define SIZE 100

void sort_by_grade(struct student studs[], int num_studs);
void show_students(struct student studs[], int num_studs, float
  avg_grd);

struct student
{
  char name[50];
  int code;
  float grd;
};

  int main()
  {
    int i;
    float sum_grd;
    struct student studs[SIZE];

    sum_grd = 0;
    for(i = 0; i < SIZE; i++)
    {
      printf("\nGrade [0-10]: ");
      scanf("%f", &studs[i].grd);
      if(studs[i].grd == -1)
        break;

      sum_grd += studs[i].grd;
      getchar();

      printf("Name: ");
      gets(studs[i].name);

      printf("Code: ");
      scanf("%d", &studs[i].code);
    }
    sort_by_grade(studs, i); /* Sort the structures in grade ascending
      order. The variable i specifies the number of students. */
    show_students(studs, i, sum_grd/i); /* The last argument is the
      average grade of all students. */
    return 0;
  }

  void sort_by_grade(struct student studs[], int num_studs)
```

```
{
  int i, j;
  struct student temp;

  for(i = 0; i < num_studs; i++)
  {
    /* In each iteration, the grd field is compared against the
       others. If it is less, the structures are swapped. */
    for(j = i+1; j < num_studs; j++)
    {
      if(studs[i].grd > studs[j].grd)
      {
        temp = studs[i];
        studs[i] = studs[j];
        studs[j] = temp;
      }
    }
  }
}

void show_students(struct student studs[], int num_studs, float
  avg_grd)
{
  int i;

  for(i = 0; i < num_studs; i++)
    if(studs[i].grd >= avg_grd)
      printf("N: %s C: %d G: %f\n", studs[i].name, studs[i].code,
      studs[i].grd);
}
```

13.11 Image-editing programs often need to rotate an image by $90°$. An image can be treated as a two-dimensional array whose elements represent the pixels of the image. For example, the rotation of the original image (i.e., $p[M][N]$) to the right produces a new image (i.e., $r[N][M]$), as shown here:

$$p = \begin{bmatrix} p_{0,0} & p_{0,1} & \cdots & p_{0,N-2} & p_{0,N-1} \\ p_{1,0} & p_{1,1} & \cdots & p_{1,N-2} & p_{1,N-1} \\ \vdots & \vdots & \vdots & \vdots & \vdots \\ p_{M-2,0} & p_{M-2,1} & \cdots & p_{M-2,N-2} & p_{M-2,N-1} \\ p_{M-1,0} & p_{M-1,1} & \cdots & p_{M-1,N-2} & p_{M-1,N-1} \end{bmatrix} \quad r = \begin{bmatrix} p_{M-1,0} & p_{M-2,0} & \cdots & p_{1,0} & p_{0,0} \\ p_{M-1,1} & p_{M-2,1} & \cdots & p_{1,1} & p_{0,1} \\ \vdots & \vdots & \vdots & \vdots & \vdots \\ p_{M-1,N-2} & p_{M-2,N-2} & \cdots & p_{1,N-2} & p_{0,N-2} \\ p_{M-1,N-1} & p_{M-2,N-1} & \cdots & p_{1,N-1} & p_{0,N-1} \end{bmatrix}$$

In particular, the first row of the original image becomes the last column of the new image, the second row becomes the last but one column, up to the last row, which becomes the first column. For example, the image

$$p = \begin{bmatrix} 1 & 2 & 3 & 4 & 5 \\ 6 & 7 & 8 & 9 & 10 \\ 11 & 12 & 13 & 14 & 15 \end{bmatrix} \text{ is transformed to } r = \begin{bmatrix} 11 & 6 & 1 \\ 12 & 7 & 2 \\ 13 & 8 & 3 \\ 14 & 9 & 4 \\ 15 & 10 & 5 \end{bmatrix}$$

The color of each pixel follows the RGB color model, in which the red, green, and blue colors are mixed together to reproduce a wide range of colors. The color is expressed as an RGB triplet (r, g, b), in which each component value varies from 0 to 255.

Define the structure type `pixel` with three integer fields named `red`, `green`, and `blue`. Write a program that creates a two-dimensional image (i.e., 3 × 5) whose elements are structures of type `pixel`. Initialize the fields of each `pixel` with random values within [0, 255]. Then, the program should display the original image, rotate the image by 90 degrees right, and display the rotated image (i.e., 5 × 3).

Hint: Use a second array to store the rotated image.

```c
#include <stdio.h>
#include <stdlib.h>
#include <time.h>

#define ROWS 3
#define COLS 5

struct pixel/* RGB format (Red-Green-Blue). */
{
  unsigned char red; /* Value in [0, 255]. */
  unsigned char green;
  unsigned char blue;
};

void rotate_right_90(struct pixel img[][COLS], struct pixel tmp[]
  [ROWS]);

int main()
{
  int i, j;
  struct pixel img[ROWS][COLS], tmp[COLS][ROWS];

  srand((unsigned int)time(NULL));
  /* Create random colors. */
  for(i = 0; i < ROWS; i++)
  {
    for(j = 0; j < COLS; j++)
    {
      img[i][j].red = rand()%256;
      img[i][j].green = rand()%256;
      img[i][j].blue = rand()%256;
    }
  }
  printf("*** Original Image ***\n\n");
  for(i = 0; i < ROWS; i++)
  {
    for(j = 0; j < COLS; j++)
    {
      printf("(%3d,%3d,%3d) ", img[i][j].red, img[i][j].green, img[i]
        [j].blue);
    }
    printf("\n");
  }
  rotate_right_90(img, tmp);

  printf("\n*** Rotated Image ***\n\n");
```

```
    for(i = 0; i < COLS; i++)
    {
      for(j = 0; j < ROWS; j++)
      {
        printf("(%3d,%3d,%3d) ", tmp[i][j].red, tmp[i][j].green, tmp[i]
          [j].blue);
      }
      printf("\n");
    }
    return 0;
}

void rotate_right_90(struct pixel img[][COLS], struct pixel tmp[]
[ROWS])
{
  int i, j, k = 0;
  for(i = ROWS-1; i >= 0; i--)
  {
    for(j = 0; j < COLS; j++)
    {
      tmp[j][i] = img[k][j];
    }
    k++;
  }
}
```

Unions

Like a structure, a union consists of one or more fields, which may be of different types. Their difference is that the fields of a structure are stored at *different* addresses, while the fields of a union are stored at the same address.

Using Unions

The definition of a union type resembles that of a structure, with the **union** word used instead of **struct**. When a union variable is declared, the compiler allocates memory to store the value of the largest field, not for all. Therefore, its fields are stored in the same space, overlaying each other.

For example, in the following program, the s variable allocates 8 bytes because its largest field is of type **double**.

```
#include <stdio.h>

union sample
{
  char ch;
  int i;
  double d;
};
```

```
int main()
{
    union sample s;

    printf("Size:%d\n", sizeof(s));
    return 0;
}
```

Since the compiler doesn't allocate memory for all the fields of a union, the main use of unions is to save memory space.

> As with structures, when you need to calculate the memory space of a union type, use the
> **sizeof** *operator.*

Since all union fields share the same memory, *only* the first field of a union variable may be initialized when it is declared. For example, the compiler would let us write

```
union sample s = {'x'};
```

but not

```
union sample s = {'x', 10, 1.23};
```

Access Union Fields

The fields of a union are accessed in the same way as the fields of a structure. However, since they are stored in the same memory, only the last assigned field has a valid value. For example, the following program assigns a value into an s field and displays the rest fields:

```
#include <stdio.h>

union sample
{
  char ch;
  int i;
  double d;
};

int main()
{
  union sample s;

  s.ch = 'a';
  printf("%c %d %f\n", s.ch, s.i, s.d);

  s.i = 64;
  printf("%c %d %f\n", s.ch, s.i, s.d);

  s.d = 12.48;
  printf("%c %d %f\n", s.ch, s.i, s.d);
  return 0;
}
```

The statement s.ch = 'a'; assigns the value 'a' into ch field. This value is stored in the common space, which has been allocated for all s fields. Therefore, printf() displays a and meaningless values for s.i and s.d.

The statement s.i = 64; assigns the value 64 into i field. Since this value is stored in the common space, it overwrites the value of s.ch. Therefore, printf() displays 64 and meaningless values for s.ch and s.d.

The statement s.d = 12.48; assigns the value 12.48 into d field. Similarly, the existing value of s.i is overwritten and the program displays 12.48 and meaningless values for s.ch and s.i.

When a union field is assigned with a value, any value previously stored in another field is overwritten.

The following program declares an array of structures of type person, reads the sex type, and stores the user's preferences in the respective union fields:

```c
#include <stdio.h>

#define SIZE   5
#define MAN    0
#define WOMAN 1

struct man
{
  char game[20];
  char movie[20];
};

struct woman
{
  char tv_show[30];
  char book[30];
};

union data
{
  struct man m;
  struct woman w;
};

struct person
{
  int type;
  union data d;
};

int main()
{
  int i, type;
  struct person pers_arr[SIZE];

  for(i = 0; i < SIZE; i++)
  {
    printf("\nSelection: 0 for man - 1 for woman: ");
    scanf("%d", &type);

    pers_arr[i].type = type;
    getchar();
    if(type == MAN)
```

```
    {
      printf("Enter favourite game: ");
      gets(pers_arr[i].d.m.game);

      printf("Enter favourite movie: ");
      gets(pers_arr[i].d.m.movie);
    }
    else if(type == WOMAN)
    {
      printf("Enter favourite TV show: ");
      gets(pers_arr[i].d.w.tv_show);

      printf("Enter favourite book: ");
      gets(pers_arr[i].d.w.book);
    }
  }
  for(i = 0; i < SIZE; i++)
  {
    if(pers_arr[i].type == MAN)
    {
      printf("\nGame: %s\n", pers_arr[i].d.m.game);
      printf("Show: %s\n", pers_arr[i].d.m.movie);
    }
    else if(pers_arr[i].type == WOMAN)
    {
      printf("\nMovie: %s\n", pers_arr[i].d.w.tv_show);
      printf("Book: %s\n", pers_arr[i].d.w.book);
    }
  }
  return 0;
}
```

Notice that the type field is used to indicate which type of structure, that is, man or woman, is stored in the union field so that the program may display correctly the user's choices.

Exercise

13.12 We often used union types when implementing protocol messages for network communications. For example, in ISDN technology, the SETUP message of Q.931 signalling protocol is sent by the calling user to an ISDN network to initiate call establishment with the called user. A simplified format of SETUP message is depicted in Figure 13.1.

The CONNECT message (Figure 13.2) is sent by the ISDN network to the calling user to indicate call acceptance by the called user.

The ALERTING message (Figure 13.3) is sent by the ISDN network to the calling user to indicate that called user alerting has been initiated.

Define the structure type isdn _ msg with three union fields to represent the SETUP, CONNECT and ALERTING messages, respectively. Write a program that reads a byte stream and parses the data according to the value of the MT field.

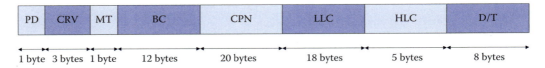

FIGURE 13.1
The SETUP message format.

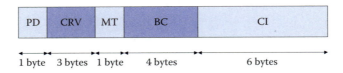

FIGURE 13.2
The CONNECT message format.

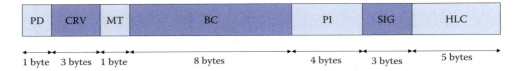

FIGURE 13.3
The ALERTING message format.

(*Note*: the value of the MT field in the SETUP, CONNECT and ALERTING messages is 5, 7 and 1, respectively.) To get the byte stream ask the user to enter up to 100 positive integers of type **char** (use -1 to stop the insertion) and store them in an array.

```
#include <stdio.h>

typedef unsigned char BYTE;

typedef struct
{
  BYTE pd;
  BYTE crv[3];
  BYTE mt;
} header;

typedef struct
{
  BYTE bc[12];
  BYTE cpn[20];
  BYTE llc[18];
  BYTE hlc[5];
  BYTE dt[8];
} setup;
```

```
typedef struct
{
  BYTE bc[4];
  BYTE ci[6];
} connect;

typedef struct
{
  BYTE bc[8];
  BYTE pi[4];
  BYTE sig[3];
  BYTE hlc[5];
} alerting;

typedef struct
{
  header hdr; /* Common header for all messages. */
  union
  {
    setup set;
    connect con;
    alerting alrt;
  };
} isdn_msg;

int main()
{
  BYTE pkt[100];
  int i;
  isdn_msg msg;

  for(i = 0; i < 100; i++)
  {
    printf("Enter octet: ");
    scanf("%d", &pkt[i]);
    if(pkt[i] == -1)
      break;
  }
  msg.hdr.pd = pkt[0];
  for(i = 0; i < 3; i++)
    msg.hdr.crv[i] = pkt[i+1];
  msg.hdr.mt = pkt[4];
  if(msg.hdr.mt == 5) /* SETUP. */
  {
    for(i = 0; i < 12; i++)
      msg.set.bc[i] = pkt[5+i];
    for(i = 0; i < 20; i++)
      msg.set.cpn[i] = pkt[17+i];
    for(i = 0; i < 18; i++)
      msg.set.llc[i] = pkt[37+i];
    for(i = 0; i < 5; i++)
      msg.set.hlc[i] = pkt[55+i];
    for(i = 0; i < 8; i++)
      msg.set.dt[i] = pkt[60+i];
  }
  else if(msg.hdr.mt == 7) /* CONNECT. */
```

```
{
  for(i = 0; i < 4; i++)
    msg.con.bc[i] = pkt[5+i];
  for(i = 0; i < 6; i++)
    msg.con.ci[i] = pkt[9+i];
}
else if(msg.hdr.mt == 1) /* ALERT. */
{
  for(i = 0; i < 8; i++)
    msg.alrt.bc[i] = pkt[5+i];
  for(i = 0; i < 4; i++)
    msg.alrt.pi[i] = pkt[13+i];
  for(i = 0; i < 3; i++)
    msg.alrt.sig[i] = pkt[17+i];
  for(i = 0; i < 5; i++)
    msg.alrt.hlc[i] = pkt[20+i];
}
  return 0;
}
```

Unsolved Exercises

13.1 Define the structure type employee with fields: first name, last name, age, and salary. Write a program that uses this type to read the data of 50 employees and store them in an array. The program should display the last name of the employee with the highest salary and the name of the older employee. (*Note*: if more than one employee has the same highest salary or greater age, the program should display the data of the first found.)

13.2 Modify the function swap() in 13.6 (Exercise) to return the book structure with the bigger price. Modify the main program to test the function.

13.3 In 13.8 (Exercise), add the function center() that should return the coordinates of the center of the rectangle as a pointer to a coord structure. Modify the main program to test the function.

13.4 Figure 13.4 depicts the status register of a printer (16 bits). Use bit fields and define the structure print_reg, with the five fields presented in Figure 13.4. Write a program that uses a structure of that type to simulate a printing job of 20 pages:

1. The low ink field is set to 3, when the 9th page is print and up to the end of the printing job.

2. The error code field is set to 10, only when the 13th page is print.

3. The paper jam field is set to 1, only when the 15th page is print.

4. The clean field is set to 1, only when the last page is print.

For each printing page, the program should display the value of the status register.

13.5 Define the structure type student with fields: name and code. Assume that the outline of a classroom can be simulated by a two-dimensional array whose elements are structures of type student. Write a program that reads the data of students and stores

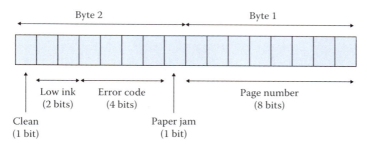

FIGURE 13.4
Printer status register.

them in a 3×5 array of structures. Then, the program should read the name of a student and his code and display his position in the array (i.e., row and column), if registered.

13.6 Define the structure type `student` with fields: name, code, and grade. Write a program that uses the type `student` to read the data of 100 students and store them sorted by decrease order of grade in an array. The sorting must be done during the data insertion.

13.7 Define the structure type `time` with fields: hours, minutes, and seconds. Write a function that takes as parameters two pointers to two structures that represent the start time and the end time of a game and returns the game's duration as a structure. Write a program that uses the type `time` to read the start time of a game and the end time and uses the function to display the game's duration. The user should enter the time in h:m:s format.

13.8 Rename the function `sort _ by _ grade()` in 13.10 (Exercise) to `sort()` and add an extra argument. If it is 0, the function should sort the students according to their codes, if it is 1 according to their names, and if it is 2 according to their grades. Modify the main program to test the function.

13.9 Define the structure type `publisher` with fields: name, address, and e-mail. Define the type `book` with fields: title, authors, code, price, and a field of type `publisher`. (*Note*: assume that the character fields are up to 100 characters.) Write a program that uses the type `book` to read the data of 100 books and store them in an array. Then, the program should read a book's code and if it is found it should display the title of the book and the publisher's name.

13.10 Define the structure type `student` with fields: name, code, and grade. Write a program that declares an array of 6 such structures and initializes the first five places with random data sorted by increase order of grade (i.e., "A.Smith, 100, 3.2", "B.Jones, 200, 4.7", "K.Lones, 175, 6.4", ...). Then the program should read the data of the sixth student and store it in the proper position, so that the array remains sorted.

13.11 Define the structure type `car` with fields: model, price, and manufacture year. Write a program that uses this type to read the data of 100 cars and store them in an array. Then, the program should provide a menu to perform the following operations:

1. *Show model.* The program should read a model and display the information about it. If the user enters `'*'`, the program should show the information about all models.

2. *Show prices.* The program should read a price and show all models that cost more.

3. *Program termination.*

13.12 Define the structure type any _ type   with an integer field (i.e., s _ type) and a field of type union. The union's fields are a structure of type time (i.e., t) as defined in 13.6 (Unsolved Exercise) and a structure of type student as defined in 13.9 (Unsolved Exercise). Write a program that reads an integer and store it into the s _ type field of a structure variable of type any _ type. Then, the program should check the value of s _ type, read data according to that value and store them in the respective structure field. For example, if its value is 1 the input data should be stored into the t field.

13.13 Define the **union** selected _ type with the fields named: u _ char (of type **char**), u _ int (of type **int**), u _ float (of type **float**), and u _ double (of type **double**). Define the structure var _ type with the fields named: type (of type **int**) and st (**union** of type selected _ type). Write a function that takes as parameter a structure of type var _ type. The function should read a number according to the value of the type field, store that number in the appropriate st field, and display the value of that field. Write a program that prompts the user to enter a number that represents the data type (i.e., 1: **char**, 2: **int**, 3: **float**, 4: **double**) and then uses the function to read a corresponding value and display it.

14

Memory Management and Data Structures

Memory management is mainly related to memory allocation and the release of allocated memory when it is no longer needed. Memory allocation can be performed either *statically* or *dynamically*. Up to this point, our program variables are stored in fixed size memory, statically allocated. In this chapter, we'll show you how to allocate memory dynamically during program execution in order to form flexible data structures, like stacks, queues, and linked lists.

Memory Blocks

When a program runs, it asks for memory resources from the operating system. Various operating systems and compilers use their own models to manage the available memory. Typically, the system memory is divided into four parts:

1. The *code* segment, which is used to store the program code.
2. The *data* segment, which is used to store the global and static variables. This segment may be also used to store literal strings (typically, in a read-only memory).
3. The *stack* segment, which is used to store function's data, like local variables.
4. The *heap* segment, which is used for dynamic memory allocation.

For example, the following program displays the memory addresses of the program data:

```c
#include <stdio.h>
#include <stdlib.h>

void test();

int global;

int main()
{
  int *ptr;
  int i;
  static int st;

  ptr = (int *) malloc(sizeof(int)); /* Allocate memory from the heap. */
  if(ptr != NULL)
  {
    printf("Code seg: %p\n", test);
    printf("Data seg: %p %p\n", &global, &st);
    printf("Stack seg: %p\n", &i);
```

```
    printf("Heap: %p\n", ptr);
    free(ptr);
  }
  return 0;
}
void test()
{
}
```

Static Memory Allocation

In static allocation, the memory is allocated from the stack. The size of the allocated memory is fixed; we must specify its size when writing the program and it can't change during program execution. For example, with the statement

```
float grades[100];
```

the compiler allocates 500*4 = 2000 bytes to store the grades of 500 students. The size of the array remains fixed; if we need to store grades for more students, we can't change it during program execution. If the students proved to be less than 500, we'd have a waste of memory. The only way to change the size of the array is to modify the program and compile it again.

Now, let's see what happens when a function is called. The compiler allocates memory in the stack to store the function's data. For example, if the function returns a value, accepts parameters, and declares local variables, the compiler allocates memory to store:

(a) The values of the parameters

(b) The local variables

(c) The return value

(d) The address of the next statement to be executed, after the function terminates

When the function terminates, the following actions take place:

(a) If the return value is assigned into a variable, it is extracted from the stack and stored into that variable.

(b) The address of the next statement is extracted from the stack and the execution of the program continues with that statement.

(c) The memory allocated to store the function's data is released; therefore, the values of the parameters and local variables may be overwritten.

For example, in the following program, when test() is called, the compiler allocates 808 bytes in the stack to store the values of i, j, and arr elements:

```
#include <stdio.h>

void test(int i, int j);
```

```
int main()
{
  float a[1000], b[10];

  test();
  return 0;
}

void test(int i, int j)
{
  int arr[200];
  ...
}
```

This memory is released when `test()` terminates. Similarly, the memory of 4040 bytes that is allocated for the local variables of `main()` is deallocated when the program terminates.

> *If the stack hasn't enough memory to store a function's data, the execution of the program would terminate abnormally and the message "Stack overflow" may appear.*

This situation may occur when a function allocates large blocks of memory and calls other nested functions that also have big memory needs. For example, a recursive function that calls itself many times may cause the exhaustion of the available stack memory.

Dynamic Memory Allocation

In dynamic allocation, the memory can be allocated during program execution. That memory is allocated from the heap and its size, unlike static allocation, isn't fixed; it may shrink or grow according to the program's needs.

Typically, the default stack size isn't very large. For example, the following program may not run because of unavailable stack memory:

```
#include <stdio.h>
int main()
{
  int arr[10000000]; /* Static allocation. */
  return 0;
}
```

On the other hand, the size of the heap is usually much larger than the stack size. For example, if we use dynamic allocation, the same memory would be successfully allocated:

```
#include <stdio.h>
#include <stdlib.h>
int main()
{
  int *arr;
  arr = (int *) malloc(10000000*sizeof(int));
```

```
  if(arr != NULL)
    free(arr);
  return 0;
}
```

As we'll see later, dynamically allocated memory should be released when no longer needed.

malloc() Function

To allocate memory dynamically, we'll need to use one of the realloc(), calloc(), or malloc() functions declared in stdlib.h. The most used is the malloc() function. It has the following prototype:

```
void *malloc(size_t size);
```

The size parameter declares the number of the bytes to be allocated. The type size _ t is a synonym of the **unsigned int** type. If the memory is allocated successfully, malloc() returns a pointer to that memory. If not, it returns NULL.

malloc() returns a **void** pointer, meaning that *any type of data* can be stored in the allocated memory. For example,

```
int *ptr;
ptr = (int *) malloc(100);
```

The (**int***) casting indicates that the allocated memory will be used for the storage of integers. Note that the casting isn't necessary; we could write

```
ptr = malloc(100);
```

The reason we prefer to cast the return pointer is to make clear which type of data will be stored in the allocated memory so that someone who reads our code won't have to look for the pointer declaration to find it.

If the allocation succeeds, ptr will point to the beginning of that memory. If not, its value would be NULL.

Always check the return value of malloc(). If it is NULL, an attempt to use a null pointer would have unpredictable results; the program may crash.

In the aforementioned example, the maximum number of stored integers is 25 because each integer needs 4 bytes. Note that we could write 25*sizeof(int) instead of 100.

In fact, the common practice is to use the **sizeof** operator in order to get a platform-independent program. For example, an **int** type may reserve two bytes in one computer and four in another. If we weren't using the **sizeof** operator, malloc() would reserve 50 integers in one case and 25 in the other.

*Always use the **sizeof** operator to specify the number of bytes to be allocated.*

However, when allocating memory for a **char** type, it is safe not to use the **sizeof** operator because the size of the **char** type is always one. For example, to allocate memory for a string of n characters, we write

```c
char *ptr;
ptr = (char *) malloc(n+1);
```

The extra byte is for the null character.

As another example, to allocate memory for 100 structures of type student, we write

```c
struct student *ptr;
ptr = (struct student *) malloc(100*sizeof(student));
```

Always remember that the size of the allocated memory is calculated in bytes.

For example, the following program is wrong because only one integer can be stored in the allocated memory, not four:

```c
#include <stdio.h>
#include <stdlib.h>
int main()
{
  int *arr, i;

  arr = (int *) malloc(4);
  if(arr != NULL)
  {
    for(i = 0; i < 4; i++)
      arr[i] = 10;

    free(arr);
  }
  return 0;
}
```

The assignment of 10 to arr[0] is correct, but the assignments to arr[1], arr[2], and arr[3] are wrong since no memory is allocated for them. Had we written

```c
arr = (int *) malloc(4*sizeof(int));
```

the program would be executed normally.

free() Function

A dynamically allocated memory must be released when it is no longer needed so that the operating system may reuse it. To release the memory, the free() function must be used. It is declared in stdlib.h, like this:

```c
void free(void *ptr);
```

The ptr must point to a block of memory, previously allocated by an allocation function (i.e., malloc()).

If `ptr` doesn't point to a dynamically allocated memory, the program may have unpredictable behavior. For example, the following program may crash when `free()` is called:

```c
#include <stdio.h>
#include <stdlib.h>
int main()
{
  int *ptr;

  free(ptr);
  return 0;
}
```

Note that `free()` is used to release dynamically allocated memory, not static. For example, if we declare `ptr` as an array (e.g., `int ptr[100]`), the program would fail again.

Also, the attempt to release an already released memory can cause unpredictable behavior. In the following example, the second call to `free()` is wrong because the memory is already released.

```c
int *ptr = (int *) malloc(100*sizeof(int));
if(ptr != NULL)
{
  free(ptr);
  free(ptr);
}
```

Don't forget to call `free()` to release a dynamically allocated memory when you don't need it any more.

memcpy() and memmove() Functions

The `memcpy()` function copies any type of data from one memory region to another. It is declared in `string.h`, like this:

```c
void *memcpy(void *dest, const void *src, size_t size);
```

`memcpy()` copies `size` bytes from the memory pointed to by `src` to the address pointed to by `dest`. If the source and destination addresses overlap, the behavior of `memcpy()` is undefined.

For example, the following program allocates memory to store the string `"ABCDE"` and displays its content:

```c
#include <stdio.h>
#include <string.h>
#include <stdlib.h>
int main()
{
  char *arr;
  /* Don't forget the null character. */
  arr = (char *) malloc(6);
```

```
  if(arr != NULL)
  {
    memcpy(arr, "ABCDE", 6);
    printf("%s\n", arr);
    free(arr);
  }
  return 0;
}
```

The main difference with `strcpy()` is that `strcpy()` stops copying when the null character is met. On the other hand, `memcpy()` doesn't stop copying until it copies the number of specified bytes. For example, if we have

```
char str2[6], str1[] = {'a', 'b', 'c', '\0', 'd', 'e'};
```

and write `memcpy(str2, str1, 6);` the content of `str2` would be equal to that of `str1`.

On the other hand, if we write `strcpy(str2, str1);` the content of `str2` would be equal to `{'a', 'b', 'c', '\0'}`.

`memmove()` is similar to `memcpy()`, except that `memmove()` guarantees that the bytes will be copied correctly even if the source and destination memory overlap. Because `memcpy()` doesn't check if the two memory regions overlap, it is executed faster than `memmove()`.

Notice that the size of the destination memory should be `size` bytes, at least. If it isn't, the extra bytes would be written in nonallocated memory, meaning that the existing data will be overwritten. For example, the next copy is not correct because the size of the destination memory is 3 bytes, while the copied bytes are 6.

```
char str1[3], str2[] = "abcde";
memcpy(str1, str2, sizeof(str2));
```

> **Tip**
>
> *`memcpy()` is often very useful because it is usually implemented in a way that copying large blocks of data is accomplished faster than an iteration loop.*

For example, the following program declares two arrays of `100000` integers, sets the values `1` to `100000` into the elements of the first array, and uses `memcpy()` to copy them into the second array:

```
#include <stdio.h>
#include <string.h>

#define SIZE 100000

int main()
{
  int i, arr1[SIZE], arr2[SIZE];

  for(i = 0; i < SIZE; i++)
    arr1[i] = i+1;

  memcpy(arr2, arr1, sizeof(arr1));
  /* We use memcpy() rather than an iteration loop:
  for(i = 0; i < SIZE; i++)
    arr2[i] = arr1[i]; */
  for(i = 0; i < SIZE; i++)
```

```
    printf("%d\n", arr2[i]);
  return 0;
}
```

memcmp() Function

The memcmp() function is used to compare the data stored in one memory region with the data stored in another. It is declared in string.h, like this:

```
int memcmp(const void *ptr1, const void *ptr2, size_t size);
```

memcmp() compares size bytes of the memory regions pointed to by ptr1 and ptr2. If they are the same, memcmp() returns 0, otherwise a nonzero value. For example, the following program declares two arrays of 100 integers, initializes them with random values, and compares them:

```
#include <stdio.h>
#include <string.h>
#include <stdlib.h>
#include <time.h>

#define SIZE 100

int main()
{
  int i, arr1[SIZE], arr2[SIZE];

  srand((unsigned int)time(NULL));
  for(i = 0; i < SIZE; i++)
  {
    arr1[i] = rand();
    arr2[i] = rand();
  }
  if(memcmp(arr1, arr2, sizeof(arr1)) == 0)
    printf("Same content\n");
  else
    printf("Different content\n");
  return 0;
}
```

If we were using an iteration loop instead, we could replace memcmp() with that:

```
for(i = 0; i < SIZE; i++)
  if(arr2[i] != arr1[i])
  {
    printf("Different content\n");
    return 0;
  }
```

The main difference with strcmp() is that strcmp() stops comparing when it encounters the null character in either string. memcmp(), on the other hand, stops comparing only when size bytes are compared; it doesn't look for the null character. For example,

```
#include <stdio.h>
#include <string.h>
#include <stdlib.h>
int main()
{
  char str1.[] = {'a', 'b', 'c', '\0', 'd', 'e'};
  char str2[] = {'a', 'b', 'c', '\0', 'd', 'f'};

  if(strcmp(str1, str2) == 0)
    printf("Same content\n");
  else
    printf("Different content\n");

  if(memcmp(str1, str2, sizeof(str1)) == 0)
    printf("Same content\n");
  else
    printf("Different content\n");
  return 0;
}
```

Because strcmp() stops comparing when the null character is met, the program displays Same content. On the contrary, memcmp() compares all bytes, and the program displays Different content.

Exercises

14.1 How many bytes does the following malloc() allocate?

```
double *ptr;
ptr = (double *) malloc(100*sizeof(*ptr));
```

Answer: Since ptr is a pointer to **double**, *ptr is a **double** number. Therefore, the sizeof(*ptr) calculates the size of the **double** type, which is 8, and malloc() allocates 800 bytes.

14.2 What is the output of the following program?

```
#include <stdio.h>
#include <string.h>
#include <stdlib.h>

void test(char *p);

int main()
{
  char *p1, *p2;

  p1 = p2 = (char *) malloc(5);
  if(p1 != NULL)
  {
    strcpy(p1, "test");
    test(p2);
    printf("%s\n", p1);
  }
```

```
    return 0;
}
void test(char *p)
{
  printf("%s\n", p);
  free(p);
}
```

Answer: When `malloc()` returns, `p1` and `p2` point to the same memory. When `test()` is called, we have `p = p2` and `test()` displays `test`. Since `free()` releases the memory pointed to by `p` (that's the same pointed to by `p1` and `p2`), `main()` displays random characters.

14.3 Write a program that reads a number of products and allocates memory to store their prices. The program should read the prices and store them into the memory.

```
#include <stdio.h>
#include <stdlib.h>
int main()
{
  double *arr;
  int i, size;

  printf("Enter size: ");
  scanf("%d", &size);

  arr = (double *) malloc(size * sizeof(double));
  if(arr == NULL)
  {
    printf("Error: Not available memory\n");
    exit(1);
  }
  for(i = 0; i < size; i++)
  {
    printf("Enter price: ");
    scanf("%lf", &arr[i]);
  }
  free(arr);
  return 0;
}
```

Comments: The `exit()` function is declared in `stdlib.h` and terminates the program. The integer argument indicates the termination status and it is made available to the operating system. For example, the value `1` indicates abnormal termination, while `0` indicates normal termination.

14.4 Write a function that resizes a dynamically allocated memory that stores integers. The function takes as parameters a pointer to the original memory, the initial memory size, the new size, and returns a pointer to the new memory. The existing data should be copied in the new memory. Write a program that allocates memory dynamically to store an array of 10 integers and sets the values 100 up to 109 to its elements. Then, the program should call the function to re-allocate new memory to store 20 integers and display its content.

```c
#include <stdio.h>
#include <stdlib.h>
#include <string.h>

int *realloc_mem(int *ptr, int old_size, int new_size);

int main()
{
  int *arr, i;

  /* Allocate memory for 10 integers. */
  arr = (int *) malloc(10 * sizeof(int));
  if(arr == NULL)
  {
    printf("Error: Not available memory\n");
    exit(1);
  }
  for(i = 0; i < 10; i++)
    arr[i] = i+100;

  arr = realloc_mem(arr, 10, 20); /* arr points to the new memory. */
  printf("\n***** Array elements *****\n");
  for(i = 0; i < 20; i++)
    printf("%d\n", arr[i]);

  free(arr); /* Release new memory. */
  return 0;
}

int *realloc_mem(int *old_mem, int old_size, int new_size)
{
  int *new_mem;

  /* Allocate memory for new_size integers. */
  new_mem = (int *) malloc(new_size * sizeof(int));
  if(new_mem == NULL)
  {
    printf("Error: Not available memory\n");
    exit(1);
  }
  /* Copy the existing data to the new memory. */
  memcpy(new_mem, old_mem, old_size * sizeof(int));
  free(old_mem); /* Release old memory. */
  return new_mem; /* Return the pointer to the new memory. */
}
```

Comments: The program displays the values 100–109 for the first ten elements and random values for the next ten since they aren't initialized.

C library provides a function similar to realloc_mem(), called realloc(). realloc() is declared in stdlib.h and it can change the size of a dynamically allocated memory. A short description is provided in Appendix C.

14.5 Write a function similar to memcmp(). The program should read two strings up to 100 characters, the number of the characters to be compared, and use the function to display the result of the comparison.

```
#include <stdio.h>
#include <stdlib.h>

int mem_cmp(const void *ptr1, const void *ptr2, size_t size);

int main()
{
  char str1[100], str2[100];
  int num;

  printf("Enter first text: ");
  gets(str1);

  printf("Enter second text: ");
  gets(str2);

  printf("Enter characters to compare: ");
  scanf("%d", &num);

  printf("%d\n", mem_cmp(str1, str2, num));
  return 0;
}

int mem_cmp(const void *ptr1, const void *ptr2, size_t size)
{
  char *p1, *p2;

  p1 = (char *)ptr1;
  p2 = (char *)ptr2;
  while(size != 0)
  {
    if(*p1 != *p2)
      return *p1 - *p2;
    p1++;
    p2++;
    size-- ;
  }
  return 0;
}
```

Comments: The **while** loop compares the characters pointed to by p1 and p2. Since we compare characters, we typecast the type **void*** to **char***. If all characters are the same, mem _ cmp() returns 0, otherwise the difference of the first two non-matching characters.

14.6 Write a function similar to strcpy(). The function should take as parameters two pointers and copy the string pointed to by the second pointer into the memory pointed to by the first pointer. The memory pointed to by the first pointer should have been allocated dynamically and its size should be equal to the length of the copied string. Write a program that reads a string up to 100 characters, calls the function to copy it in the dynamically allocated memory, and displays the content of that memory.

```
#include <stdio.h>
#include <string.h>
#include <stdlib.h>

char *str_cpy(char *trg, const char *src);
```

```
int main()
{
  char *trg, src[100];

  printf("Enter text: ");
  gets(src);
  /* Allocate memory to store the input string and the null
     character. */
  trg = (char *) malloc(strlen(src)+1);
  if(trg == NULL)
  {
    printf("Error: Not available memory\n");
    exit(1);
  }
  printf("Copied text:%s\n", str_cpy(trg, src));
  free(trg);
  return 0;
}

char *str_cpy(char *trg, const char *src)
{
  int i = 0;
  while(*(src+i) != '\0')/* Equivalent to while(src[i] != '\0')*/
  {
    *(trg+i) = *(src+i); /* Equivalent to trg[i] = src[i]; */
    i++;
  }
  *(trg+i) = '\0'; /* Add the null character. */
  return trg;
}
```

Comments: We could avoid the use of i; nevertheless, we'd get a more complicated code. Here is an example:

```
char *str_cpy(char *trg, const char *src)
{
  char *ptr = trg;
  while(*trg++= *src++);
  return ptr;
}
```

The loop executes until the null character is copied into the memory pointed to by trg. For example, in the first iteration, we have trg[0] = src[0] and the pointers are increased to point to the next elements. In fact, the statement *trg++= *src++; is equivalent to

```
*trg = *src;
trg++;
src++;
```

14.7 Write a function that takes as parameters three pointers to strings and stores the last two strings into the first one. The memory for the first string should have been allocated dynamically. Write a program that reads two strings up to 100 characters and uses the function to store them into a dynamically allocated memory.

```
#include <stdio.h>
#include <string.h>
#include <stdlib.h>

void str_cat(char *fin, char *str1, char *str2);

int main()
{
  char *fin, str1[100], str2[100];

  printf("Enter first text: ");
  gets(str1);

  printf("Enter second text: ");
  gets(str2);
  /* Allocate memory to store both strings and the null character. */
  fin = (char *) malloc(strlen(str1)+strlen(str2)+1);
  if(fin == NULL)
  {
    printf("Error: Not available memory\n");
    exit(1);
  }
  str_cat(fin, str1, str2);
  printf("Merged text: %s\n", fin);
  free(fin);
  return 0;
}

void str_cat(char *fin, char *str1, char *str2)
{
  while(*str1 != '\0')/* Equivalent to while(*str1) */
    *fin++= *str1++;

  /* Copy the second string right after the first one. */
  while(*str2 != '\0')/* Equivalent to while(*str2) */
    *fin++ = *str2++;

  *fin = '\0'; /* Add the null character. */
}
```

Comments: The first loop copies the characters of the string pointed to by str1 into the memory pointed to by fin. Similarly, the next loop adds the characters of the second string.

Instead of **while** loops, we could use **for** loops, as follows:

```
void str_cat(char *fin, char *str1, char *str2)
{
  for(; *str1; *fin++ = *str1++);
  for(; *str2; *fin++ = *str2++);
  *fin = '\0';
}
```

Notice the semicolon ; at the end of both **for** statements.
How about a more complex solution with a single **for** loop?

```
void str_cat(char *fin, char *str1, char *str2)
```

```
{
  for(; *str2; *str1 ? *fin++ = *str1++ : *fin++ = *str2++);
  *fin = '\0';
}
```

That's another example of what we consider bad-written code. Don't forget our advice. Write code having simplicity in mind.

14.8 Write a function that takes as parameters two arrays of doubles (i.e., a1 and a2) and their number of elements (note that number is the same), and allocates memory to store the elements of a1 that are not contained in a2 and returns a pointer to that memory. Write a program that reads pairs of doubles and stores them into two arrays of 100 positions (i.e., p1 and p2). If the user enters –1, the insertion of numbers should end. The program should use the function to display the elements of p1 that are not contained in p2.

```
#include <stdio.h>
#include <stdlib.h>

#define SIZE 100

double *find_diff(double a1[], double a2[],int size,int *items); /*
The parameter items indicates how many elements are stored in the
memory. A pointer is passed, so that the function may change its
value. */

int main()
{
  int i, elems;
  double *p3, j, k, p1[SIZE], p2[SIZE];

  for(i = 0; i < SIZE; i++)
  {
    printf("Enter numbers: ");
    scanf("%lf%lf", &j, &k);
    if((j == -1) || (k == -1))
      break;
    p1[i] = j;
    p2[i] = k;
  }
  elems = 0;
  p3 = find_diff(p1, p2, i, &elems);
  if(elems == 0)
    printf("\n***** No different elements *****\n");
  else
  {
    for(i = 0; i < elems; i++)
      printf("%f\n", p3[i]);
  }
  free(p3);
  return 0;
}

double *find_diff(double a1[], double a2[], int size,int *items)
{
  int i, j, found;
  double *mem;
```

```
  mem = (double *) malloc(size * sizeof(double));
  if(mem == NULL)
  {
    printf("Error: Not available memory\n");
    exit(1);
  }
  for(i = 0; i < size; i++)
  {
    found = 0; /* This variable indicates whether an element of the
      first array exists in the second, or not. The value 0 means
      that it doesn't exist. */
    for(j = 0; j < size; j++)
    {
      if(a2[j] == a1[i])
      {
        found = 1;
        break; /* Since this element exists, we stop searching. */
      }
    }
    /* If it doesn't exist, it is stored in the memory. */
    if(found == 0)
    {
      mem[*items] = a1[i];
      (*items)++;
    }
  }
  return mem;
}
```

14.9 Write a program that declares an array of five pointers to strings and allocates memory to store strings up to 100 characters. Then, the program should use these pointers to read five strings and display the larger one (*Note*: If more than one string has the same maximum length, the program should display the one found first.)

```
#include <stdio.h>
#include <string.h>
#include <stdlib.h>

#define SIZE 5

int main()
{
  char *ptr[SIZE];
  int i, pos, len, max_len; /* The variable pos indicates the ptr
    element, which points to the larger string. The variable max_len
    holds its length. */

  pos = max_len = 0;
  /* Each pointer points to a dynamically allocated memory of 100
  bytes. */
  for(i = 0; i < SIZE; i++)
  {
    ptr[i] = (char *) malloc(100);
    if(ptr[i] == NULL)
```

```
      {
        printf("Error: Not available memory\n");
        exit(1);
      }
      printf("Enter text: ");
      gets(ptr[i]);
      /* We compare the length of each string against max_len and if a
         larger string is found, we store its position and length. */
      len = strlen(ptr[i]);
      if(len > max_len)
      {
        pos = i;
        max_len = len;
      }
    }
    printf("Larger string: %s\n", ptr[pos]);
    for(i = 0; i < SIZE; i++)
      free(ptr[i]);
    return 0;
}
```

14.10 Write a program that reads its command line arguments and allocates memory to store their characters in reverse order. For example, if the arguments are next and time, the program should store into the memory txen and emit.

```
#include <stdio.h>
#include <stdlib.h>
#include <string.h>

int main(int argc, char *argv[])
{
  char *rvs_str;
  int i, j, len;

  if(argc == 1)
  {
    printf("Missing string arguments…\n");
    exit(1);
  }
  for(i = 1; i < argc; i++)
  {
    len = strlen(argv[i]);

    rvs_str = (char *) malloc(len+1); /* Allocate one extra place for
      the null character. */
    if(rvs_str == NULL)
    {
      printf("Error: Not available memory\n");
      exit(1);
    }
    for(j = 0; j < len; j++)
      rvs_str[j] = argv[i][len-1-j]; /* The last character is stored
        in position len-1. */

    rvs_str[j] = '\0'; /* Terminate the string. */
```

```c
      printf("Reverse of %s is: %s\n", argv[i], rvs_str);
      free(rvs_str);
    }
  return 0;
}
```

14.11 Write a program that reads continuously the N×M dimensions of a two-dimensional array of doubles, allocates memory for it, then it reads, and stores numbers into it. If any input dimension is less or equal to zero, the program should terminate.

```c
#include <stdio.h>
#include <stdlib.h>

int main()
{
  int i, j, rows, cols;
  double **arr; /* We handle the two-dimensional array as pointer to
    pointer. */

  while(1)
  {
    printf("\nEnter dimensions of array[N][M] (zero or less to
      terminate): ");
    scanf("%d%d", &rows, &cols);

    if(rows <= 0 || cols <= 0)
      return 0;

    arr = (double**) malloc(rows * sizeof(double*)); /* We allocate
      memory for 'rows' pointers to doubles. For example, if rows is
      3, we allocate memory for arr[0], arr[1] and arr[2] pointers.
      */
    if(arr == NULL)
    {
      printf("Error: Not available memory\n");
      exit(1);
    }
    for(j = 0; j < rows; j++)
    {
      arr[j] = (double*) malloc(cols * sizeof(double)); /* Allocate
        memory to store the elements of each row. Each row contains
        'cols' elements. */
      if(arr[j] == NULL)
      {
        printf("Error: Not available memory\n");
        exit(1);
      }
    }
    for(i = 0; i < rows; i++)
      for(j = 0; j < cols; j++)
      {
        printf("Set arr[%d][%d]: ", i, j);
        scanf("%lf", &arr[i][j]);
        printf("arr[%d][%d] = %f\n", i, j, arr[i][j]);
      }
```

```
    for(j = 0; j < rows; j++)
      free(arr[j]);
    free(arr);
  }
  return 0;
}
```

14.12 Define the structure type `publisher` with fields: name, address, and phone number. Then, define the structure type `book` with fields: title, author, field, code, price, and a pointer to a structure of type `publisher`. Except the price, all other fields must be pointers. Assume that the maximum string length is 100 characters.

Write a program that reads the number of books and allocates memory to store their data and the data of the publishers as well. Then, the program should read a book's code and, if it is registered, the program should display its title and the name of its publisher.

```
#include <stdio.h>
#include <string.h>
#include <stdlib.h>
typedef struct
{
  char *name;
  char *addr;
  char *phone;
} pub;

typedef struct
{
  char *title;
  char *auth;
  char *code;
  pub *pub_ptr;
  float prc;
} book;
/* For the sake of brevity, we assume that all malloc() calls are
   successful. */
int main()
{
  book *books_ptr;
  char str[100];
  int i, num;

  printf("Enter number of books: ");
  scanf("%d", &num);
  getchar();

  books_ptr = (book *) malloc(num * sizeof(book));
  for(i = 0; i < num; i++)
  {
    printf("\nTitle: ");
    gets(str);
    books_ptr[i].title = (char *)malloc(strlen(str)+1);
    strcpy(books_ptr[i].title, str);

    printf("Authors: ");
```

```
    gets(str);
    books_ptr[i].auth = (char *) malloc(strlen(str)+1);
    strcpy(books_ptr[i].auth, str);

    printf("Code: ");
    gets(str);
    books_ptr[i].code = (char *) malloc(strlen(str)+1);
    strcpy(books_ptr[i].code, str);

    printf("Price: ");
    scanf("%f", &books_ptr[i].prc);

    getchar();
    /* Allocate memory to store the data of the publishing firm. */
    books_ptr[i].pub_ptr = (pub *) malloc(sizeof(pub));

    printf("Name: ");
    gets(str);
(books_ptr[i].pub_ptr)->name = (char *) malloc(strlen(str)+1);
    strcpy((books_ptr[i].pub_ptr)->name, str);

    printf("Address: ");
    gets(str);
    (books_ptr[i].pub_ptr)->addr = (char *) malloc(strlen(str)+1);
    strcpy((books_ptr[i].pub_ptr)->addr, str);

    printf("Phone: ");
    gets(str);
    (books_ptr[i].pub_ptr)->phone = (char *) malloc(strlen(str)+1);
    strcpy((books_ptr[i].pub_ptr)->phone, str);
  }
  printf("\nEnter code to search: ");
  gets(str);

  for(i = 0; i < num; i++)
  {
    if(strcmp(books_ptr[i].code, str) == 0)
    {
      printf("\nTitle: %s\tPublisher: %s\n\n", books_ptr[i].title,
        (books_ptr[i].pub_ptr)->name);
      break;
    }
  }
  if(i == num)
    printf("\nCode '%s' isn't registered\n", str);

  for(i = 0; i < num; i++)
  {
    free((books_ptr[i].pub_ptr)->name);
    free((books_ptr[i].pub_ptr)->addr);
    free((books_ptr[i].pub_ptr)->phone);

    free(books_ptr[i].title);
    free(books_ptr[i].auth);
    free(books_ptr[i].code);
    free(books_ptr[i].pub_ptr);
  }
  free(books_ptr);
  return 0;
}
```

Dynamic Data Structures

The data structures we've met so far are used for data storing and processing in an easy and fast way. For example, arrays are data structures, which are used for the storage of the same type of data. Structures and unions are also data structures, which can be used for the storage of any type of data.

These data structures are *static* because the allocated memory is *fixed* and can't be modified during program execution. However, in many applications, it'd be more efficient to use a dynamic data structure, a structure whose size may grow or shrink as needed.

The next sections describe how to create some simple forms of dynamic data structures, like linked lists, queues, and stacks.

Linked List

A linked list consists of a chain of linked elements, called *nodes*. Each node is a structure, which contains its data and a pointer to the next node in the chain, as depicted in Figure 14.1.

The first node is the head of the list and the last one its tail. The value of the pointer field in the last node must be equal to NULL to indicate the end of the list.

Unlike an array whose size remains fixed, a linked list is more flexible because we can insert and delete nodes as needed. On the other hand, any array element can be accessed very fast; we just use its position as an index. The time to access a node depends on the position of the node in the list. It is fast if the node is close to the beginning, and slow if it is near the end.

Insert a Node

A new node can be inserted at any point in the list. To insert a new node, we check the following cases:

1. If the list is empty, the node is inserted at the beginning and becomes the head and the tail of the list. The value of its pointer field is set to NULL since there is no next node in the list.
2. If the list isn't empty, we check the following subcases:
   (a) If the node is inserted at the beginning of the list, it becomes the new head of the list and its pointer field points to the previous head, which now becomes the second node of the list.
   (b) If the node is inserted at the end of the list, it becomes the new tail of the list and the value of its pointer field is set to NULL. The old tail becomes the second

FIGURE 14.1
Linked list.

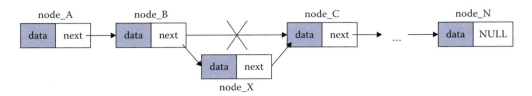

FIGURE 14.2
Insert a new node in linked list.

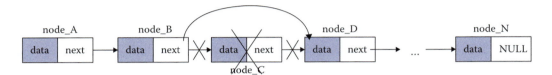

FIGURE 14.3
Delete an existing node from linked list.

to last node and the value of its pointer field changes from NULL to point to the
new node.

(c) If the node is inserted after an existing node, the pointer field of the new node
must point to the node after the current node and the pointer field of the current node must point to the new node. Figure 14.2 depicts how the new node X
is inserted between the nodes B and C.

Delete a Node

To delete a node from the list, we check the following cases:

1. If it is the head of the list, we check the following subcases:
   (a) If there is a next node, this node becomes the new head of the list.
   (b) If there is no next node, the list becomes empty.
2. If it is the tail of the list, the previous node becomes the new tail and its pointer
   field is set to NULL.
3. If it is an intermediate node, the pointer field of the previous node must point to
   the node after the one to be deleted. This operation is shown here with the deletion
   of node C (Figure 14.3).

Examples

Before creating a linked list, we'll implement two special cases of a linked list, a stack and
a queue. Note that you can implement these data structures in several ways; we tried to
implement them in a simple and comprehensible way.

In the following examples, each node is a structure of type student. If you need to
develop similar data structures, the most part of the code can be reused as is.

Implementing a Stack

In this section, we'll create a LIFO (*Last In First Out*) stack, where, as its name declares, the last inserted node it is the first to get extracted. It is a special case of a linked list with the following restrictions:

1. A new node can be inserted only at the beginning of the stack and becomes its new head.
2. Only the head can be deleted.

Exercise

14.13 Define the structure type `student` with fields: code, name, and grade. Create a stack whose nodes are structures of type `student`. Write a program that displays a menu to perform the following operations:

1. Insert a student. The program should read the student's data and store them in a node, which becomes the new head of the stack.
2. Display the data of the stored students.
3. Display the data of the last inserted student.
4. Delete the last inserted student.
5. Display the total number of the stored students.
6. Program termination.

To handle the stack, we use a global pointer. This pointer always points to the head of the stack.

```c
#include <stdio.h>
#include <stdlib.h>

typedef struct student
{
  char name[100];
  int code;
  float grd;
  struct student *next; /* Pointer to the next node. */
} student;

student *head; /* Global pointer that always points to the head of
  the stack. */

void add_stack(const student *stud_ptr);
void show_stack();
void pop();
int size_stack();
void free_stack();
int main()
```

```
{
  int sel;
  student stud;

  head = NULL; /* This initial value indicates that the stack is
    empty. */
  while(1)
  {
    printf("\nMenu selections\n");
    printf("- - - - - - -\n");

    printf("1. Add student\n");
    printf("2. View all students\n");
    printf("3. View top student\n");
    printf("4. Delete top student\n");
    printf("5. Number of students\n");
    printf("6. Exit\n");

    printf("\nEnter choice: ");
    scanf("%d", &sel);
    switch(sel)
    {
      case 1:
        getchar();

        printf("Name: ");
        gets(stud.name);
        printf("Code: ");
        scanf("%d", &stud.code);

        printf("Grade: ");
        scanf("%f", &stud.grd);

        add_stack(&stud);
      break;

      case 2:
        if(head != NULL)
          show_stack();
        else
          printf("\nThe stack is empty\n");
      break;

      case 3:
        if(head != NULL)
          printf("\nData:%d %s %.2f\n\n",
          head->code,head->name,head->grd);
        else
          printf("\nThe stack is empty\n");
      break;

      case 4:
        if(head != NULL)
          pop();
        else
          printf("\nThe stack is empty\n");
      break;

      case 5:
```

```
      if(head != NULL)
        printf("\n%d students exist in stack\n", size_stack());
      else
        printf("\nThe stack is empty\n");
      break;

      case 6:
        if(head != NULL)
          free_stack();
      return 0;
      default:
        printf("\nWrong choice\n");
      break;
    }
  }
  return 0;
}
void add_stack(const student *stud_ptr)
{
  student *new_node;
  /* Allocate memory to create a new node. */
  new_node = (student *) malloc(sizeof(student));
  if(new_node == NULL)
  {
    printf("Error: Not available memory\n");
    exit(1);
  }
  *new_node = *stud_ptr; /* Copy the student's data into the new
    node. */
  new_node->next = head; /* The new node is inserted at the
    beginning of the stack. For example, when the first node is
    inserted the value of new_node->next becomes NULL, because that
    is the initial value of the head pointer. */
  head = new_node; /* head points to the new node, therefore that
    new node becomes the new head of the stack. */
}
void show_stack()
{
  student *ptr;

  ptr = head;
  printf("\n***** Student Data *****\n\n");
  while(ptr != NULL)
  {
    printf("C:%d N:%s G:%.2f\n\n", ptr->code, ptr->name, ptr->grd);
    ptr = ptr->next; /* In each iteration, ptr points to the next
      node. When its value becomes NULL means that there is no other
      node in the stack and the loop terminates. */
  }
}
void pop()
{
  student *ptr;
```

```
  ptr = head->next; /* ptr points to the node after the head. This
    node will become the new head of the stack. */
  printf("\nStudent with code = %d is deleted\n",head->code);
free(head); /* Release the allocated memory. The information for
  which is the next node is not lost, because we saved it in ptr. */
  head = ptr; /* head points to the new head of the stack. */
}
int size_stack()
{
  student *ptr;
  int num;

  num = 0;
  ptr = head;
  while(ptr != NULL)
  {
  ptr = ptr->next;
  num++; /* This variable counts the nodes, until we reach the last
    one. */
  }
  return num;
}

void free_stack()
{
  student *ptr, *next_node;

  ptr = head;
  while(ptr != NULL)
  {
    next_node = ptr->next; /* next_node always points to the node
      after the one to be deleted. */
    free(ptr); /* Release the allocated memory. The information for
      which is the next node is not lost, because we saved it in
      next_node. */
    ptr = next_node; /* ptr points to the next node. */
  }
}
```

Comments:

1. In add _ stack() we pass a pointer and not the structure itself to avoid the creation of a structure's copy. The word **const** is added to prevent the function from modifying the values of the structure's fields.

2. To display immediately the number of the existing students, without traversing the whole stack, we could remove the size _ stack() function and declare a global variable that should be increased each time a student is inserted and reduced when a student is deleted. The reason we are using the size _ stack() is to show you how to traverse the nodes of the stack.

3. If the variable head had been declared locally in main(), we should pass its memory address in the functions that need it. For example, the add _ stack() would change to

```
void add_stack(const student *stud_ptr, student **head_ptr)
{
  student *new_node;
  new_node = (student *) malloc(sizeof(student));
  if(new_node == NULL)
  {
    printf("Error: Not available memory\n");
    exit(1);
  }
  *new_node = *stud_ptr;
  new_node->next = *head_ptr;
  *head_ptr = new_node;
}
```

To call it we'd write add _ stack(&stud, &head);

Since the address of head is passed to add _ stack(), the function may modify its value.

Because we think that this code is more complicated, at least for a beginner, we preferred to declare head as a global variable and use it directly when needed.

Implementing a Queue

In this section, we'll create a FIFO (*First In First Out*) queue, where, as its name declares, the first inserted node it is the first to get extracted. It is a special case of a linked list with the following restrictions:

1. A new node can be inserted only at the end of the queue and becomes its new tail.
2. Only the head can be deleted.

Exercise

14.14 Define the structure type student with fields: code, name, and grade. Create a queue whose nodes are structures of type student. Write a program that displays a menu to perform the following operations:

1. Insert a student. The program should read the student's data and store them in a node, which becomes the new tail of the queue.
2. Display the data of the stored students.
3. Display the data of the last inserted student.
4. Delete the last inserted student.
5. Program termination.

To handle the queue, we use two global pointers. The first one always points to the head of the queue and the second one to its tail.

```
#include <stdio.h>
#include <stdlib.h>
```

```c
typedef struct student
{
  char name[100];
  int code;
  float grd;
  struct student *next;
} student;
```

```c
student *head; /* Global pointer that always points to the head of
the queue. */
student *tail; /* Global pointer that always points to the tail of
the queue. */
```

```c
void add_queue(const student *stud_ptr);
void show_queue();
void pop();
void free_queue();
```

```c
int main()
{
  int sel;
  student stud;

  head = NULL;
  while(1)
  {
    printf("\nMenu selections\n");
    printf("- - - - - - -\n");
    printf("1. Add student\n");
    printf("2. View all students\n");
    printf("3. View last student\n");
    printf("4. Delete top student\n");
    printf("5. Exit\n");

    printf("\nEnter choice: ");
    scanf("%d", &sel);

    switch(sel)
    {
      case 1:
        getchar();

        printf("Name: ");
        gets(stud.name);

        printf("Code: ");
        scanf("%d", &stud.code);

        printf("Grade: ");
        scanf("%f", &stud.grd);

        add_queue(&stud);
      break;

      case 2:
        if(head != NULL)
          show_queue();
        else
```

```
        printf("\nThe queue is empty\n");
      break;

      case 3:
       if(head != NULL)
         printf("\nData:%d %s %.2f\n\n",
         tail->code,tail->name,tail->grd);
        else
         printf("\nThe queue is empty\n");
      break;

      case 4:
        if(head != NULL)
          pop();
        else
          printf("\nThe queue is empty\n");
      break;

      case 5:
        if(head != NULL)
          free_queue();
      return 0;

      default:
        printf("\nWrong choice\n");
      break;
    }
  }
  return 0;
}
void add_queue(const student *stud_ptr)
{
  student *new_node;
  new_node = (student *) malloc(sizeof(student));
  if(new_node == NULL)
  {
    printf("Error: Not available memory\n");
    exit(1);
  }
*new_node = *stud_ptr;
new_node->next = NULL;

if(head == NULL)
  head = tail = new_node; /* If the queue is empty, both pointers
    point to the new node. */
  else
  {
    tail->next = new_node; /* The new node is inserted at the end of
      the queue. */
    tail = new_node; /* Now, tail points to the last node. */
  }
}
```

Comments: The code of the show _ queue(), pop(), and free _ queue() functions
is the same as the code of show _ stack(), pop(), and free _ stack() functions
in the previous exercise.

Implementing a Linked List

The following program implements a linked list whose nodes correspond to students. Like before, we use one global pointer to point to the head of the list and another one to point to its tail.

Exercises

14.15 Define the structure type `student` with fields: code, name, and grade. Create a list whose nodes are structures of type `student`. Write a program that displays a menu to perform the following operations:

1. Insert a student at the end of the list. The program should read the student's data and store them in a node, which becomes the new tail of the list.

2. Insert a student in another point. The program should read the code of a student, locate the respective node in the list, and create a new node after it to insert the data of the new student.

3. Display the data of the stored students.

4. Find a student. The program should read the code of a student and, if it is found in the list, it should display the student's data.

5. Modify the grade of a student. The program should read the code of a student and the new grade and modify the existing grade.

6. Delete a student. The program should read the code of a student and remove the node that corresponds to the student.

7. Program termination.

```c
#include <stdio.h>
#include <stdlib.h>

typedef struct student
{
  char name[100];
  int code;
  float grd;
  struct student *next;
} student;

student *head; /* Global pointer that always points to the head of
  the list. */
student *tail; /* Global pointer that always points to the tail of
  the list. */

void add_list_end(const student *stud_ptr);
void add_list(const student *stud_ptr, int code);
void show_list();
student *find_node(int code);
void del_node(int code);
void free_list();
```

```c
int main()
{
  int sel, code;
  float grd;
  student stud, *ptr;

  head = NULL;
  while(1)
  {
    printf("\nMenu selections\n");
    printf("- - - - - - -\n");

    printf("1. Add student at the end\n");
    printf("2. Add student\n");
    printf("3. View all students\n");
    printf("4. View student\n");
    printf("5. Modify student\n");
    printf("6. Delete student\n");
    printf("7. Exit\n");

    printf("\nEnter choice: ");
    scanf("%d", &sel);

    switch(sel)
    {
      case 1:
      case 2:/* To avoid the repetition of the same code we use the
        same case. Then, the if statement checks the user's choice
        and calls the respective function. */
        getchar();

        printf("Name: ");
        gets(stud.name);

        printf("Code: ");
        scanf("%d", &stud.code);

        printf("Grade: ");
        scanf("%f", &stud.grd);

        if(sel == 1)
          add_list_end(&stud);
        else
        {
          printf("\nEnter student code after which the new student
            will be added: ");
          scanf("%d", &code);
          add_list(&stud, code);
        }
      break;

      case 3:
        if(head == NULL)
          printf("\nThe list is empty\n");
        else
          show_list();
      break;
```

```c
    case 4:
      if(head == NULL)
        printf("\nThe list is empty\n");
      else
      {
        printf("\nEnter student code to search: ");
        scanf("%d", &code);
        ptr = find_node(code);
        if(ptr == NULL)
          printf("\nStudent with code = %d does not exist\n", code);
        else
          printf("\nData: %s %.2f\n\n", ptr->name, ptr->grd);
      }
    break;

    case 5:
      if(head == NULL)
        printf("\nThe list is empty\n");
      else
      {
        printf("\nEnter student code to modify: ");
        scanf("%d", &code);

        printf("Enter new grade: ");
        scanf("%f", &grd);
        ptr = find_node(code);
        if(ptr != NULL)
          ptr->grd = grd;
        else
          printf("\nStudent with code = %d does not exist\n", code);
      }
    break;

    case 6:
      if(head == NULL)
        printf("\nThe list is empty\n");
      else
      {
        printf("\nEnter student code to delete: ");
        scanf("%d", &code);
        del_node(code);
      }
    break;

    case 7:
      if(head != NULL)
        free_list();
    return 0;

    default:
      printf("\nWrong choice\n");
    break;
    }
  }
  return 0;
}
```

```c
/* For a better understanding of the add_list_end(), read the
   comments of the add_queue() in the previous exercise. */
void add_list_end(const student *stud_ptr)
{
  student *new_node;

  new_node = (student *) malloc(sizeof(student));
  if(new_node == NULL)
  {
    printf("Error: Not available memory\n");
    exit(1);
  }
  *new_node = *stud_ptr;
  new_node->next = NULL;

  if(head == NULL)
    head = tail = new_node;
  else
  {
    tail->next = new_node;
    tail = new_node;
  }
}

void add_list(const student *stud_ptr, int code)
{
  student *new_node, *ptr;

  ptr = head;
  /* We traverse the list, until the node with the indicated code is
     found. If it is found, the new node is added after it and the
     function terminates. */
  while(ptr != NULL)
  {
    if(ptr->code == code)
    {
      new_node = (student *)malloc(sizeof(student));
      if(new_node == NULL)
      {
        printf("Error: Not available memory\n");
        exit(1);
      }
      *new_node = *stud_ptr; /* Copy the student's data. */
      new_node->next = ptr->next; /* Now, the new node is linked to
        the node after the current node. */
      ptr->next = new_node; /* Now, the current node is linked to the
        new node. */
      if(ptr == tail)/* Check if the new node is added at the end of
        the list. If it is, it becomes the new tail. */
        tail = new_node;
      return;
    }
    ptr = ptr->next; /* Check the next node. */
  }
  /* If the execution reaches this point, means that the input code
     doesn't correspond to an existing student. */
```

```c
    printf("\nStudent with code = %d does not exist\n", code);
}

void show_list()
{
  student *ptr;

  ptr = head;
  printf("\n***** Student Data *****\n\n");
  while(ptr != NULL)
  {
    printf("C:%d N:%s G:%.2f\n\n", ptr->code, ptr->name, ptr->grd);
    ptr = ptr->next;
  }
}
student *find_node(int code)
{
  student *ptr;
  ptr = head;
  while(ptr != NULL)
  {
    if(ptr->code == code)
      return ptr;
      ptr = ptr->next;
  }
  return NULL;
}

void del_node(int code)
{
  student *ptr, *prev_node; /* prev_node always points to the
    previous node from the one that is going to be deleted. */

  ptr = prev_node = head;
  while(ptr != NULL)
  {
    if(ptr->code == code)
    {
      if(ptr == head)
        head = ptr->next; /* If the node is the head of the list, the
          next node becomes the new head. If there is no other node,
          the list becomes empty and the value of head becomes
          NULL. */
      else
      {
        /* Now, ptr points to the node that will be deleted and prev_
          node points to the previous one. This statement links the
          previous node with the node after the one that will be
          deleted. */
        prev_node->next = ptr->next;
        if(ptr == tail)/* Check if the deleted node is the tail of
          the list. If it is, the previous node becomes the new
          tail. */
          tail = prev_node;
      }
      free(ptr); /* Delete the node. */
```

```
      printf("\nStudent with code = %d is deleted\n", code);
      return;
    }
    prev_node = ptr; /* Now, prev_node points to the node that was
       just checked and found that it doesn't correspond to the node
       having the input code. */
    ptr = ptr->next; /* Now, ptr points to the next node. Note that
       prev_node points always to the previous node from the one that
       is going to be checked. */
  }
  printf("\nStudent with code = %d does not exist\n", code);
}

void free_list()
{
  student *ptr, *next_node;

  ptr = head;
  while(ptr != NULL)
  {
    next_node = ptr->next;
    free(ptr);
    ptr = next_node;
  }
}
```

Comments: To change the first operation and insert the data of the new student at the beginning of the list and not at its end, replace the add _ list _ end() with the add _ stack() presented in the stack implementation. In that case, the tail pointer is not needed.

14.16 Consider the linked list of the previous exercise. Write a function that takes as parameters the codes of two students and, if they are stored in the list, the function should swap their grades and return 0. If not, the function should return –1.

```
int swap(int code_a, int code_b)
{
  student *ptr, *tmp1, *tmp2;
  float grd;

  ptr = head;
  tmp1 = tmp2 = NULL;

  while(ptr != NULL)
  {
    if(ptr->code == code_a)
      tmp1 = ptr;
    else if(ptr->code == code_b)
      tmp2 = ptr;

    if(tmp1 && tmp2)
    {
      grd = tmp1->grd;
      tmp1->grd = tmp2->grd;
      tmp2->grd = grd;
      return 0;
    }
```

```
    ptr = ptr->next;
  }
  return -1;
}
```

Unsolved Exercises

14.1   Complete the following program to read an integer and a **double** number and display their sum. Don't use any other variable.

```
int main()
{
  int *p1;
  double *p2;

  ...
}
```

14.2   Write a program that uses a pointer variable to read three integers and display the greatest. Don't use any other variable.

14.3   Use the two pointer variables and complete the following function to return how many products cost less than $20 and how many more than $100.

```
void find(double *arr, int size, int *l20, int *m100);
```

Write a program that prompts the user to enter the number of the products (i.e., `size`), their prices, and stores them in a dynamically allocated memory (i.e., `arr`). Then, the program should use the function to display how many products cost less than $20 and how many more than $100.

14.4   Write a program that

(a) Allocates memory to store a number of integers. The program should prompt the user to enter that number.

(b) The program should read those integers and store them in the allocated memory.

(c) If the user enters −5, the program should release the memory and continue from the first step.

14.5   Use the `ptr` pointer and complete the following program to read and display the data of one student.

```
struct student
{
  char *name;
  int code;
  float grd;
};
int main()
{
  struct student *ptr;

  ...
}
```

Then, the program should read a number of students and use `ptr` to allocate memory and store the data of the students whose name begin with an `'A'`.

14.6 Write a program that reads 10 **double** numbers and stores them in a dynamically allocated memory. Then, the program should allocate an extra memory of the same size and prompt the user to enter a number, as follows:

(a) If it is `0`, the program should store the numbers in that new memory in reverse order.

(b) If it is `1`, the program should store first the negatives and then the positives.

Use pointer arithmetic to handle the memories.

For example, assume that the first memory contains the numbers:

```
-3.2 4 3 -9.1 7 6 -2 15 9 -37
```

If the user enters `0`, the numbers should be stored in the second memory in reverse order, as follows:

```
-37 9 15 -2 …
```

If the user enters `1`, it should be stored in that order:

```
-3.2 -9.1 -2 -37 4 3 …
```

14.7 Use the `ptr` pointer and complete the following program to read and display the data of one book. Don't use any other variable.

```
struct student
{
  char *title; /* Assume that the characters fields are less than
    100 characters. */
  char *authors;
  int *code;
  double *prc;
};

int main()
{
  struct book *ptr;
  …
}
```

14.8 Write a program that reads integers continuously and displays only those which are not repeated. If the user enters –1, the insertion of numbers should terminate. See an example of program execution:

```
Enter number: -20
Output: -20
Enter number: 345
Output: 345
Enter number: -20
Enter number: 432
Output: 432
```

As you see, the number `-20` doesn't appear twice.

14.9 Consider the linked list of 14.15 (Exercise). Write a function that takes as a parameter the code of a student and, if it is stored in the list, the function should return how many nodes are left up to the end of the list. If not, the function should return -1. Don't use the `tail` pointer variable.

14.10 Consider the linked list of 14.15 (Exercise). Write a **void** function that returns a pointer to the node with the maximum grade in the list and a pointer to the node with the minimum grade, as well. Write a sample program to show how to test the function.

14.11 Consider the queue of 14.14 (Exercise). Add a pointer field (i.e., `prev`) to the type `student` to point to the previous node. Modify the `add_queue()` and `pop()` functions and add the function `show_reverse()` to traverse the nodes reversely, from the tail to the head.

14.12 Write a program that generates 100 random integers and creates a linked list with those numbers.

15

Files

Real-world programs often need to perform access operations on files. This chapter introduces C's most important library functions designed specifically for use with files. We won't describe them in full detail, but we'll give you the material needed in order to perform file operations. In particular, you'll learn how to open and close a file, as well as how to read and write data in *text* and *binary* files.

Files in C

C supports access operations on two kinds of files: *text* and *binary*.

A *text* file consists of one or more lines that contain readable characters according to a standard format, like the ASCII code. Each line ends with special character(s) the operating system uses to indicate the end of line.

In *Windows* systems, for example, the pair of '\r' (Carriage Return) and '\n' (Line Feed) characters, that is, CR/LF, with ASCII codes 13 and 10 respectively, indicate the end of line. Therefore, the new line character '\n' is replaced by '\r' and '\n' when written in the file. The reverse replacement happens when the file is read. On the other hand, this replacement does not take place in *Unix* systems because the '\n' character indicates the end of line.

Unlike the *text* files, the bytes of a *binary* file don't necessarily represent readable characters. For example, an executable C program is stored in a *binary* file. If you open it, you'll probably see some unintelligible characters. A *binary* file isn't divided into lines and no character translation takes place. In *Windows*, for example, the new line character isn't expanded to \r\n when written in the file.

Another difference between *text* and *binary* files is that the operating system may add a special character at the end of a *text* file to mark its end. In *Windows*, for example, the control-Z (CTRL-Z) character marks the end of a *text* file. On the other hand, no character has a special significance in a *binary* file. They are all treated the same.

Storing data in a *binary* file can save space compared to a *text* file. For example, suppose that we are using the ASCII character set to write the number 47654 in a *text* file. Since this number is represented with 5 characters, the size of the file would be 5 bytes, as shown in Figure 15.1.

On the other hand, if this number is stored in *binary*, the size of the file would be 2 bytes, as shown in Figure 15.2.

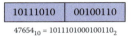

00110100	00110111	00110110	00110101	00110100
'4'	'7'	'6'	'5'	'4'

FIGURE 15.1
Storing a five-digit number in a text file.

10111010	00100110

$47654_{10} = 1011101000100110_2$

FIGURE 15.2
Storing a five-digit number in a binary file.

Open a File

To open a file, we use the fopen() function. It is declared in stdio.h, like this:

```
FILE *fopen(const char *filename, const char *mode);
```

The first argument points to the name of the file to be opened. The "name" may include path information. The second argument specifies the operation to perform on the file, according to Table 15.1.

To open a *text* file, we'd choose one of these modes. To open a *binary* file, add the letter b. For example, the "rb" mode opens a binary file for reading, while the "w+b" mode opens a binary file for reading and writing.

If the file is opened successfully, fopen() returns a pointer to a structure of type FILE. This file pointer can be used to perform file operations. The FILE structure is defined in stdio.h and holds information concerning the opened file. If the file can't be opened, fopen() returns NULL. For example, fopen() fails when the mode is set to "r" and the specified file doesn't exist.

Always check the return value of fopen() against NULL to see if the file was opened successfully or not.

If the program runs in the same directory where you are looking for the file, just put the file name in double quotes. Here's some examples of typical fopen() calls:

- fopen("test.txt", "r"); opens the test.txt text file for reading.
- fopen("test.dat", "a+b"); opens the test.dat binary file for reading and appending. If it doesn't exist, it will be created.

If the operating system uses \ to specify the path, write \\ because C treats \ as the beginning of an escape sequence.

For example, if your program runs in *Windows* and you intend to open the test.txt file in d:\dir1\dir2, you should write

```
fopen("d:\\dir1\\dir2\\test.txt", "r");
```

TABLE 15.1

File Open Modes

Mode	Operation
r	Open file for reading.
w	Open file for writing. If the file exists, it will be truncated and its data will be lost. If it doesn't exist, it will be created.
a	Open file for appending. If the file exists, the existing data will be preserved and the new data will be added at its end. If it doesn't exist, it will be created.
r+	Open file for reading and writing.
w+	Open file for reading and writing. If the file exists, it will be truncated and its data will be lost. If it doesn't exist, it will be created.
a+	Open file for reading and appending. If the file exists, the existing data will be preserved and the new data will be added at its end. If it doesn't exist, it will be created.

However, if the program obtains the file name from the command line, it isn't needed to add an extra \. For example, type d:\dir1\dir2\test.txt with one \.

The following program reads a file name and displays a message to indicate if it is opened successfully or not:

```
#include <stdio.h>
#include <stdlib.h>
int main()
{
  FILE *fp;
  char fname[100];

  printf("Enter file name: ");
  gets(fname);

  fp = fopen(fname, "r"); /* Open file for reading. */
  if(fp == NULL)
  {
    printf("Error: File can not be opened\n");
    exit(1); /* Program termination. */
  }
  printf("File is opened successfully\n");
  fclose(fp); /* Close file. */
  return 0;
}
```

We could check the return value of fopen() against NULL in one line, like this:

```
if((fp = fopen(fname, "r")) == NULL)
```

The inner parentheses are necessary for reasons of priority.

Close a File

The `fclose()` function is used to close an open file. It is declared in `stdio.h`, like this:

```c
int fclose(FILE *fp);
```

It takes as argument a file pointer associated with an open file.

If the file is closed successfully, `fclose()` returns 0, EOF otherwise. The EOF (*End Of File*) is a special constant value, which indicates that either the end of the file is reached or a file operation failed. It is defined in `stdio.h` equal to –1.

> *Although an open file is closed automatically when the program terminates, it'd be better to call* `fclose()` *when you no longer use it. A good reason is that if your program crashes the file will be intact.*

Process a File

As discussed, the file pointer returned from a successful call to `fopen()` is associated with the opened file and points to a structure of type `FILE`. This structure holds information concerning the file. For example, it keeps the file position where the next read or write operation can be performed.

When a file is opened for reading or writing, the file position is set at the beginning of the file. If it is opened for appending, it is set at the end of the file. When a write or read operation is performed, the file position advances automatically. For example, if a file is opened for reading and the program reads 50 characters, the file position advances 50 bytes from the beginning of the file. Similarly, in a write operation, the file position advances a number of places equal to the number of the written bytes.

As we'll see, the `fseek()` function can be used to set the file position anywhere in the file.

Write Data in a Text File

The most common functions for writing data in a text file are `fprintf()`, `fputs()`, and `fputc()`. These functions are mostly used with text files, although you can use them with binary files.

`fputs()` Function

The `fputs()` function writes a string in an output file. It is declared in `stdio.h`, like this:

```c
int fputs(const char *str, FILE *fp);
```

The first argument contains the string written in the file indicated by `fp`.

If `fputs()` is executed successfully, it returns a non-negative value, EOF otherwise.

fprintf() Function

The fprintf() function is more generic than fputs() because it can write a variable number of different data items to an output file. Like printf(), fprintf() uses a format string to define the order and the type of data written in the file.

```
int fprintf(FILE* fp, const char *format,…);
```

The only difference between them is that printf() always writes to stdout, while fprintf() writes to the file indicated by fp. In fact, if fp is replaced by stdout, a call of fprintf() is equivalent to a call of printf().

The stdin and stdout streams are declared in stdio.h and they are ready to use; it is not needed to open or close them. By default, the stdout output stream is associated with the screen, while the stdin input stream is associated with the keyboard.

If fprintf() is executed successfully, it returns the number of bytes written in the file, otherwise a negative value.

For example, the following program shows how to use fputs() and fprintf() for writing some data in a text file:

```c
#include <stdio.h>
#include <stdlib.h>
int main()
{
  FILE *fp;
  int i;

  fp = fopen("test.txt", "w"); /* Open file for writing. */
  if(fp == NULL)
  {
    printf("Error: File can't be created\n");
    exit(1);
  }
  fputs("Hello_1\n", fp);
  for(i = 0; i < 3; i++)
    fprintf(fp, "%d.%s\n", i+1, "Hello_2"); /* Use fprintf() to write a
      string along with an increasing number. */
  fclose(fp);
  return 0;
}
```

Exercise

15.1 Suppose that the following program runs in Windows. What would be the size of the output file?

```c
#include <stdio.h>
#include <stdlib.h>
int main()
{
  FILE *fp;
```

```
    fp = fopen("test.txt", "w");
    if(fp == NULL)
    {
      printf("Error: File can't be created\n");
      exit(1);
    }
    fprintf(fp, "%d\n", 123);
    fclose(fp);
    return 0;
}
```

Answer: Since the number 123 consists of three digits, fprintf() writes the '1', '2', '3' characters in the file. In *Windows*, the '\n' is replaced by '\r' and '\n', therefore the file size would be 5 bytes.

What would be its size if we write %c instead of %d, that is, fprintf(fp, "%c\n", 123);

In that case, fprintf() would write only the character with ASCII value 123 ('{'). Therefore, the file size would be 3 bytes.

fputc() Function

The fputc() function writes one character in an output file. It is declared in stdio.h, like this:

```
int fputc(int ch, FILE *fp);
```

The first argument specifies the character written in the file indicated by fp. Notice that any integer can be passed to fputc(), but only the lower 8 bits will be written.

If fputc() is executed successfully, it returns the written character, EOF otherwise.

A function similar to fputc() is the putc(). Their difference is that putc() is usually implemented as a macro, while fputc() is a function. Since macros tend to be executed faster, it is preferable to use putc() instead of fputc().

The following program writes one by one the characters of a string in a text file:

```
#include <stdio.h>
#include <stdlib.h>
int main()
{
  FILE *fp;
  char str[] = "This text will be saved in the file";
  int i;

  fp = fopen("test.txt", "w");
  if(fp == NULL)
  {
    printf("Error: File can't be created\n");
    exit(1);
  }
  for(i = 0; str[i] != '\0'; i++)
    putc(str[i], fp);

  fclose(fp);
  return 0;
}
```

Exercises

15.2 Write a program that reads products' prices continuously and store in a text file those that cost more than $10 and less than $20. If the user enters –1, the insertion of prices should terminate.

```c
#include <stdio.h>
#include <stdlib.h>
int main()
{
  FILE *fp;
  double prc;

  fp = fopen("test.txt", "w");
  if(fp == NULL)
  {
    printf("Error: File can't be created\n");
    exit(1);
  }
  while(1)
  {
    printf("Enter price: ");
    scanf("%lf", &prc);
    if(prc == -1)
      break;
    if(prc > 10 && prc < 20)
      fprintf(fp, "%.2f\n", prc);
  }
  fclose(fp);
  return 0;
}
```

15.3 Write a program that reads strings continuously (assume less than 100 characters) and appends in a user-selected file those with less than 10 characters and begin with an 'a'. If the user enters "final", the input of the strings should terminate and the program should display how many strings were written in the file.

```c
#include <stdio.h>
#include <stdlib.h>
#include <string.h>
int main()
{
  FILE *fp;
  char str[100];
  int i, cnt;

  printf("Enter file name: ");
  gets(str);
  fp = fopen(str, "a"); /* Open file for appending. */
  if(fp == NULL)
  {
    printf("Error: File can't be created\n");
    exit(1);
  }
```

```
    cnt = 0;
    while(1)
    {
      printf("Enter text: ");
      gets(str);
      if(strcmp(str, "final") == 0)
        break;
      if((str[0] == 'a') && (strlen(str) < 10))
      {
        cnt++;
        fputs(str, fp);
      }
    }
    printf("\n%d strings were written\n", cnt);
    fclose(fp);
    return 0;
}
```

15.4 What would be written in the output file?

```
#include <stdio.h>
int main()
{
  FILE *fp[2];

  if((fp[0] = fp[1] = fopen("test.txt", "w")) != NULL)
  {
    fputs("One", fp[0]);
    fclose(fp[0]);
    fputs("Two", fp[1]);
    fclose(fp[1]);
  }
  return 0;
}
```

Answer: The fp[0] and fp[1] pointers become equal to the return value of fopen() and they are both associated with the test.txt file. The first call of fputs() writes One in the file, while the second call fails because the file is closed with the statement fclose(fp[0]). Therefore, only the string One will be written in the file. Actually, had we checked the return value of the second fputs(), we would see that it is EOF.

15.5 Write a program that intends to write some data into a file. The program reads a file name and, if it does not exist, the program should create it and write the string "One" in it. If it exists, the program should ask the user to overwrite it. If the answer is positive, the program should write the string "One" in it, otherwise the user should be able to enter another file name in order to repeat the aforementioned procedure.

```
#include <stdio.h>
#include <stdlib.h>

FILE *open_file(char name[], int *f);
```

```c
int main()
{
  FILE *fp;
  char name[100];
  int flag, sel;

  flag = 0;
  do
  {
    printf("Enter file name: ");
    scanf("%s", name);

    fp = fopen(name, "r"); /* Check whether the file exists or not.
      If not, we create and open it for writing. Otherwise, we close
      the file and ask the user. */
    if(fp == NULL)
      fp = open_file(name, &flag);
    else
    {
      fclose(fp);
      printf("Would you like to overwrite existing file (Yes:1
        - No:0)? ");
      scanf("%d", &sel);
      if(sel == 1)/* Overwrites the existing file. */
        fp = open_file(name, &flag);
    }
} while(flag == 0);
  fputs("One", fp);
  fclose(fp);
  return 0;
}

FILE *open_file(char name[], int *f)
{
  FILE *fp;

  fp = fopen(name, "w");
  if(fp == NULL)
  {
    printf("Error: File can't be created\n");
    exit(1);
  }
  *f = 1;
  return fp;
}
```

Read Data from a Text File

The most common functions for reading data from a text file are fscanf(), fgets(), and fgetc(). These functions are mostly used with text files, but we can use them with binary files as well.

`fscanf()` Function

The `fscanf()` function is used to read a variable number of different data items from an input file. Like `scanf()`, `fscanf()` uses a format string to define the order and the type of data that will be read from the file.

```
int fscanf(FILE* fp, const char *format,…);
```

The only difference between them is that `scanf()` always reads from `stdin`, while `fscanf()` reads from the file indicated by `fp`.

fscanf() returns the number of data items that were successfully read from the file and assigned to program variables. If either the end of file is reached or the read operation failed, it returns EOF. Later, we'll see that we can use `feof()` to determine which one happened.

The following program assumes that the `test.txt` text file contains the annual temperatures in an area. It reads them and displays those within [−5, 5].

```
#include <stdio.h>
#include <stdlib.h>
int main()
{
  FILE *fp;
  int ret;
  double temp;

  fp = fopen("test.dat", "r");
  if(fp == NULL)
  {
    printf("Error: File can't be loaded\n");
    exit(1);
  }
  while(1)
  {
    ret = fscanf(fp, "%lf", &temp); /* Since fscanf() reads one item, the
      return value 1 implies that the value was successfully read and
      assigned to temp. */
    if(ret != 1)/* We could omit ret and write if(fscanf(fp, "%lf", &temp)
      != 1) */
      break;
    if(temp >= -5 && temp <= 5)
      printf("%f\n", temp);
  }
  fclose(fp);
  return 0;
}
```

We could also put the `fscanf()` into the **while** statement and write it like this:

```
while(fscanf(fp, "%lf", &temp) == 1)
```

As a matter of style, we prefer the **while**(1) statement to make clearer the terminating condition of the loop.

We also suggest to test the return value of `fscanf()` against the number of the assigned items, rather than EOF. For example, assume that the following code reads the floats contained in a text file:

```c
float i;
while(fscanf(fp, "%d", &i) != EOF)
{
  /* Do some work with the i variable. */
}
```

Although the variable i is not assigned successfully because of the wrong specifier "%d", `fscanf()` doesn't return EOF and the loop continues. Had we written the **while** statement like

```c
while(fscanf(fp, "%d", &i) != 1)
```

the wrong assignment would have been traced and the loop would terminate.

Before using `fscanf()`, we should know the type of the stored data and their order in the file. For example, in the following program, we should know in advance that each line of the `test.txt` file contains a string of up to 100 characters, an **int** and a **double** number, in order to pass the right arguments in `fscanf()`.

```c
#include <stdio.h>
#include <stdlib.h>
int main()
{
  FILE *fp;
  char str[100];
  int i;
  double d;

  fp = fopen("test.txt", "r");
  if(fp == NULL)
  {
    printf("Error: File can't be loaded\n");
    exit(1);
  }
  while(1)
  {
    if(fscanf(fp, "%s%d%lf", str, &i, &d) != 3)
      break;
    printf("%s%d%f\n", str, i, d);
  }
  fclose(fp);
  return 0;
}
```

Like `scanf()`, `fscanf()` uses the space character to distinguish the read values.

Like `scanf()`, *be careful when using* `%s` *in* `fscanf()` *to read strings. If the string consists of multiple words, only the first word will be read.*

If the first argument of fscanf() is replaced by stdin, a call of fscanf() is equivalent to a call of scanf(). For example, the following program uses fscanf() and fprintf() instead of scanf() and printf(), respectively, to read an integer and a double number and display them:

```c
#include <stdio.h>
int main()
{
  int i;
  double d;

  fprintf(stdout, "Enter an integer and a double: ");
  if(fscanf(stdin, "%d%lf", &i, &d) == 2)
    fprintf(stdout, "%d%f\n", i, d);
  return 0;
}
```

Exercises

15.6 Write a program that reads products' codes (assume less than 20 characters) and prices and stores them in a text file like this:

```
C101    17.5
C102    32.8
...
```

If the user enters –1 for price, the insertion of products should terminate. Then, the program should read a product's code and search the file to find and display its price.

```c
#include <stdio.h>
#include <stdlib.h>
#include <string.h>
int main()
{
  FILE *fp;
  char flag, str[20], prod[20];
  double prc;

  fp = fopen("test.txt", "w+"); /* Open file for reading and
    writing. */
  if(fp == NULL)
  {
    printf("Error: File can't be created\n");
    exit(1);
  }
  while(1)
  {
    printf("Enter price: ");
    scanf("%lf", &prc);
```

```
    if(prc == -1)
      break;
    getchar();
    printf("Enter product code: ");
    gets(str);
    fprintf(fp, "%s %f\n", str, prc);
  }
  getchar();
  printf("Enter product code to search for: ");
  gets(prod);

  flag = 0;
  fseek(fp, 0, SEEK_SET);
  while(1)
  {
    if(fscanf(fp, "%s%lf", str, &prc) != 2)
      break;
    if(strcmp(str, prod) == 0)
    {
      flag = 1;
      break; /* Since the product is found exit from the loop. */
    }
  }
  if(flag == 0)
    printf("The %s product is not listed\n", prod);
  else
    printf("The price for product %s is %f\n", prod, prc);
  fclose(fp);
  return 0;
}
```

Comments: The fseek() function is discussed later, so—for now—you can bypass the statement fseek(fp, 0, SEEK _ SET); in the aforementioned code. You just need to know that we used this statement to rewind the file pointer at the beginning of the file and start reading from there.

15.7 Suppose that each line of the students.txt file contains the names of the students and their grades (read them as **double** variables) in two lessons like this:

```
John       Morne       7        8.12
Jack       Lommi       4.50     9
Peter      Smith       2        5.75
...
```

Write a program that reads each line of students.txt and stores in suc.txt the names and grades of the students with average grade greater or equal to 5, while in fail.txt the students with average grade less than 5. The program should display the total number of students written in each file.

```
#include <stdio.h>
#include <stdlib.h>
int main()
{
  FILE *fp_in, *fp_suc, *fp_fail;
  char fnm[100], lnm[100];
```

```
    int suc_stud, fail_stud;
    double grd1, grd2;

    fp_in = fopen("students.txt", "r");
    if(fp_in == NULL)
    {
      printf("Error: File can't be loaded\n");
      exit(1);
    }
    fp_suc = fopen("suc.txt", "w");
    if(fp_suc == NULL)
    {
      printf("Error: File_1 can't be created\n");
      exit(1);
    }
    fp_fail = fopen("fail.txt", "w");
    if(fp_fail == NULL)
    {
      printf("Error: File_2 can't be created\n");
      exit(1);
    }
    suc_stud = fail_stud = 0;
    while(1)
    {
      if(fscanf(fp_in,"%s%s%lf%lf", fnm, lnm, &grd1, &grd2) != 4)
        break;
      if((grd1 + grd2)/2 >= 5)
      {
        fprintf(fp_suc,"%s %s %f %f\n", fnm, lnm, grd1, grd2);
        suc_stud++;
      }
      else
      {
        fprintf(fp_fail,"%s %s %f %f\n", fnm, lnm, grd1, grd2);
        fail_stud++;
      }
    }
    printf("Failed:%d Succeeded:%d\n", fail_stud, suc_stud);
    fclose(fp_suc);
    fclose(fp_fail);
    fclose(fp_in);
    return 0;
}
```

fgets() Function

The fgets() function reads a number of characters from an input file. It is declared in stdio.h, like this:

```
char *fgets(char *str, int size, FILE *fp);
```

The first argument points to the memory allocated to store the read characters.

The second argument declares the maximum number of characters that will be read from the file indicated by fp. Its value shouldn't be greater than the size of the allocated memory; otherwise, a memory overflow may occur. `fgets()` adds a null character at the end of the string.

`fgets()` stops reading characters when a new line character is read or `size-1` characters have been read, whichever comes first.

If `fgets()` is executed successfully, the read characters are stored in the memory pointed by `str` and this pointer is returned. If either the end of file is reached or a read error occurs, `fgets()` returns `NULL`.

As discussed in Chapter 10, the risk with `gets()` is that if you read more characters than the size of the allocated memory the program would have unpredictable behavior.

Since `fgets()` *stops reading when the maximum number of characters is stored, it is much safer to write*

```
fgets(str, sizeof(str), stdin);
```

instead of

```
gets(str);
```

In that case, memory overflow can't happen because `fgets()` *won't store more characters in* str *than its size. Once more, use* gets() *only if you are absolutely sure that the user won't insert a string with more characters than the size of the allocated memory.*

For sake of consistency with the rest of the book's examples, we'd still use gets() *to read strings from the keyboard, assuming that the user input won't exceed the maximum length.*

The following program reads 50 names (we assume less than 100 characters each) and stores them in different lines of a text file. Then, the program reads a character, reads the names from the file, and displays those that begin with the input character.

```c
#include <stdio.h>
#include <stdlib.h>
int main()
{
  FILE *fp;
  char ch, str[100];
  int i, times;

  fp = fopen("test.txt", "w+");
  if(fp == NULL)
  {
    printf("Error: File can't be loaded\n");
    exit(1);
  }
  for(i = 0; i < 50; i++)
  {
    printf("Enter name: ");
    gets(str);
    fprintf(fp, "%s\n", str);
  }
  printf("Enter char: ");
  ch = getchar();

  fseek(fp, 0, SEEK_SET);
  times = 0;
```

```c
  while(1)
  {
    if(fgets(str, sizeof(str), fp) == NULL)
      break;
    if(str[0] == ch)
    {
      printf("Name:%s\n", str);
      times++;
    }
  }
  printf("Total occurrences = %d\n", times);
  fclose(fp);
  return 0;
}
```

Exercise

15.8 Write a program that reads the names of two files, compares them line by line (assume that each line contains less than 100 characters), and displays their first common line. If the two files have no common line, the program should display an informative message.

```c
#include <stdio.h>
#include <string.h>
#include <stdlib.h>
int main()
{
  FILE *fp1, *fp2;
  char flag, str1[100], str2[100];

  printf("Enter first file: ");
  gets(str1);

  fp1 = fopen(str1, "r");
  if(fp1 == NULL)
  {
    printf("Error: File can't be loaded\n");
    exit(1);
  }
  printf("Enter second file: ");
  gets(str1);

  fp2 = fopen(str1, "r");
  if(fp2 == NULL)
  {
    printf("Error: File can't be loaded\n");
    exit(1);
  }
  flag = 0;
  while(1)
  {
    if((fgets(str1, sizeof(str1), fp1) == NULL)
```

```
       || (fgets(str2, sizeof(str2), fp2) == NULL))
        break; /* We check if a read error occurred or the end of a
          file is reached. In either case, the loop terminates. */
      if(strcmp(str1, str2) == 0)
      {
        printf("The same line is:%s\n", str1);
        flag = 1;
        break; /* Since a common line is found exit from the loop. */
      }
    }
    if(flag == 0)
      printf("There is no common line\n");

    fclose(fp1);
    fclose(fp2);
    return 0;
}
```

fgetc() Function

The fgetc() function reads a character from an input file. It is declared in stdio.h, like this:

```
int fgetc(FILE *fp);
```

If fgetc() is executed successfully, it returns the read character. If either the end of file is reached or the read operation failed, it returns EOF.

A function similar to fgetc() is the getc(). Like putc() and fputc(), their difference is that getc() is usually implemented as a macro; therefore, it'd be executed faster.

Like getchar(), *always store the return value of* getc() *and* fgetc() *into an* **int** *variable, not a* **char**. *For example, suppose that we use* getc() *to read characters from a binary file which contains the value 255. When 255 is read, the value that would be stored in a signed* **char** *variable is –1. Testing this value against EOF is true, therefore the program would stop reading more characters.*

Exercises

15.9 Write a program that reads the name of a file and displays its second line.

```
#include <stdio.h>
#include <stdlib.h>
int main()
{
  FILE *fp;
  char fname[100];
  int ch, lines;

  printf("Enter file name: ");
  gets(fname);
```

```
    fp = fopen(fname,"r");
    if(fp == NULL)
    {
      printf("Error: File can't be loaded\n");
      exit(1);
    }
    printf("\nLine contents\n");
    lines = 1;
    while(1)
    {
      ch = getc(fp);
      if((ch == EOF) || (lines > 2))
        break;
      if(ch == '\n')/* Increase the line counter. */
        lines++;
      if(lines == 2)/* Only the characters of the second line are
        displayed. */
        printf("%c", ch);
    }
    fclose(fp);
    return 0;
}
```

15.10 Assume that the "test.txt" is a text file. What would be the output of the following program?

```
#include <stdio.h>
int main()
{
  FILE *fp;
  char ch;
  if((fp = fopen("test.txt", "r")) != NULL)
  {
    while(ch = getc(fp) != EOF)
      putc(ch, stdout);
    fclose(fp);
  }
  return 0;
}
```

Answer: If the **while** statement were written correctly, the program would have displayed the file's characters. Its correct form is

```
while((ch = getc(fp)) != EOF)
```

The inner parentheses are needed for reasons of priority. Since they are missing, the expression getc(fp) != EOF is first executed and its value is 1, as long as the end of file isn't reached. Therefore, ch becomes 1 and the program displays the respective character continuously. When getc() returns EOF, ch becomes 0 and the loop terminates.

15.11 A simple way to encrypt data is to XOR them with a secret key. Write a program that reads a key character and the name of a text file (assume that it contains readable characters using the ASCII set) and encrypts its content by XORing each character with the key. The encrypted characters should be stored in a second file selected by the user.

```c
#include <stdio.h>
#include <stdlib.h>
int main()
{
  FILE *fp_in, *fp_out;
  char fname[100];
  int ch, key_ch;

  printf("Enter input file: ");
  gets(fname);

  fp_in = fopen(fname, "r");
  if(fp_in == NULL)
  {
      printf("Error: Input file can't be loaded\n");
      exit(1);
  }
  printf("Enter output file: ");
  gets(fname);

  fp_out = fopen(fname, "w");
  if(fp_out == NULL)
  {
      printf("Error: Output file can't be created\n");
      exit(1);
  }
  printf("Enter key char: ");
  key_ch = getchar();
  while(1)
  {
    ch = getc(fp_in);
    if(ch == EOF)
        break;
    putc(ch ^ key_ch, fp_out);
  }
  fclose(fp_in);
  fclose(fp_out);
  return 0;
}
```

Comments: If you rerun the program and give as an input the encrypted file and the same key, the output file would be the same with the original file because according to the Boolean algebra we have (a ^ b) ^ b = a.

15.12 Define a structure of type `country` with fields: name, capital, and population. Suppose that a text file contains the data of several countries. The first line contains the number of countries and the following lines store the country's data in the form:

```
name capital population
```

Write a program that reads the name of the file and stores in an array of such structures the countries' data. Then, the program should read a number and display the countries with higher population than this number.

```c
#include <stdio.h>
#include <stdlib.h>

typedef struct {
  char name[50];
  char capital[50];
  int pop;
} country;

int main()
{
  FILE *fp;
  country *cntr;
  char fname[100];
  int i, num_cntr, pop;

  printf("Enter file name: ");
  gets(fname);

  fp = fopen(fname, "r");
  if(fp == NULL)
  {
    printf("Error: File can't be loaded\n");
    exit(1);
  }
  if(fscanf(fp, "%d", &num_cntr) != 1)
  {
    printf("Error: fscanf() failed\n");
    exit(1);
  }
  /* Dynamic memory allocation to store the countries' data. */
  cntr = (country *)malloc(num_cntr * sizeof(country));
  if(cntr == NULL)
  {
    printf("Error: Not available memory\n");
    exit(1);
  }

  for(i = 0; i < num_cntr; i++)
    if(fscanf(fp, "%s%s%d", cntr[i].name, cntr[i].capital, &cntr[i].
      pop) != 3)
    {
      printf("Error: fscanf() read error\n");
      exit(1);
    }
  fclose(fp);
  printf("Enter population: ");
  scanf("%d", &pop);
```

```
     for(i = 0; i < num_cntr; i++)
       if(cntr[i].pop >= pop)
         printf("%s %s\t%d\n", cntr[i].name, cntr[i].capital, cntr[i].
           pop);

     free(cntr);
     return 0;
   }
```

End of File

As discussed, the operating system may add a special character at the end of a *text* file to mark its end, whereas none character marks the end of a *binary* file. In *Windows*, for example, the CTRL-Z character with ASCII value 26 marks the end of a *text* file.

For example, the following program writes some characters in a *text* file and then it reads and displays them:

```
#include <stdio.h>
#include <stdlib.h>
int main()
{
  FILE *fp;
  int ch;

  fp = fopen("test.txt", "w+");
  if(fp == NULL)
  {
    printf("Error: File can't be created\n");
    exit(1);
  }
  fprintf(fp, "%c%c%c%c%c\n", 'a', 'b', 26, 'c', 'd');
  fseek(fp, 0, SEEK_SET);
  while(1)
  {
    ch = getc(fp);
    if(ch == EOF)
      break;
    printf("%c", ch);
  }
  fclose(fp);
  return 0;
}
```

If this program runs on *Windows*, the loop will display only the characters 'a' and 'b' and the next call of getc() ends it because the read character with ASCII value 26 marks the end of file. On the other hand, had we used the "w+b" mode, the program would have displayed all the characters since none character has a special significance in a *binary* file.

`fseek()` Function

The `fseek()` function moves the file pointer to a specific location. It is declared in `stdio.h`, like this:

```
int fseek(FILE *fp, long int offset, int origin);
```

The `fseek()` function moves the file pointer indicated by `fp` to a new location `offset` bytes from the point indicated by `origin`. If `offset` is negative, the file pointer moves back.

The value of the `origin` must be one of the following constants, defined in `stdio.h`:

- `SEEK_SET`. The file pointer is moved `offset` bytes from the beginning of the file.
- `SEEK_CUR`. The file pointer is moved `offset` bytes from its current position.
- `SEEK_END`. The file pointer is moved `offset` bytes from the end of the file.

For example, to move to the end of the file, we would write

```
fseek(fp, 0, SEEK_END);
```

To move 20 bytes from the beginning of the file, we would write

```
fseek(fp, 20, SEEK_SET);
```

If `fseek()` is executed successfully, it returns 0. Otherwise, it returns a nonzero value.

When `fseek()` is used with text files, care is required with the new line character(s). For example, suppose that the following listing writes some text in the first two lines of a text file. If the operating system expands the new line character, the value of `offset` should be 6 and not 5 in order to move to the beginning of the second line.

```
fputs("text\n", fp);
fputs("another text\n", fp);
/* move to the beginning of the second line. */
fseek(fp, 6, SEEK_SET);
```

`ftell()` Function

The `ftell()` function gets the current position of a file pointer. It is declared in `stdio.h`, like this:

```
long int ftell(FILE *fp);
```

The `ftell()` function returns the current position of the file pointer indicated by `fp`. The position is expressed as the number of bytes from the beginning of the file.

`ftell()` may be used together with `fseek()` to return to a previous file position, like this:

```
long int old_pos;
/*… open file and do some work with it. */
old_pos = ftell(fp);
/*… do some other work. */
fseek(fp, old_pos, SEEK_SET); /* return back to the old position. */
```

It is safer to use fseek() *and* ftell() *functions with binary and not text files because new line character translations can produce unexpected results.*
fseek() *is guaranteed to work with text files, only if*
    a) offset *is* 0 *or*
    b) offset *is obtained from a previous call to* ftell() *and* origin *is set to* SEEK_SET.

Write and Read Data from a Binary File

The fwrite() and fread() functions are used to write and read data from a binary file. Both are often used to write and read large blocks of data in a single step. These functions are used primarily with binary files, although with some care we can use them with text files as well.

fwrite() Function

The fwrite() function is very useful for writing large blocks of data in a single step. It is declared in stdio.h, like this:

```
int fwrite(const void *buf, size_t size, size_t count, FILE *fp);
```

The size_t type is a synonym for **unsigned int**. The first argument points to the memory that holds the data to be written in the file indicated by fp. Since the pointer is declared as **void**, any kind of data can be written in the file.

The third argument specifies how many elements will be written in the file, while the second argument specifies the size of one element in bytes. The product of the second and third arguments should be equal to the number of the bytes to be written in the file.

For example, to write an array of 1000 integers, the second argument should be equal to the size of one integer, that is, 4, while the third argument should be equal to 1000.

```
int arr[1000];
fwrite(arr, 4, 1000, fp);
```

It is best to use the **sizeof** operator to specify the size of one element in order to make your program platform independent. For example,

```
fwrite(arr, sizeof(int), 1000, fp);
```

As another example, the following listing writes one **double** number.

```
double a = 1.2345;
fwrite(&a, sizeof(double), 1, fp);
```

The fwrite() function returns the number of the elements actually written in the file. If the return value equals the third argument, it implies that the fwrite() was executed successfully. If not, a write error occurred.

fread() Function

Like fwrite(), fread() is very useful for reading large blocks of data in a single step. It is declared in stdio.h, like this:

```
int fread(void *buf, size_t size, size_t count, FILE *fp);
```

The first argument points to the memory in which the read data will be stored. Like fwrite()'s arguments, the second argument specifies the size of one element in bytes, while the third argument specifies how many elements will be read from the file.

The fread() function returns the number of the elements actually read from the file. Like fwrite(), this value should be tested against the third argument. If they are equal, fread() was executed successfully. If not, either the end of file is reached or a read error occurred.

In the following example, a **double** number is read and stored into a.

```
double a;
fread(&a, sizeof(double), 1, fp);
```

As another example, 1000 integers are read and stored into array arr.

```
int arr[1000];
fread(arr, sizeof(int), 1000, fp);
```

> *It is safer to use* fwrite() *and* fread() *with binary and not text files because new line character translations can produce unexpected results.*

For example, suppose that we use fwrite() to write a string of 50 characters in a text file and the value of count is set to 50. If the program runs in Windows and the string contains new line character(s), the replacement(s) with the \r\n pair would make its size more than 50, therefore the fwrite() won't write the entire string.

Exercises

15.13 Write a program that declares an array of 5 integers with values 10, 20, 30, 40, and 50 and writes them in a binary file. Next, the program should read an integer and replace the third stored integer with that. The program should read and display the file's content before it ends.

```
#include <stdio.h>
#include <stdlib.h>

#define SIZE 5
```

```
int main()
{
  FILE *fp;
  int i, arr[SIZE] = {10, 20, 30, 40, 50};

  fp = fopen("test.bin", "w+b");
  if(fp == NULL)
  {
    printf("Error: File can't be created\n");
    exit(1);
  }
  fwrite(arr, sizeof(int), SIZE, fp);

  printf("Enter new value: ");
  scanf("%d", &i);
  fseek(fp, 2*sizeof(int), SEEK_SET); /* Since each integer is 4
    bytes, fseek() moves the file pointer 2*sizeof(int) = 8 bytes
    from the beginning of the file to get to the third integer. */
  fwrite(&i, sizeof(int), 1, fp); /* The second and third arguments
    specify that one element of sizeof(int) = 4 bytes will be
    written in the file. The first argument specifies the memory
    address of the written number. */
  fseek(fp, 0, SEEK_SET);
  fread(arr, sizeof(int), SIZE, fp);
  printf("\n***** File contents *****\n");
  for(i = 0; i < SIZE; i++)
    printf("%d\n", arr[i]);

  fclose(fp);
  return 0;
}
```

15.14 Write a program that reads 10 book titles (assume less than 100 characters) and writes them in a user-selected binary file (first write the size of the title and then the title). Next, the program should read a title and display a message to indicate if it is contained in the file or not.

```
#include <stdio.h>
#include <string.h>
#include <stdlib.h>
int main()
{
  FILE *fp;
  char found, str[100], tmp_str[100];
  int i, len;

  printf("Enter file name: ");
  gets(str);

  fp = fopen(str, "w+b");
  if(fp == NULL)
  {
    printf("Error: File can't be created\n");
    exit(1);
  }
```

```
for(i = 0; i < 10; i++)
{
  printf("Enter text: ");
  gets(str);

  len = strlen(str);
  fwrite(&len, sizeof(int), 1, fp);
  fwrite(str, 1, len, fp);
}
printf("Enter title to search: ");
gets(tmp_str);

found = 0;
fseek(fp, 0, SEEK_SET);
while(1)
{
  if(fread(&len, sizeof(int), 1, fp) != 1)
    break;
  if(fread(str, 1, len, fp) != len)
    break;
  str[len] = '\0';
  if(strcmp(str, tmp_str) == 0)
  {
    found = 1;
    break;
  }
}
if(found == 0)
  printf("\n%s isn't found\n", tmp_str);
else
  printf("\n%s is found\n", tmp_str);

fclose(fp);
return 0;
}
```

15.15 Suppose that the purpose of the following program is to write a string in a binary file, read it from the file and display it. Is this code error-free?

```
#include <stdio.h>
#include <stdlib.h>
int main()
{
  FILE* fp;
  char str1[5], str2[] = "test";
  if((fp = fopen("text.bin", "w+b")) != NULL)
  {
    fwrite(str2, 1, 4, fp);
    fread(str1, 1, 4, fp);
    printf("%s\n", str1);
    fclose(fp);
  }
  return 0;
}
```

Answer: No, it isn't.

The first bug is that the file pointer hasn't been moved to the beginning of the file before calling fread().

> *To switch safely from reading to writing mode and vice versa, call* fseek().

Therefore, this bug is eliminated by adding the fseek() between the fwrite() and fread() calls, like this:

```
fseek(fp, 0, SEEK_SET);
```

Now, fread() reads successfully the four characters and stores them into str1.

The second bug is that str1 isn't null terminated, therefore printf() won't work properly. To eliminate this bug, add this statement before printf():

```
str1[4] = '\0';
```

or initialize str1 like **char** str1[5] = '\0';

Exercises

15.16   Suppose that the test.bin binary file contains a student's grades. Write a program that reads the grades (**float** type) from the binary file, then reads a number, and displays those with a greater value than this.

```c
#include <stdio.h>
#include <stdlib.h>
int main()
{
  FILE *fp;
  int i, grd_num;
  float grd, *grd_arr;

  fp = fopen("test.bin", "rb");
  if(fp == NULL)
  {
    printf("Error: File can't be loaded\n");
    exit(1);
  }
  fseek(fp, 0, SEEK_END);
  grd_num = ftell(fp)/sizeof(float); /* Since the file pointer is
    at the end of file, ftell() returns the size of the file in
    bytes. Since each grade is stored as float, their division
    calculates the number of grades stored in the file. */
  fseek(fp, 0, SEEK_SET);
  grd_arr = (float *) malloc(grd_num * sizeof(float)); /* Allocate
  memory to store the grades. */
  if(grd_arr == NULL)
  {
    printf("Error: Not available memory\n");
    exit(1);
  }
}
```

```
  /* Read all grades and check if they are read successfully. */
  if(fread(grd_arr, sizeof(float), grd_num, fp) == grd_num)
  {
    printf("Enter grade: ");
    scanf("%f", &grd);
    for(i = 0; i < grd_num; i++)
      if(grd_arr[i] > grd)
        printf("%f\n", grd_arr[i]);
  }
    else
    printf("Error: fread() failed\n");

  free(grd_arr);
  fclose(fp);
  return 0;
}
```

15.17 A common method that antivirus software uses to identify viruses is the signature-based detection. The signature is a sequence of bytes that identify a particular virus. For example, a signature could be (in hex): F3 BA 20 63 7A 1B ... When a file is scanned, the antivirus software searches the file for signatures that identify the presence of viruses.

Write a program that reads a virus signature (e.g., 5 integers) and checks if it is contained in the binary file test.dat.

```
#include <stdio.h>
#include <stdlib.h>
#include <string.h>

#define SIZE 5

int main()
{
  FILE *fp;
  int i, found, len, buf[SIZE], pat[SIZE];

  fp = fopen("test.dat", "rb");
  if(fp == NULL)
  {
    printf("Error: File can't be loaded\n");
    exit(1);
  }
  printf("Enter virus signature (%d integers)\n", SIZE);
  for(i = 0; i < SIZE; i++)
  {
    printf("Enter number: ");
    scanf("%d", &pat[i]);
  }
  len = sizeof(pat);
  found = 0;
  while(1)
  {
    if(fread(buf, sizeof(int), SIZE, fp) != SIZE)
      break;
```

```
      if(memcmp(buf, pat, len) == 0)
      {
        found = 1;
        break;
      }
      else
        fseek(fp, -(len-sizeof(int)), SEEK_CUR); /* Go back to check
          the next group of five. */
    }
    if(found == 1)
      printf("SOS: Virus found\n");
    else
      printf("That virus signature isn't found\n");

    fclose(fp);
    return 0;
}
```

15.18 Define a structure of type `employee` with fields: first name, last name, tax number, and salary (assume that the fields contain less than 30 characters). Write a program that uses this type to read the data of 100 employees and store them in an array of such structures. If the user enters `"fin"` as the employee's first name, the data insertion should terminate and the program should write the data structures in a binary file.

```
#include <stdio.h>
#include <string.h>
#include <stdlib.h>

#define SIZE 100
#define LEN 30

struct employee
{
  char name[LEN];
  int tax_num;
  int salary;
};

int main()
{
  FILE *fp;
  int i, num_empl;
  struct employee empl[SIZE];

  fp = fopen("test.bin", "wb");
  if(fp == NULL)
  {
    printf("Error: File can't be created\n");
    exit(1);
  }
  num_empl = 0;
  for(i = 0; i < SIZE; i++)
  {
    printf("\nEnter full name: ");
```

```
      gets(empl[i].name);
      if(strcmp(empl[i].name, "fin") == 0)
        break;

      printf("Enter tax number: ");
      scanf("%d", &empl[i].tax_num);

      printf("Enter salary: ");
      scanf("%d", &empl[i].salary);

      num_empl++;
      getchar();
    }
    /* Write the data structures in a single step. */
    fwrite(empl, sizeof(struct employee), num_empl, fp);
    fclose(fp);
    return 0;
}
```

15.19 Suppose that the test.bin binary file contains structures of the type employee defined in 15.18 (Exercise). Write a program that reads them and copies the employees' data whose salary is more than an input amount in the data.bin binary file. The program should also display the average salary of the employees stored in data.bin.

```
#include <stdio.h>
#include <stdlib.h>

#define LEN 30

struct employee
{
char name[LEN];
int tax_num;
int salary;
};

int main()
{
  FILE *fp_in, *fp_out;
  int count, amount, sum_sal;
  struct employee tmp_emp;

  fp_in = fopen("test.bin", "rb");
  if(fp_in == NULL)
  {
    printf("Error: Input file can't be loaded\n");
    exit(1);
  }
  fp_out = fopen("data.bin", "wb");

    if(fp_out == NULL)
  {
    printf("Error: Output file can't be created\n");
    exit(1);
  }
```

```
    printf("Enter amount: ");
    scanf("%d", &amount);
    count = sum_sal = 0;
    while(1)
    {
      if(fread(&tmp_emp, sizeof(employee), 1, fp_in) != 1)
        break;
      if(tmp_emp.salary > amount)
      {
        fwrite(&tmp_emp, sizeof(employee), 1, fp_out);
        sum_sal += tmp_emp.salary;
        count++;
      }
    }
    if(count)
      printf("Avg = %.2f\n", (float)sum_sal/count);
    else
      printf("No employee gets more than %d\n", amount);

    fclose(fp_in);
    fclose(fp_out);
    return 0;
}
```

15.20 Define a structure of type band with fields: name, category, singer, and records (assume that the fields contain less than 30 characters). Suppose that the test.bin binary file contains such structures. Write a program that reads them, then it reads the name of a band, a new singer, and replaces the current singer with the new one.

```
#include <stdio.h>
#include <stdlib.h>
#include <string.h>

#define LEN 30

typedef struct {
  char name[LEN];
  char category[LEN];
  char singer[LEN];
  int records;
} band;

int main()
{
  FILE *fp;
  band *band_arr;
  char found, name[LEN], singer[LEN];
  int i, band_num;

  fp = fopen("test.bin", "r+b");
  if(fp == NULL)
  {
    printf("Error: File can't be loaded\n");
    exit(1);
  }
```

```c
    fseek(fp, 0, SEEK_END);
    band_num = ftell(fp)/sizeof(band); /* We divide the return value
      of ftell() with the size of one structure to calculate the
      number of the stored structures. */
    fseek(fp, 0, SEEK_SET);

    band_arr = (band *) malloc(sizeof(band) * band_num);
    if(band_arr == NULL)
    {
      printf("Error: Not available memory\n");
      exit(1);
    }
    if(fread(band_arr, sizeof(band), band_num, fp) == band_num)
    {
      printf("Enter band name: ");
      gets(name);

      printf("Enter new singer: ");
      gets(singer);

      found = 0;
      for(i = 0; i < band_num; i++)
        if(strcmp(band_arr[i].name, name) == 0)
        {
          fseek(fp, i*sizeof(band), SEEK_SET); /* If the band is
            found, move the file pointer to the beginning of this
            structure. */
          strcpy(band_arr[i].singer, singer); /* Change the singer
            field and write the data structure in the current
            position. */
          fwrite(&band_arr[i], sizeof(band), 1, fp);
          printf("\n% Singer of band %s is changed to %s\n", name,
            singer);
          found = 1;
          break;
        }
    }
    else
      printf("Error: fread() failed !!!\n");

    if(found == 0)
      printf("\n %s band isn't found\n\n", name);

    free(band_arr);
    fclose(fp);
    return 0;
}
```

15.21 Suppose that each line of the test.bin binary file contains a student's grades in 5 lessons. Write a program that reads the file and displays the average grade of each student.

```c
#include <stdio.h>
#include <stdlib.h>

#define LESSONS 5

int main()
```

```
{
  FILE *fp;
  int i, j, stud_num, grd_num;
  float sum_grd, **grd_arr;

  fp = fopen("test.bin", "rb");
  if(fp == NULL)
  {
    printf("Error: File can't be loaded\n");
    exit(1);
  }
  fseek(fp, 0, SEEK_END);
  grd_num = ftell(fp)/sizeof(float);
  fseek(fp, 0, SEEK_SET);

  stud_num = grd_num/LESSONS; /* Since grd_num indicates the total
    number of grades stored in the file and each line contains
    LESSONS grades, their division calculates the number of rows,
    that is the number of students. */
    /* We use the grd_arr as a two-dimensional array of 'stud_num'
      rows and 'LESSONS' columns, meaning that each line holds a
      student's grades in 'LESSONS' courses. */
  grd_arr = (float **) malloc(stud_num * sizeof(float *));

  if(grd_arr == NULL)
  {
    printf("Error: Not available memory\n");
    exit(1);
  }
  for(i = 0; i < stud_num; i++)
  {
    grd_arr[i] = (float *) malloc(LESSONS * sizeof(float));
    if(grd_arr[i] == NULL)
    {
      printf("Error: Not available memory\n");
      exit(1);
    }
  }
  for(i = 0; i < stud_num; i++)
  {
    sum_grd = 0;
    if(fread(grd_arr[i], sizeof(float), LESSONS, fp) == LESSONS)
    {
      for(j = 0; j < LESSONS; j++)
        sum_grd += grd_arr[i][j];

      printf("%d.%f\n", i+1, sum_grd/LESSONS);
    }
    else
    {
      printf("Error: fread() failed\n");
      break; /* Stop reading. */
    }
  }
  for(i = 0; i < stud_num; i++)
    free(grd_arr[i]);
```

```
      free(grd_arr);
      fclose(fp);
      return 0;
   }
```

Comments: We could present a simpler solution similar to 15.16 (Exercise). That is, we could declare the `grd_arr` variable as a pointer, store the grades in the allocated memory, and read the grades from this memory in groups of five in order to calculate the average grade of each student. The reason to choose the aforementioned solution is to show you how to handle the allocated memory as a two-dimensional array.

`feof()` Function

The `feof()` function is used to determine whether the end of file is reached. It is declared in `stdio.h`, like this:

```
int feof(FILE *fp);
```

If a read operation attempts to read beyond the end of file indicated by `fp`, `feof()` returns a nonzero value, `0` otherwise.

When we used a read function in our programs, we checked its return value to determine if it was completed successfully or not. If the operation is unsuccessful, we can use `feof()` to determine whether the failure was due to an end of file condition or for another reason.

For example, the following program reads the contents of a text file (we assume that each line has less than 100 characters) and if `fgets()` fails, we use `feof()` to see why.

```
#include <stdio.h>
#include <stdlib.h>
int main()
{
  FILE *fp;
  char str[100];

  fp = fopen("test.txt", "r");
  if(fp == NULL)
  {
    printf("Error: File can't be loaded\n");
    exit(1);
  }
  while(1)
  {
    if(fgets(str, sizeof(str), fp) != NULL)
      printf("%s", str);
    else
    {
      if(feof(fp))
        printf("End of file\n");
      else
```

```
      printf("Failed for another reason\n");
    break;
    }
  }
  fclose(fp);
  return 0;
}
```

Exercise

15.22 Assume that each line of `test.c` text file contains less than 100 characters. What does the following "badly-written" program do?

```c
#include <stdio.h>
int main()
{
  FILE *fp;
  char str[100];

  for(fp = fopen("test.c", "r"); fp && !feof(fp);
    fgets(str,sizeof(str),fp) ? printf("%s", str) : 1);
  return fp ? fclose(fp) : 0;
}
```

Answer: The program uses `fgets()` to read and display each line of `test.c`, while the **for** loop is executed as long as the end of file isn't reached.

If `fopen()` fails, `fp` would be equal to `NULL`, the **for** loop won't be executed and the program returns 0. Otherwise, the `fclose()` closes the file and the program returns the return value of `fclose()`.

Unsolved Exercises

15.1 Suppose that the text file `grades.txt` contains the grades of a number of students. Write a program that reads the file and displays the minimum and the maximum grades, and the average grade of those who failed (grade < 5) and the average grade of those who succeeded (grade >= 5). (*Note*: suppose that the maximum grade is 10 and the minimum 0.)

15.2 Write a program that finds the sequential doubled words of a user-selected text file (i.e., "In this this chapter we we present...") and writes them into another text file.

15.3 Suppose that each line of the text file `students.txt` contains the grades of the students in three courses. Write a program that reads the file and displays the average grade of each course. For example, if the file content is

```
5      3.5     9

9      6       4.5
```

the program should display 7 4.75 6.75.

15.4   Write a program that converts the uppercase letters of a user selected text file to lowercase letters and vice versa. (*Hint*: use `fseek()` between successive read and write operations.)

15.5   Write a program that checks if the content of two user-selected text files is the same.

15.6   Write a program that reads the names of two text files from the command line and appends the content of the second file into the first one. Then, the program should display the content of the first file. (*Note*: assume that each line contains than 100 characters.)

15.7   Write a program that copies in reverse order (from the last character up to the first one) the content of a user-selected text file into another text file. (*Hint*: open the source file as binary.)

15.8   Suppose that each line of the text file `students.txt` contains the name of a student and his grades in three courses. Write a program that reads the file and displays the name(s) of the student(s) with the maximum average grade. (*Note*: assume that the students are less than 300.)

15.9   Suppose that the binary file `grades.dat` contains the grades of a number of students. Write a program that reads the file and writes the grades sorted in the file `grd_sort.dat`.

15.10   Define the structure type `book` with fields: title, authors, and price. Suppose that the binary file `book.dat` contains 100 of those structures. Write a program that reads the number of a book entry (i.e., `25`), the new data, and replace the existing data with the new data. Then, the program should read from the file the data of that entry and display them in order to verify that it was written correctly.

15.11   Consider the file `book.dat` of the previous exercise. Write a program that reads the existing entries and writes them in the file `book_rvs.dat` in reverse order, meaning that the last entry should be written first, the last but one written second, and so on.

16

Preprocessor Directives and Macros

Preprocessor is the part of the compiler that processes a C program *before* its compilation. In this chapter, we'll discuss how to define and use macros in a C program and then we'll present preprocessor directives that support the conditional compilation of a C program.

Simple Macros

In previous chapters, we have used the **#define** directive to define a simple macro, which is a symbolic name associated with a constant value. To define a simple macro, we write

```
#define macro_name replacement_characters
```

Typically, most programmers choose capital letters to name a macro in order to distinguish them from program variables. The usual practice is to define all macros with global scope at the top of the program, or in another file, which must be included in the program with the **#include** directive. However, a macro can be defined anywhere in a program, even inside a function. For example, in the following program,

```
#include <stdio.h>

#define NUM 200

int main()
{
    int i, arr[NUM];
    for(i = 0; i < NUM; i++)
        arr[i] = i + NUM;
    return 0;
}
```

the preprocessor replaces each appearance of NUM with its defined value, that is, 200.

A common mistake is to add the = symbol in a macro definition. For example, if we write

```
#define NUM = 200
```

the program won't compile because the definition of `arr` would expand to `arr[= 200]`.

Another common error is to add a semicolon at the end of the macro definition. For example, if we write

```
#define NUM 200;
```

the definition of `arr` would expand to `arr[200;]` and the compiler would raise an error message.

Although simple macros are primarily used to define names for numbers, it can be used for other purposes as well. For example, in the following program

```c
#include <stdio.h>

#define test printf("example")

int main()
{
  test;
  return 0;
}
```

the preprocessor replaces the `test` macro with `printf("example")` and the program displays `example`.

Notice that it is legal to define a macro with no replacement value. For example,

```c
#define LABEL
```

As we'll see later, this kind of macros is used for controlling conditional compilation.

The name of a macro can be used in the definition of another macro. For example,

```c
#define SIZE 200
#define NUM SIZE
```

Remember that first the preprocessor replaces the macro names with the corresponding values and then the program is compiled.

To extend a macro in several lines, add the backslash character (\) at the end of each line. For example, the following program displays 60.

```c
#include <stdio.h>

#define NUM \
    10 + \
    20 + \
    30
int main()
{
  printf("%d\n", NUM);
  return 0;
}
```

As discussed in Chapter 2, the main advantage of using a macro is that we can change the value of a constant throughout a program by making a single modification; the definition of the macro.

Macros with Parameters

Besides its simple form, a macro can take parameters and behave like a function. For example, in the following program

```
#include <stdio.h>
#define MIN(a, b)  ((a) < (b) ? (a) : (b))
int main()
{
  int i, j, min;

  printf("Enter numbers: ");
  scanf("%d%d", &i, &j);
  min = MIN(i, j);
  printf("Min = %d\n", min);
  return 0;
}
```

the MIN macro takes two parameters. Once it is met, the preprocessor replaces a with i, b with j, and expands this line to

```
min = ((i) < (j) ? (i) : (j));
```

A macro may be more generic than a function since it can take parameters of any data type. For example, the MIN macro may be also used to find the minimum of **int**, **char**, **double**, and other data types.

Because of operators' precedence, always enclose each macro parameter in parentheses.

For example, see what happens if we omit the parentheses:

```
#define MUL(a, b)  (a*b)/* Instead of ((a)*(b)). */
```

A statement like this

```
j = MUL(9+1, 8+2);
```

is expanded to

```
j = 9+1*8+2;
```

and j becomes 19 and not 100 because the multiplication is performed before the addition.

Do not leave a whitespace after the name of a parameterized macro because the preprocessor would handle the left parenthesis as the beginning of a simple macro definition.

For example, if we write

```
#define MIN (a,b)  ((a) < (b) ? (a) : (b))
```

the compilation would fail because the preprocessor starts the replacement of MIN from the left parenthesis.

Using a parameterized macro instead of a function that does the same may execute faster because a function call imposes some run-time overhead due to the storage of context information (e.g., the memory address of the function) and copy of the arguments. On the other hand, a macro invocation doesn't impose any delay because the preprocessor expands it before the execution of the program.

However, using a parameterized macro instead of a function has several disadvantages. For example, because macros rely on text substitution, you can get into serious trouble when the macro contains complex code, like control-flow constructs. Needless to say, that is much harder to read, maintain and debug a macro which contains multiple statements. Also, it is not possible to declare a pointer to a macro, because macros are removed during preprocessing, so there is no "memory" for a pointer to point to.

When a function is called, the compiler checks the types of the arguments. If an argument has a wrong type and the compiler can't convert it to the proper type, it produces an error message. In contrast, the preprocessor doesn't check the types of the macro arguments, so non-desirable values may pass to it.

Typically, a parameterized macro is used in place of a function, when its code is not complicated and it doesn't extend in many lines.

See some more examples why writing a macro might be error prone.

When passing an argument, don't apply operators that may change its value.

For example, the following assignment

```
x = MIN(i++, j);
```

is expanded to

```
x = ((i++) < (j) ? (i++) : (j));
```

Therefore, if i is less than j, it will be unintentionally incremented twice and a wrong value will be assigned to x.

If a macro contains expressions with operators, enclose them in parentheses.

For example, see what happens if we omit the parentheses:

```
#define NUM 2*5 /* Instead of (2*5). */
```

A statement like this

```
float j = 3.0/NUM;
```

is expanded to

```
float j = 3.0/2*5;
```

which assigns the value 7.5 to j and not 0.3 because the division is performed before the multiplication.

A macro should be written with great care in order to avoid unexpected results.

and ## Preprocessor Operators

The # operator in front of an argument instructs the preprocessor to create a string literal having the name of that argument. For example,

```
#include <stdio.h>
#define f(s) printf("%s = %s\n", #s, s);
int main()
{
  char *str = "text";
  f(str);
  return 0;
}
```

When the preprocessor expands the f macro, it replaces #s with the name of the argument, that is, str. Therefore, the program displays str = text.
   Let's see another example:

```
#include <stdio.h>

#define sum(a, b) printf(#a "+" #b " = %d\n", a+b)

int main()
{
  int i, j;

  printf("Enter numbers: ");
  scanf("%d%d", &i, &j);
  sum(i, j);
  sum(i*j, i*j);
  return 0;
}
```

The first invocation of sum is translated to

```
printf("i" "+" "j" " = %d\n", i+j);
```

Since the consecutive strings literals can be concatenated, it is equivalent to

```
printf("i+j = %d\n",i+j);
```

Similarly, the second invocation of sum is translated to

```
printf("i*j" "+" "i*j" " = %d\n", i*j+i*j);
```

which is equivalent to

```
printf("i*j+i*j = %d\n", i*j+i*j);
```

For example, if the user enters 2 and 5, the program displays

```
i+j = 7
i*j+i*j = 20
```

The ## operator is used to merge identifiers together. If the identifier is a macro parameter,
the preprocessor first replaces it with the value of the argument and then pasting occurs.
For example,

```
#include <stdio.h>
```

```
#define f(a) s##u##m##a
```

```
int sum1(int a, int b);
int main()
{
  int i, j;

  printf("Enter numbers: ");
  scanf("%d%d", &i, &j);
  printf("%d\n", f(1)(i, j));
  return 0;
}
int sum1(int a, int b)
{
  return a+b;
}
```

When the preprocessor expands the f macro, it replaces a with the value of its argument, that is, 1, and then merges the s, u, m characters. Therefore, f(1)(i,j) expands to sum1(i,j) and the program calls sum1() to display the sum of the two input numbers.

Preprocessor Directives and Conditional Compilation

A preprocessor directive begins with the symbol # and instructs the preprocessor to do something. For example, the #include directive instructs the preprocessor to open a particular file and add its content into the program.

This section describes the preprocessor directives that allow the conditional compilation of a section of the program. The conditional compilation is very useful when multiple versions of the same program must be maintained.

#if, #else, #elif, and #endif Directives

The #if and #endif directives are used to define which parts of a program will be compiled, depending on the value of an expression. The general form is

```
#if expression
   ... /* block of statements */
#endif
```

If the value of expression is true, the preprocessor keeps the block of the statements to be processed by the compiler. If not, this block will be removed from the program.

Notice that if the expression is an undefined macro the #if directive evaluates to false. For example,

```
#if NUM
```

If the NUM identifier is not defined as a macro, the outcome is 0.

The #else directive is used in conjunction with the #if directive to define a block of statements that will be compiled if the value of the expression is false.

```
#if expression
   ... /* block of statements A */
```

```
#else
   ... /* block of statements B */
#endif
```

For example, in the following program, the preprocessor will remove the first `printf()` from the program because `NUM` is greater than `0`.

```
#include <stdio.h>

#define NUM 10

int main()
{
  #if NUM < 0
    printf("Seg_1\n");
  #else
    printf("Seg_2\n");
  #endif
  return 0;
}
```

Therefore, the **#else** part will be compiled and the program displays `Seg _ 2`.

The **#elif** directive is used to define multiple compilation paths. The general form is

```
#if expression_A
   ... /* block of  statements A */
#elif expression_B
   ... /* block of  statements B */
   .
   .
#else
   ... /* block of statements N */
#endif
```

For example, the second block of statements will be compiled only if the value of expression _ A is false and the value of expression _ B is true. The last block of statements will be compiled only if the previous expressions are false. Consider the following program:

```
#include <stdio.h>

#define VER_2 1

int main()
{
  int cnt;

  #if VER_1
        cnt = 1;
        printf("Version_1\n");
  #elif VER_2
        cnt = 2;
        printf("Version_2\n");
  #else
```

```
        cnt = 3;
        printf("Version_3\n");
  #endif

  printf("Cnt = %d\n", cnt);
  return 0;
}
```

Since the VER _ 1 macro is not defined and the value of VER _ 2 is true, cnt becomes 2 and the program displays

```
Version_2
Cnt = 2
```

If we change the value of VER _ 2 to 0, the **#elif** expression becomes false and the program would display

```
Version_3
Cnt = 3
```

#ifdef, #ifndef, and #undef Directives

The **#ifdef** directive is used to check if an identifier is defined as a macro. The general form is

```
#ifdef macro_name
  ... /* block of statements */
#endif
```

The difference with the **#if** directive is that the **#ifdef** directive only checks if the identifier is defined as a macro and it does not evaluate its value. For example,

```
#include <stdio.h>

#define VER_1

int main()
{
  #ifdef VER_1
    printf("Version_1\n");
  #else
    printf("Version_2\n");
  #endif
  return 0;
}
```

Since VER _ 1 is defined, the program displays Version _ 1.

The **#ifndef** directive is used to check whether an identifier is not defined as a macro. The **#ifndef** directive is often used to force the single compilation of a header file.

For example, suppose that a program consists of several source files and some of them include the same file, for example, **#include** "test.h". If the test.h contains declarations of variables, the multiple compilation of the test.h would fail because re-declaration of the same variables is not allowed. A technique to avoid multiple compilation of the same file (e.g., test.h) is to add the following lines at its beginning.

```
ifndef ANY_TAG
#define ANY_TAG/* Just define a macro. */
/* Contents of test.h */
#endif
```

When the first file that includes the **#include** "test.h" is compiled, the preprocessor will include the contents of the test.h because the ANY _ TAG macro is not defined yet. However, the next directives won't re-include the test.h because the ANY _ TAG macro is now defined and the **#ifndef** directive will fail.

The **#undef** directive is used to cancel the definition of a macro. For example,

```
#include <stdio.h>

#define NUM 100

int main()
{
  int arr[NUM];
#undef NUM

  printf("Array contains %d elements\n", NUM);
  return 0;
}
```

When printf() is compiled, the compiler will produce an error message because the **#undef** directive cancels the definition of NUM.

defined **Operator**

An alternative way to check if an identifier is defined as a macro is to use the **defined** operator. The **defined** operator is usually used together with the **#if** directive. For example,

```
#if defined(VER_1)/* Equivalent to #ifdef VER_1 */
...
#endif
```

Similarly, the following

```
#if !defined(VER_1)
```

is equivalent to

```
#ifndef VER_1.
```

Exercises

16.1 Write a macro that calculates the absolute value of a number. Write a program that reads an integer and uses the macro to display its absolute value.

```
#include <stdio.h>

#define abs(a) ((a) >= 0 ? (a) : -(a))

int main()
{
  int i;

  printf("Enter number: ");
  scanf("%d", &i);
  printf("abs = %d\n", abs(i));
  return 0;
}
```

16.2 Write a macro that checks whether a number is odd or even. Write a program that reads an integer and uses the macro to display whether it is odd or even.

```
#include <stdio.h>

#define odd_even(a) (((a) & 1) == 0)

int main()
{
  int i;

  printf("Enter number: ");
  scanf("%d", &i);
  if(odd_even(i))
      printf("Even\n");
  else
    printf("Odd\n");
  return 0;
}
```

16.3 What is the output of the following program?

```
#include <stdio.h>

#define TEST

#ifdef TEST
#define f() printf("One ");
#undef TEST
#endif

int main()
{
  f();
#ifdef TEST
  f();
#endif

  f();
  return 0;
}
```

Answer: The preprocessor replaces the first appearance of f with the printf() statement and then cancels the definition of TEST due to the **#undef** directive. Since TEST is not defined, the preprocessor doesn't expand the second f and continues with the third one. Therefore, the program displays One One.

16.4 Write a program that reads double numbers continuously and counts either the positives or the negatives depending on the definition of a macro. For example, if the CNT _ POS macro is defined, the program should count the positives, otherwise the negatives. If the user enters 0, the insertion of numbers should terminate.

```c
#include <stdio.h>

#define CNT_POS

int main()
{
  int cnt = 0;
  double num = 1;

  while (num != 0)
  {
    printf("Enter number: ");
    scanf("%lf", &num);
    #ifdef CNT_POS
      if (num > 0)
        cnt++;
    #else
      if (num < 0)
        cnt++;
    #endif
  }
  printf("Cnt =%d\n", cnt);
  return 0;
}
```

16.5 Write a program that displays One if both VER _ 1 and VER _ 2 macros are not defined. Otherwise, if either VER _ 3 or VER _ 4 macro is defined, the program should display Two. If nothing from the above happens, it should display Three.

```c
#include <stdio.h>

int main()
{
#if !defined(VER_1) && !defined(VER_2)
  printf("One\n");
#elif defined(VER_3) || defined(VER_4)
  printf("Two\n");
#else
  printf("Three\n");
#endif
  return 0;
}
```

Comments: Notice how the ! operator is used to check whether a macro is defined or not.

16.6 What is the output of the following program?

```c
#include <stdio.h>
#include <string.h>

#define f(text) printf(text); if(strlen(text) < 5) return 0;
```

```
int main()
{
  f("One");
  f("Two");
  f("Three");
  return 0;
}
```

Answer: The preprocessor expands the first appearance of f to the following lines:

```
printf("One");
if(strlen("One") < 5)
return 0;
```

When the program runs, the **return** statement terminates the program because the length of the "One" is 3, less than 5. Therefore, the program displays One.

16.7 Write a macro that calculates the greatest of three numbers. Write a program that reads three double numbers and uses the macro to display the greatest.

```
#include <stdio.h>

#define max(a, b, c) ((a) >= (b) && (a) >= (c) ? (a) : \
(b) >= (a) && (b) >= (c) ? (b) : (c))

int main()
{
  double i, j, k;

  printf("Enter numbers: ");
  scanf("%lf%lf%lf", &i, &j, &k);
  printf("Max = %f\n", max(i, j, k));
  return 0;
}
```

16.8 Write a macro that calculates the greater of two numbers. Write a program that reads four integers and uses the macro to display the greater.

```
#include <stdio.h>

#define max(a, b) ((a) > (b) ? (a) : (b))

int main()
{
  int i, j, k, l;

  printf("Enter numbers: ");
  scanf("%d%d%d%d", &i, &j, &k, &l);
  printf("Max = %d\n", max(max(max(i, j), k), l));
  return 0;
}
```

Comments: This program is an example of nested macros. The preprocessor expands the nested macros from the inner to the outer.

16.9 Write a macro that may read a character, an integer, or a float and display it. Write a program that uses this macro to read and display the values of a character, an integer, and a float.

```
#include <stdio.h>
#define f(type, text, a) printf(text); scanf(type, &a);
  printf(type"\n", a);

int main()
{
  char ch;
  int i;
  float fl;

  f("%c", "Enter character: ", ch);
  f("%d", "Enter integer: ", i);
  f("%f", "Enter float: ", fl);
  return 0;
}
```

16.10 What is the output of the following program?

```
#include <stdio.h>
#define hide(t, r, a, p, i, n) p##a##r##t(i, n)
double show(int a, int b);
int main()
{
  printf("%d\n", (int)hide(w, o, h, s, 1, 2));
  return 0;
}
double show(int a, int b)
{
  return (a+b)/2.0;
}
```

Answer: The preprocessor replaces one by one the arguments of the hide macro. Therefore, the preprocessor substitutes t with w, r with o, and so on. Since the p,a,r,t are replaced by the s,h,o,w arguments and the **##** operator merges the operands together, the p##a##r##t(i,n) is expanded to show(i,n).

Therefore, the program calls the show() function with arguments 1 and 2, which returns their average, that is, 1.5. Since the return value casts to **int**, the program displays 1.

16.11 What is the output of the following program?

```
#include <stdio.h>
#define no_main(type, name, text,num) type name() {printf(text);
  return num;}
no_main(int, main, "No main() program", 0)
```

Answer: Here is a weird program with no main() included. The preprocessor substitutes the type with **int**, name with main, and so on. Therefore, the preprocessor expands the no _ main macro to

```
int main() {printf("No main() program\n"); return 0;}
```

and the program displays No main() program.

Note that if we were using **void** instead of **int** the compilation would fail because a **void** function can't return a value.

Unsolved Exercises

16.1 Write a macro that checks whether a number is between two other numbers. Write a program that reads three numbers (i.e., x, a, and b) and uses the macro to check if x belongs in [a, b].

16.2 The following macro calculates the minimum of two numbers, but it contains an error.

```
#define MIN(a, b) (a) < (b) ? (a) : (b)
```

Give an example to show why the macro won't operate as expected and fix that error.

16.3 Write a macro that takes two arguments and calculates their average. Write a program that reads four integers and uses the macro to display their average.

16.4 Write a macro that takes one argument and if it is a lowercase letter it expands to the respective uppercase letter. Write a program that reads a string up to 100 characters and uses the macro to display its lowercase letters in uppercase.

16.5 Complete the following macro to display the integer result of the math operation specified by the sign.

```
#define f(a, sign, b)…
```

Assume that the a and b arguments are integers. Write a program that reads two integers and a math sign and uses the macro to display the result of the math operation.

16.6 Modify the macro of the previous exercise in order to display the result of math operations applied on both integers and floats.

```
#define f(a, sign, b, fmt)…
```

The fmt argument can be either %d or %f. Write a program that reads an integer and a **double** number and a math sign and uses the macro to display the result of the math operation.

16.7 What would be the output of the following program when the macro ONE:

(a) Is defined?

(b) Not defined?

Explain why.

```
#include <stdio.h>

#define ONE 1
#define myprintf(a) printf("x" #a " = %d\n", x##a)

int main()
{
#ifdef ONE
#define TWO ONE+ONE
#else
#define ONE 2
#define TWO (ONE+ONE)
#endif
```

```
    int x1 = 3*(ONE+ONE), x2 = 3*TWO;

    myprintf(1);
    myprintf(2);
    return 0;
}
```

16.8 Complete the macro `SET_BIT` to set the bit in position `pos`, the macro `CLEAR_BIT` to clear the bit in position `pos` and the macro `CHECK_BIT` to check the value of the bit in position `pos`. Write a program that reads an integer and a bit position and tests the operation of the three macros.

```
#define SET_BIT(a, pos)
#define CLEAR_BIT(a, pos)
#define CHECK_BIT(a, pos)
```

16.9 Write a program that reads a string up to 100 characters and, based on the definition of a macro, it displays either the number of its lowercase letters or the number of its uppercase letters or the number of its digits. For example, if the macro `UL` is defined, the program should display the number of the uppercase letters.

17

Review Exercises

17.1 Write a program that reads its command line arguments and allocates memory to store them.

```c
#include <stdio.h>
#include <stdlib.h>
#include <string.h>

int main(int argc, char *argv[])
{
  char *tot_str;
  int i, tot_chars;

  if(argc == 1)/* Check if there is only one argument. */
  {
    printf("Missing arguments…\n");
    exit(1);
  }
  tot_chars = 0; /* It counts the characters of all arguments. */
  for(i = 1; i < argc; i++)/* Remember that argv[1] points to the
    first argument, argv[2] to the second one, and so forth. argv[0]
    points to the name of the program. */
    tot_chars += strlen(argv[i]);

  tot_str = (char *) malloc(tot_chars+1); /* Allocate an extra place
    for the null character. */
  if(tot_str == NULL)
  {
    printf("Error: Not available memory\n");
    exit(1);
  }
  *tot_str = '\0'; /* Initialize the allocated memory with the null
    character. */
  for(i = 1; i < argc; i++)
    strcat(tot_str, argv[i]);

  printf("The merged string is: %s\n", tot_str);
  free(tot_str);
  return 0;
}
```

17.2 What is the output of the following program?

```c
#include <stdio.h>

void test(int **tmp, int i);

int main()
```

```
{
  int *ptr;
  int **tmp;
  int i, arr[] = {10, 20, 30};

  ptr = arr;
  tmp = &ptr;
  for(i = 0; i < 3; i++)
  {
    test(tmp, i);
    printf("%d ", arr[i]);
  }
  return 0;
}

void test(int **tmp, int i)
{
  *(*tmp + i) += 50;
}
```

Answer: The variable tmp is declared as a pointer to a pointer to an integer. The statement tmp = &ptr; makes tmp to point to the address of ptr, which points to the address of the first element of the array arr.

Since *tmp is equal to ptr, we have *(*tmp+i) = *(ptr+i) = *(arr+i) = arr[i]. Therefore, each call to test() increases the respective element of the array arr by 50, and the program displays 60 70 80.

17.3 Define the structure type city with fields: city name, country name, and population. Write a program that uses this type to read the data of 100 cities and store them in an array of such structures. Then, the program should read the name of a country and a number and it should display the cities of that country whose population is greater than the input number.

```
#include <stdio.h>
#include <string.h>

#define SIZE 100

struct city
{
  char name[50];
  char country[50];
  int pop;
};

int main()
{
  char country[50];
  int i, pop, flag;
  struct city cities[SIZE];

  for(i = 0; i < SIZE; i++)
  {
    printf("\nCity: ");
    gets(cities[i].name);

    printf("Country: ");
    gets(cities[i].country);
```

```
   printf("Population: ");
   scanf("%d", &cities[i].pop);

   getchar();
 }
 printf("\nEnter country to search: ");
 gets(country);

 printf("Population: ");
 scanf("%d", &pop);

 flag = 0;
 for(i = 0; i < SIZE; i++)
 {
   if((strcmp(cities[i].country, country) == 0) && (cities[i].pop >
     pop))
   {
     flag = 1;
     printf("C:%s P:%d\n", cities[i].name, cities[i].pop);
   }
 }
 if(flag == 0)
   printf("\nNone city is found\n");
 return 0;
}
```

17.4 To connect to an Internet address, the network card must transmit an IP packet that encapsulates a special TCP segment. The IPv4 header format is depicted in Figure 17.1.

The TCP header format is depicted in Figure 17.2.

Write a program that reads the source IP address in x.x.x.x format (each x is an integer in [0, 255]), the destination IP address, the TCP destination port (integer in [1, 65535]) and creates an IP packet that encapsulates the required TCP segment. The program must store the content of the IP packet (in hexadecimal format) in a user-selected text file, in which each line must contain 16 bytes.

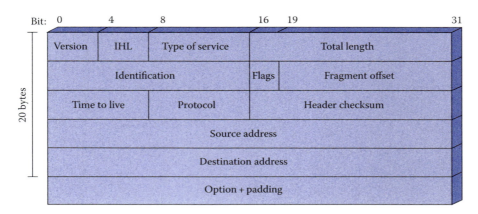

FIGURE 17.1
IPv4 header format.

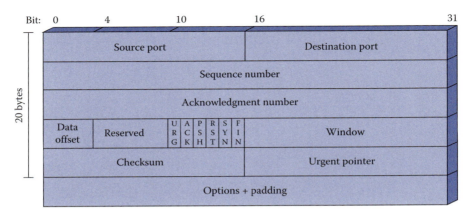

FIGURE 17.2
TCP header format.

Set the following values in the IPv4 header:

(a) `Version = 4`

(b) `IHL = 5`

(c) `Total Length =` total length of the IP packet, including the TCP data

(d) `Protocol = 6`

(e) `Time to Live = 255`

(f) `Destination Address =` destination IP address

(g) `Source Address =` source IP address

Set the following values in the TCP header:

(a) `Destination Port =` the destination port

(b) `Source Port = 1500`

(c) `Window =` the maximum value for this field

(d) `SYN bit = 1`

Set the rest fields to 0 and assume that there are no `Options` fields.
    The program has one restriction; if the destination IP address starts from `130.140` or `160.170` and the TCP destination port is `80`, do not create the IP packet.

```
#include <stdio.h>
#include <stdlib.h>

typedef unsigned char BYTE;

void Build_Pkt(int IP_src[], int IP_dst[], int port);
void Save_Pkt(BYTE pkt[], int len);

int main()
{
  int IP_src[4], IP_dst[4], TCP_dst_port;

  do
```

```c
{
  printf("Enter dst port [1-65535]: ");
  scanf("%d", &TCP_dst_port);
} while(TCP_dst_port < 1 || TCP_dst_port > 65535);
printf("Enter dst IP (x.x.x.x): ");
scanf("%d.%d.%d.%d", &IP_dst[0], &IP_dst[1], &IP_dst[2],
  &IP_dst[3]);
if(TCP_dst_port == 80)
{
  if(IP_dst[0] == 130 && IP_dst[1] == 140)
  {
    printf("It isn't allowed to connect to network 130.140.x.x\n");
    return 0;
  }
  else if(IP_dst[0] == 160 && IP_dst[1] == 170)
  {
    printf("It isn't allowed to connect to network 160.170.x.x\n");
    return 0;
  }
}
printf("Enter src IP (x.x.x.x): ");
scanf("%d.%d.%d.%d", &IP_src[0], &IP_src[1], &IP_src[2],
  &IP_src[3]);
Build_Pkt(IP_src, IP_dst, TCP_dst_port);
return 0;
}

void Build_Pkt(int IP_src[], int IP_dst[], int port)
{
  BYTE pkt[40] = {0}; /* Initialize all fields to 0. */
  int i, j;
  pkt[0] = 0x45; /* Version, IHL. */
  pkt[8] = 255; /* Time to Live. */
  pkt[9] = 6; /* Protocol = TCP. */
  for(i = 12, j = 0; i < 16; i++, j++)
    pkt[i] = IP_src[j]; /* IP Source. */
  for(i = 16, j = 0; i < 20; i++, j++)
    pkt[i] = IP_dst[j]; /* IP Destination. */
  pkt[20] = 1500 >> 8; /* TCP Source Port. */
  pkt[21] = 1500 & 0xFF;
  pkt[22] = port >> 8; /* TCP Destination Port. */
  pkt[23] = port & 0xFF;
  pkt[33] = 2; /* SYN bit. */
  pkt[34] = 0xFF; /* To get the max value of the Window field set
  its bits to 1. */
  pkt[35] = 0xFF;
  /* The values of the CheckSum and Urgent Pointer fields are set in
    positions 36-40, therefore the total length of the IP packet is
    40 bytes. */
  pkt[2] = 0; /* IP Total Length. */
  pkt[3] = 40;

  Save_Pkt(pkt, 40);
}
```

```
void Save_Pkt(BYTE pkt[], int len)
{
  FILE *fp;
  char name[100];
  int i;

  printf("Enter file name: ");
  scanf("%s", name);

  fp = fopen(name, "w");
  if(fp == NULL)
  {
    printf("Error: File can't be created\n");
    exit(1);
  }
  for(i = 0; i < len; i++)
  {
    if((i > 0) && (i%16 == 0))
      putc('\n', fp);
    fprintf(fp, "%02X ", pkt[i]);
  }
  fclose(fp);
}
```

Comments: In a real application, this IP packet is encapsulated in a MAC frame (yes, we know that these concepts may look strange to you, but this is what really happens in computer communications), which is transmitted to the IP destination address through the computer's network card.

Do you have any idea about what this program really does? This program is a simplified version of a common application, probably installed in your computer, called *firewall*. Like this program, a firewall may prevent communication to specific IP addresses and specific applications (e.g., the communication port for the transfer of web pages is set to 80). In fact, a firewall is nothing more than a sequence of if-else statements.

17.5 Write a program that provides a menu to perform the following operations:

1. *Intersection*. The program reads the common size of two arrays (i.e., A and B) and allocates memory dynamically to store their elements. Then, it reads integers and stores them into A only if they are not already stored. Then, it does the same for the array B. The program should display the intersection of the two arrays, meaning their common elements.

2. *Union*. The program should store the union of the two arrays (the elements that belong either to array A or to array B) into a third array.

3. *Program termination*.

```
#include <stdio.h>
#include <stdlib.h>
#include <string.h>

void fill_table(int arr[], int size);
void find_inter(int arr1[], int arr2[], int size);
void find_union(int arr1[], int arr2[], int size);
```

```c
int main()
{
  int *arr1, *arr2, sel, size;

  arr1 = arr2 = NULL;
  while(1)
  {
    printf("\nMenu selections\n");
    printf("- - - - - - -\n");

    printf("1. Intersection\n");
    printf("2. Union\n");
    printf("3. Exit\n");

    printf("\nEnter choice: ");
    scanf("%d", &sel);

    switch(sel)
    {
      case 1:
        do
        {
          printf("\nEnter size: ");
          scanf("%d", &size);
        } while(size <= 0);

        /* Free the allocated memory. */
        if(arr1 != NULL)
          free(arr1);
        if(arr2 != NULL)
          free(arr2);

        arr1 = (int *) malloc(size * sizeof(int));
        arr2 = (int *) malloc(size * sizeof(int));
        if(arr1 == NULL || arr2 == NULL)
        {
          printf("Not available memory");
          exit(1);
        }
        printf("\nArray_1 elements\n");
        fill_table(arr1, size);

        printf("\nArray_2 elements\n");
        fill_table(arr2, size);

        find_inter(arr1, arr2, size);
      break;

      case 2:
        if(arr1 && arr2)/* Check that memory has been allocated. */
          find_union(arr1, arr2, size);
      break;

      case 3:
        if(arr1 != NULL)
          free(arr1);
        if(arr2 != NULL)
          free(arr2);
```

```
        return 0; /* Program termination. */
      }
    }
    return 0;
}

void fill_table(int arr[], int size)
{
  int i, j, num, found;

  i = 0; /* It counts how many numbers are stored in the array. */
  while(i < size)
  {
    printf("Enter number: ");
    scanf("%d", &num);

    found = 0;
    /* Check if the input number is already stored in the array. If
       it is, the loop terminates. */
    for(j = 0; j < i; j++)
    {
      if(arr[j] == num)
      {
        printf("Error: %d is in array\n", num);
        found = 1;
        break;
      }
    }
    /* If the number isn't found, store it. */
    if(found == 0)
    {
      arr[i] = num;
      i++;
    }
  }
}

void find_inter(int arr1[], int arr2[], int size)
{
  int *arr3, i, j, k;

  arr3 = (int *) malloc(size * sizeof(int));
  if(arr3 == NULL)
  {
    printf("Error: Not available memory\n");
    exit(1);
  }

  k = 0;
  for(i = 0; i < size; i++)
  {
    /* If a common element is found, store it in arr3 and increase k,
       which counts the number of stored items. */
    for(j = 0; j < size; j++)
    {
      if(arr1[i] == arr2[j])
```

```
      {
        arr3[k] = arr1[i];
        k++;
        break; /* This loop terminates and the external loop
          continues with the next arr1 element. */
      }
    }
  }
  if(k == 0)
    printf("\nResult: There is no common elements\n");
  else
  {
    printf("\nIntersection: ");
    for(i = 0; i < k; i++)
      printf("%d ", arr3[i]);
    printf("\n");
  }
  free(arr3);
}

void find_union(int arr1[], int arr2[], int size)
{
  int *arr3, i, j, k, found;

  /* The maximum memory size that may be needed is 2*size, which
    covers the worst case that the arrays have no common elements. */
  arr3 = (int *)malloc(2*size*sizeof(int));
  if(arr3 == NULL)
  {
    printf("Error: Not available memory\n");
    exit(1);
  }
  k = size;
  memcpy(arr3, arr1, size * sizeof(int)); /* First, store the arr1
    elements. Other elements may be stored after the k-th place. */
  for(j = 0; j < size; j++)
  {
    found = 0;
    /* Check if arr2[j] exists in arr3. If it doesn't, it is stored
      in arr3 and k, which counts the number of the stored items, is
      increased. */
    for(i = 0; i < size; i++)
    {
      if(arr2[j] == arr3[i])
      {
        found = 1;
        break;
      }
    }
    if(found == 0)
    {
      arr3[k] = arr2[j];
      k++;
    }
  }
```

```
    printf("\nUnion: ");
    for(i = 0; i < k; i++)
      printf("%d ", arr3[i]);

    printf("\n");
    free(arr3);
}
```

17.6   What is the output of the following program?

```
#include <stdio.h>
void test(int **tmp);

int main()
{
  int *ptr, arr[] = {5, 10, 15};

  ptr = arr;
  test(&ptr);

  printf("%d ", *ptr);
  return 0;
}
void test(int **tmp)
{
  int i;

  i = *(*tmp)++;
  printf("%d ", i);

  i = (**tmp)++;
  printf("%d ", i);
}
```

Answer: When test() is called, we have tmp = &ptr and *tmp is equivalent to ptr.
   Therefore, the expression i = *(*tmp)++; is equivalent to i = *ptr++;. This
statement first assigns to i the value of *ptr, that is, *ptr = *arr = arr[0] = 5,
and then ptr is increased to point to arr[1].
   Then, we have i = (**tmp)++ = (*ptr)++ = arr[1]++;. Like before, this state-
ment first assigns to i the value of arr[1], that is, 10, and then arr[1] becomes 11.
   After the execution of test(), ptr points to arr[1].
   Therefore, the program displays 5 10 11.

17.7   Write a program that requires two command line arguments (integers) that correspond to the dimensions (rows and columns) of a two-dimensional array of doubles.
The program should allocate memory to create the array and assign random values
to its elements so that the sum of the elements in each row is equal to 1. The program
should display the elements of the array before it terminates.
   Note: To convert the command line arguments to integers, use the atoi() function. To generate random values, use the rand() and srand() functions. You can
find their prototypes in Appendix C.

```
#include <stdio.h>
#include <stdlib.h>
#include <time.h>
```

```
int main(int argc, char *argv[])
{
  int i, j, rows, cols;
  double **arr, sum_line;

  if(argc < 3)
    printf("Error: missing arguments...\n");
  else if(argc == 3)
  {
    rows = atoi(argv[1]); /* Convert the second argument to integer.*/
    cols = atoi(argv[2]); /* Do the same for the third argument. */

    srand((unsigned)time(NULL));
    arr = (double **) malloc(rows * sizeof(double*));
    if(arr == NULL)
    {
      printf("Error: Not available memory\n");
      exit(1);
    }
    for(j = 0; j < rows; j++)
    {
      arr[j] = (double *) malloc(cols * sizeof(double));
      if(arr[j] == NULL)
      {
        printf("Error: Not available memory\n");
        exit(1);
      }
    }
    for(i = 0; i < rows; i++)
    {
      sum_line = 0;
      for(j = 0; j < cols-1; j++)
      {
        arr[i][j] = (rand()%101/(cols-1))/100.0;
        sum_line += arr[i][j];
```
/* As an example, assume that the number of columns is 5. For the
first four row's elements, the expression rand()%101 generates an
integer in [0, 100]. Dividing it by the number of the columns
minus 1 (i.e. 4) we get an integer in [0, 25] and dividing that
with 100.0 we get a number in [0, 0.25]. In this way, the sum of
the first four elements is constrained in [0, 1]. */
```
      }
      arr[i][j] = 1-sum_line; /* To satisfy the condition that the
        sum of the elements of each row must be equal to 1, the value
        of the last element is calculated by subtracting from 1 the
        sum of the rest elements. */
    }
    for(i = 0; i < rows; i++)
    {
      for(j = 0; j < cols; j++)
        printf("%6.2f", arr[i][j]);

      printf("\n");
    }
    for(j = 0; j < rows; j++)
```

```
      free(arr[j]);
    free(arr);
  }
  else
    printf("Error: too many arguments...\n");
  return 0;
}
```

17.8 Write a program that simulates a simple electronic roulette that allows the player to bet on whether the winning number of the next spin of the roulette ball will be odd or even. The numbers of the roulette are from 0 up to 36. If the winning number is 0, the player loses because 0 is counted neither as odd nor as even. The program should display a menu to perform the following operations:

1. *Bet on odd numbers*
2. *Bet on even numbers*
3. *Spin the ball*
4. *Statistics*
5. *Termination*

After selecting the kind of bet (i.e., odd or even), the player should specify the bet. When the third option is chosen, the program should generate a random integer in [0, 36] and display a message to indicate if the player won or lost. The fourth option should display some statistics, that is, how many times the player won/lost, and how much money the player wins or loses.

We left the best for the end: the program should be written in such a way that the player should be definitely *lost*, at the end of the game.

```
#include <stdio.h>
#include <stdlib.h>
#include <time.h>

#define LOSS 0
#define WIN 1

void unfair_play(int sel, int bet, int *lost, int *times);
int fair_play(int sel, int bet, int *lost);

int main()
{
  int sel, last_sel, flag, tmp, bet, sum_lost, win_times,
    lost_times;

  flag = 1;
  sum_lost = win_times = lost_times = 0;

  while(1)
  {
    printf("\nRoulette Game\n");
    printf("- - - - - - \n");

    printf("1. Odd\n");
    printf("2. Even\n");
```

```c
    printf("3. Play\n");
    printf("4. Stats\n");
    printf("5. Exit\n");

    printf("\nEnter choice: ");
    scanf("%d", &sel);

    srand((unsigned)time(NULL));

    switch(sel)
    {
      case 1:
      case 2:
        last_sel = sel;
        do
        {
          printf("\nPlace your bet: ");
          scanf("%d", &bet);
        } while(bet <= 0);
      break;

      case 3:
        if(bet == 0)
        {
          printf("No bet is placed\n");
          break;
        }
        if(flag == 1)/* We force the program to "play unfair" and
          make the player to lose the first bet. */
        {
          flag = 0;
          unfair_play(last_sel, bet,
            &sum_lost, &lost_times);
        }
        else
        {
          if(bet >= sum_lost)/* If the player bets a larger amount
            than the money he loses, the program behaves "unfair" and
            the player loses. */
            unfair_play(last_sel, bet,
            &sum_lost, &lost_times);
          else
          {
            tmp = fair_play(last_sel, bet, &sum_lost);
            if(tmp == LOSS)
            {
              printf("Sorry, you lost...\n");
              lost_times++;
            }
            else
            {
              printf("Yeaaaaah, you won...\n");
              win_times++;
            }
          }
        }
    }
```

```c
        bet = 0; /* Initialize the bet to zero. */
      break;
      case 4:
        printf("\nWin_Times: %d\tLost_Times: %d\tLost_Money: %d
          euro\n", win_times, lost_times, sum_lost);
      break;

      case 5:
      return 0;

      default:
        printf("\nWrong choice\n");
      break;
    }
  }
  return 0;
}

/* The program generates a random number, so that the player loses.
   For example, if the player bets on odd numbers the program
   generates an even number, and vice versa. */
void unfair_play(int sel, int bet, int *lost, int *times)
{
  int num;

  *lost += bet;
  (*times)++;

  while(1)
  {
    num = rand()% 37;
    if(sel == 1)/* The player bets on odd numbers. */
    {
      if((num & 1) == 0)
      {
        printf("\nThe ball goes to %d. Sorry, you lost …\n", num);
        return;
      }
    }
    else/* The player bets on even numbers. */
    {
      if((num & 1) == 1)
      {
        printf("\nThe ball goes to %d. Sorry, you lost …\n", num);
        return;
      }
    }
  }
}
int fair_play(int sel, int bet, int *lost)
{
  int num;

  num = rand()% 37;
  printf("\nThe ball goes to %d. ", num);

  if(num == 0)/* If zero comes out, the player loses. */
```

```
{
  *lost += bet;
  return LOSS;
}
if(sel == 1)/* The player bets on odd numbers. */
{
  if((num & 1) == 1)
  {
    *lost -= bet; /* If the player wins, the total lost amount is
    reduced. */
    return WIN;
  }
  else
    *lost += bet; /* If the player loses, the total lost amount is
    increased. */
}
else/* The player bets on even numbers. */
{
  if((num & 1) == 0)
  {
    *lost -= bet;
    return WIN;
  }
  else
    *lost += bet;
}
return LOSS;
}
```

Comments: Since the program controls the bets, the player eventually *loses* even if he won more times than he lost. In other words, the program creates the feeling of "misfortune" to the player, whereas it actually cheats him.

However, the main purpose of this exercise is instructive. With this simple simulation example, we want to show you that gambling games' software can be written in a way that creates the feeling of "bad luck," whereas the player is just a victim of fraud.

Therefore, *stay away* from online gaming and any kind of electronic "lucky machines" (e.g., slot machines, fruit machines, etc.). The big profits they promise are not for you, but for their owners.

17.9 What is the output of the following program?

```
#include <stdio.h>

void test(int val, int *tmp);

int main()
{
  int *ptr, i, arr[] = {5, 10, 15};

  ptr = arr;
  for(i = 0; i < 2; i++)
  {
    test(*ptr, ptr);
    ptr++;
  }
```

```
  while(ptr >= arr)
  {
    printf("%d ", *ptr);
    ptr-- ;
  }
  return 0;
}
void test(int val, int *tmp)
{
  printf("%d%d\n", ++val, (*tmp)++);
}
```

Answer:

First iteration of the **for** *loop*: When test() is called, we have val = *ptr =
*arr = arr[0] = 5. The expression ++val first increases val by one, and
the program displays 6.

Also, we have tmp = ptr = arr. Therefore, the expression (*tmp)++ is
equivalent to (*arr)++ = arr[0]++. Now, the program first displays the
value of arr[0], that is, 5, and then it is increased by one.

Second iteration of the **for** *loop*: When test() is called, we have val = *ptr =
*(arr+1) = arr[1] = 10. Like before, the program displays 11.

Also, we have tmp = ptr = arr+1 = &arr[1]. Therefore, the expression
(*tmp)++ is equivalent to arr[1]++. The program first displays 10 and then
the value of arr[1] is increased.

Execution of the **while** *loop*: The first iteration displays the value of arr[2], that
is, 15, and then ptr is decreased by one and points to arr[1]. Therefore, the
second iteration displays the value of arr[1], that is, 11, and the last one the
value of arr[0], that is, 6.

To sum up, the program displays

```
6   5
10  11
15  11  6
```

17.10 A car parking station charges $6.00 for the first 3 h. Each extra hour is charged $1.50
(even for one extra minute the whole hour is charged). The whole day is charged
$12.00 and the maximum parking time is 24 h. Write a program that reads the num-
ber of the parked cars and the respective parking hours and displays the charges in
the following form:

```
Car     Hours  Charge
1        2.5    6.00
2        4.5    9.00
3        5.25  10.50
4        4      7.50
5        8     12.00
TOTAL   24.25  45.00
```

```c
#include <stdio.h>
#include <stdlib.h>
int main()
{
  int i, cars;
  float *hours_arr, rem, bill, sum_hours, total_bill;

  do
  {
    printf("Enter number of total cars: ");
    scanf("%d", &cars);
  } while(cars <= 0);

  /* Allocate memory to store the parking hours of all cars. */
  hours_arr = (float *) malloc(cars * sizeof(float));
  if(hours_arr == NULL)
  {
    printf("Error: Not available memory\n");
    exit(1);
  }
  for(i = 0; i < cars; i++)
  {
    do
    {
      printf("Enter hours for car_%d [max = 24h]: ", i+1);
      scanf("%f", &hours_arr[i]);
    } while(hours_arr[i] > 24 || hours_arr[i] < 0);
  }
  sum_hours = total_bill = 0;

  printf("\nCar\tHours\tCharge\n");
  for(i = 0; i < cars; i++)
  {
    bill = 6; /* For the first 3 hours. */
    rem = hours_arr[i] - 3; /* Find the extra hours. */
    if(rem > 0)
    {
      /* We check whether rem has a decimal part or not. For example,
         if rem = 3.2, the statement if(rem - (int)rem) = if(3.2 - 3)
         = if(0.2 != 0) is true, so rem is increased. In fact, rem
         should be increased, because the 3.2 hours are charged as if
         it were 4. */
      if(rem - (int)rem != 0)
        rem++;

      bill += (int)rem * 1.5; /* Use typecasting to remove the
        decimal part of rem. */
      if(bill > 12)
        bill = 12;
    }
    printf("%d\t%.2f\t%.2f\n", i+1, hours_arr[i], bill);

    sum_hours += hours_arr[i];
    total_bill += bill;
  }
```

```
    printf("SUM\t%.2f\t%.2f\n", sum_hours, total_bill);
    free(hours_arr);
    return 0;
}
```

17.11 Because a C program executes rather fast, the C language is often used for implementing cipher algorithms. As an example, we'll describe one of the most simple
and famous cipher algorithms.

The Caesar algorithm is one of the oldest cipher methods used by Julius Caesar
to encrypt his messages. According to this algorithm, each character is substituted
by the one located in the next three places. For example, if we apply the Caesar
algorithm in the English alphabet, the message "Watch out for Ovelix !!!"
is encrypted as "Zdwfk rxw iru Ryhola !!!". Notice that the character 'x' is
encrypted as 'a' since the substitution continues from the beginning of the alphabet. Similarly, 'y' is replaced by 'b' and 'z' by 'c'. The recipient decrypts the
message by substituting each character with the one located three places before it.

Write a program that provides a menu to perform the following operations:

1. *File Encryption.* The program should read the name of a file and the key number that will be used to encrypt the content of the file. For example, in the case
   of Caesar algorithm, the value of the key number is 3. The program should
   encrypt only the lowercase and uppercase characters.

2. *File Decryption.* The program should read a file name and the key number that
   will be used to decrypt the content of the file.

3. *Program termination.*

```c
#include <stdio.h>
#include <stdlib.h>

void cipher(FILE *fp_in, FILE *fp_out, int key);
void decipher(FILE *fp_in, FILE *fp_out, int key);
int main()
{
  FILE *fp_in, *fp_out;
  char str[100];
  int sel, key;

  while(1)
  {
    printf("\nMenu selections\n");
    printf("- - - - - - - -\n");
    printf("1. Cipher\n");
    printf("2. Decipher\n");
    printf("3. Exit\n");
    printf("\nEnter choice: ");
    scanf("%d", &sel);

    switch(sel)
    {
      case 1:
      case 2:
        getchar();
```

```
      /* Check whether the input key is valid or not. Since we are
         using the English alphabet, the valid values are between 1
         and 25. */
      do
      {
        printf("Enter key size: ");
        scanf("%d", &key);
      } while(key < 1 || key > 25);

      getchar();
      printf("Input file: ");
      gets(str);
      fp_in = fopen(str, "r");
      if(fp_in == NULL)
      {
        printf("Error: File can't be loaded\n");
        exit(1);
      }
      printf("Output file: ");
      gets(str);
      fp_out = fopen(str, "w");
      if(fp_out == NULL)
      {
        printf("Error: File can't be created\n");
        exit(1);
      }
      if(sel == 1)
        cipher(fp_in, fp_out, key);
      else
        decipher(fp_in, fp_out, key);

      fclose(fp_in);
      fclose(fp_out);
    break;

    case 3:
    return 0;

    default:
      printf("\nWrong choice\n");
    break;
    }
  }
  return 0;
}
void cipher(FILE *fp_in, FILE *fp_out, int key)
{
  int ch;

  while(1)
  {
    ch = getc(fp_in);
    if(ch == EOF)
      return;
    /* Only the lower and upper case characters are encrypted. */
    if(ch >= 'A' && ch <= 'Z')
```

```
      {
        ch += key;
        if(ch > 'Z')
          ch -= 26;
      }
      else if(ch >= 'a' && ch <= 'z')
      {
        ch += key;
        if(ch > 'z')
          ch -= 26;
      }
      putc(ch, fp_out);
    }
}
void decipher(FILE *fp_in, FILE *fp_out, int key)
{
  int ch;

  while(1)
  {
    ch = getc(fp_in);
    if(ch == EOF)
      return;
    /* Only the lower and upper case characters are decrypted. */
    if(ch >= 'A' && ch <= 'Z')
    {
      ch -= key;
      if(ch < 'A')
        ch += 26;
    }
    else if(ch >= 'a' && ch <= 'z')
    {
      ch -= key;
      if(ch < 'a')
        ch += 26;
    }
    putc(ch, fp_out);
  }
}
```

Comments: In cipher(), the value of key is added to each character. If the new value exceeds the last character of the alphabet ('Z' or 'z'), we subtract 26 to go back to the beginning of the alphabet. In decipher(), the reverse actions take place.

17.12 What is the output of the following program?

```
#include <stdio.h>
#include <string.h>

void test(char a, char str[], char *ptr);

int main()
{
  char *tmp, txt[20] = "abcde";

  tmp = txt;
  test(*(tmp+1), txt+3, &txt[1]);
```

```
  printf("%s\n", txt);
  return 0;
}
void test(char a, char str[], char *ptr)
{
  strcpy(str, "1234");
  str[0] = a;
  ptr[2] = *str + 5;
}
```

Answer: When test() is called, we have

(a) a = *(tmp+1) = *(txt+1) = txt[1] = 'b'
(b) str = txt+3, so, str[0] is equal to txt[3].
(c) ptr = &txt[1] = txt+1. Since ptr points to txt[1], ptr[0] is equal to txt[1], ptr[1] is equal to txt[2], and ptr[2] is equal to txt[3].

The statement strcpy(str, "1234"); copies the string "1234" into the memory that str points to. Since str points to txt[3], its content becomes "abc1234".
    The statement str[0] = a; is equivalent to txt[3] = a. Since the value of a is equal to 'b', txt[3] becomes 'b' and the content of txt changes to "abcb234".
    As said, ptr[2] is equal to txt[3] and str[0] is equal to txt[3]. Therefore, the value of *str is equal to 'b' and the statement ptr[2] = *str+5; is equivalent to txt[3] = 'b'+5; meaning that txt[3] becomes equal to the character located 5 places after 'b', that is, 'g'.
    Therefore, the program displays abcg234.

17.13 Suppose that we have created subnets in a Class C IP network (an IP network is specified as Class C when the first octet (byte) of its IP address is within [192, 223]). Write a program that reads the last octet of an IP address, the subnet mask, and displays the addresses of all subnets, the broadcast address of each subnet and the subnet in which the IP address belongs to.
    For example, suppose that we have a Class C IP network (e.g., x.x.x.x) and the user enters 74. Therefore, we have to find out in which subnet the IP address x.x.x.74 belongs to.
    To find the subnet, the user must enter the subnet mask in one of the following ways:

(a) To enter the last octet of the subnet mask, the valid values are 252, 248, 240, 224, 192, 128.
(b) To enter the number of the network bits, which must be an integer in [25, 30].

To find out the subnets, we calculate the distance between them:

(a) In the first case, the distance is equal to $256-x$, where x is the input number.
(b) In the second case, the distance is equal to 2^{32-x}, where x is the input number of bits. For example, if the user enters 26, the distance is equal to $2^{32-26} = 2^6 = 64$.

Starting from the IP address x.x.x.0, the IP address of each subnet starts from a number multiple of the distance, while the broadcast IP address is the last address of each subnet.

For example, assume that the distance is 64. Each subnet starts from an IP address multiple of 64. The broadcast address is always the number right before the next subnet. The program should display the last octet of each IP address, like this:

```
Network:    .0   .64   .128    .192
Broadcast: .63 .127 .191    .255
```

The valid host range is the numbers between the subnet address and the broadcast address. Therefore, the IP address x.x.x.74 is a member of the second subnet because its last octet falls in [64, 127].

Although the length of this exercise may discourage you to deal with it, it'd be useful for those who take "Computer Networks" courses and need a simple method to create subnets fast.

```c
#include <stdio.h>
#include <stdlib.h>
int main()
{
  int i, j, flag, sel, num, dist, host_byte;

  do
  {
    printf("Enter last host byte [0-255]: ");
    scanf("%d", &host_byte);
  } while(host_byte < 0 || host_byte > 255);
  do
  {
    flag = 0;
    printf("Enter mask (0: 255.255.255.x form or 1:/bits form): ");
    scanf("%d", &sel);
    if(sel == 0)
    {
      printf("Enter last mask octet 255.255.255.");
      scanf("%d", &num);

      if(num != 252 && num != 248 && num != 240
        && num != 224 && num != 192 && num != 128)
      {
        printf("Last octet should be one of {128, 192, 224, 240, 248,
          252}\n");
        flag = 1;
      }
      else
        dist = 256-num;
    }
    else if(sel == 1)
    {
      printf("Enter network bits:/");
      scanf("%d", &num);

      if(num < 25 || num > 30)
      {
        printf("Enter valid mask/25-/30\n");
        flag = 1;
      }
    }
```

```
      else
      {
        num = 32-num;
        dist = 1;
        for(i = 0; i < num; i++)
          dist = dist * 2;
        /* For faster implementation we could write:
   dist = 1 << num; instead of the loop. */
      }
    }
  } while(flag == 1);

  printf("\nThe mask 255.255.255.%d produces %d subnets, each with
    %d hosts\n", 256-dist, 256/dist, dist-2);

  printf("\nNetwork : ");
  for(i = 0; i < 256; i+=dist)
    printf(".%d\t", i);
  printf("\nBroadcast: ");
  for(i = dist-1; i < 256; i+=dist)
    printf(".%d\t", i);

  for(i = j = 0; i < 256; i+=dist, j++)
  {
    if(host_byte >= i && host_byte < i+dist)
    {
      if(host_byte == i)
        printf("\n\nThe x.x.x.%d address is the network address of
          subnet_%d\n", host_byte, j+1);
      else if(host_byte == i+dist-1)
        printf("\n\nThe x.x.x.%d address is the broadcast address of
          subnet_%d\n", host_byte, j+1);
      else
        printf("\n\nThe x.x.x.%d address belongs to subnet_%d\n",
        host_byte, j+1);
      break;
    }
  }
  return 0;
}
```

17.14 Write a program that reads the order of preference of 6 answers that 100 tourists gave in the question: "What did you enjoy the most in Greece?"

1. *The food*
2. *The climate*
3. *The islands*
4. *The night life*
5. *The ancient places*
6. *The people*

Each answer takes 1–6 points according to its rank order. The first answer takes 6 points, the second one takes 5 points, and the last one takes 1 point. For example, if two tourists answered the following:

First tourist   Second tourist
5 *(6p.)*   3 *(6p.)*
4 *(5p.)*   6 *(5p.)*
6 *(4p.)*   1 *(4p.)*
3 *(3p.)*   4 *(3p.)*
1 *(2p.)*   2 *(2p.)*
2 *(1p.)*   5 *(1p.)*

the program should display

First answer gets 6 points

Second answer gets 3 points

Third answer gets 9 points

Fourth answer gets 8 points

Fifth answer gets 7 points

Sixth answer gets 9 points

The program should read valid answers in [1, 6] and check if the answer is already given. If it does, the program should display a message and prompt the user to enter a new one.

```c
#include <stdio.h>
#include <string.h>

#define TOURISTS 100
#define ANSWERS 6

int main()
{
  int i, j, sel, pnts[ANSWERS] = {0}; /* This array holds the points
    of each answer. For example, pnts[0] holds the points of the
    first answer, pnts[1] holds the points of the second answer and
    so forth. */

  int given_ans[ANSWERS] = {0}; /* This array is used to check
    whether an answer is already given or not. If an element's value
    is 1, means that the respective answer is selected. For example,
    if the user selects the third answer the value of given_ans[2]
    becomes 1. */
  for(i = 0; i < TOURISTS; i++)
  {
    printf("\nEnter answers of tourist_%d:\n", i+1);

    memset(given_ans, 0, sizeof(given_ans)); /* The values of the
      given_ans array must be zeroed before reading the answers of a
      new tourist. See memset() in Appendix C. */
    for(j = 0; j < ANSWERS; j++)
    {
      while(1)/* Infinite loop until the user enters a valid answer
        in [1,6], not already given. */
      {
        printf("Answer_%d [1-%d]: ", j+1, ANSWERS);
        scanf("%d", &sel);
```

```
      if(sel < 1 || sel > ANSWERS)
        printf("Wrong answer …\n");
      else if(given_ans[sel-1] == 1)
        printf("Error: This answer is already given …\n");
      else
        break;
    }
    pnts[sel-1] += ANSWERS - j; /* For example, if the first answer
      (j = 0) is the fifth one, then pnts[sel-1] = pnts[5-1] =
      pnts[4] += 6-0 = 6; meaning that 6 more points will be added
      to the points of the fifth choice. */
    given_ans[sel-1] = 1;
  }
}
printf("\n***** Answer Results *****\n");
for(i = 0; i < ANSWERS; i++)
  printf("Answer #%d gets %d points\n", i+1, pnts[i]);
return 0;
}
```

17.15 What is the output of the following program?

```
#include <stdio.h>
int main()
{
  char *arr[] = {"TEXT", "SHOW", "OPTIM", "DAY"};
  char **ptr1;

  ptr1 = arr; /* Equivalently, ptr1 = &arr[0]. */
  printf("%s ", *++ptr1);
  printf("%s", *++ptr1+2);
  printf("%c\n", **++ptr1+1);
  return 0;
}
```

Answer: The arr array is declared as an array of pointers to strings. In particular, arr[0] points to the first character of "TEXT", arr[1] points to "SHOW", and so on.

The statement ptr1 = arr; makes ptr1 to point to the address of arr[0].

The expression ++ptr1 makes ptr1 to point to arr[1]. Since *++ptr1 is equivalent to arr[1], the program displays "SHOW".

Similarly, the expression ++ptr1+2 makes first ptr1 to point to arr[2]. Since *++ptr1 is equivalent to arr[2], the statement printf("%s",*++ptr1+2); is equivalent to printf("%s",ptr[2]+2); and the program displays the characters of "OPTIM" following the first two, that is, "TIM".

Like before, the expression ++ptr1+1 makes first ptr1 to point to arr[3]. Since *++ptr1 is equivalent to arr[3], **++ptr1 is equivalent to *arr[3].

What is the value of *arr[3]? Since arr[3] points to the first character of "DAY", *arr[3] is equal to 'D'. Therefore, the value of *ptr[3]+1 is 'E' and the program displays 'E'.

To sum up, the program displays SHOW TIME.

17.16 Write a program that can be used as a book library management application. Define the structure type book with fields—title, authors, and book code—and suppose that the `test.bin` binary file contains structures of type `book`. If the `test.bin` file doesn't exist, it must be created. Write a program that provides a menu to perform the following operations:

1. *Add a new book.* The program should read the details of a new book (i.e., title, authors, and code) and add it in the file.
2. *Search for a book.* The program should read the title of a book and display its details. If the user enters *, the program should display the details of all books.
3. *Modify a book.* The program should read the title of a book and its new details and it should replace the existing details with the new ones.
4. *Delete a book.* The program should read the title of a book and set its code equal to –1.
5. *Program termination.*

```c
#include <stdio.h>
#include <string.h>
#include <stdlib.h>

#define LEN 30

typedef struct
{
  char title[LEN];
  char auth[LEN];
  int code;
} book;

void read_data(book *ptr);
void find_book(FILE *fp, char title[]);
void modify_book(FILE *fp, book *ptr, int flag);
void show_books(FILE *fp);

int main()
{
  FILE *fp;
  char title[LEN];
  int sel;
  book b;

  fp = fopen("test.bin", "r+b");
  if(fp == NULL)
  {
    /* If the file doesn't exist, it is created. */
    fp = fopen("test.bin", "w+b");
    if(fp == NULL)
    {
      printf("Error: File can't be created\n");
      exit(1);
    }
  }
  while(1)
  {
    printf("\n\nMenu selections\n");
    printf("- - - - - - - -\n");
```

```c
      printf("1. Add Book\n");
      printf("2. Find Book\n");
      printf("3. Modify Book \n");
      printf("4. Erase Book\n");
      printf("5. Exit\n");

      printf("\nEnter choice: ");
      scanf("%d", &sel);
      getchar();

      switch(sel)
      {
        case 1:
          read_data(&b);
          fseek(fp, 0, SEEK_END); /* Add the details of the new book at
            the end of the file. */
          fwrite(&b, sizeof(book), 1, fp);
        break;

        case 2:
          printf("\nTitle to search: ");
          gets(title);
          if(strcmp(title, "*") != 0)
            find_book(fp, title);
          else
            show_books(fp);
        break;

        case 3:
          read_data(&b);
          modify_book(fp, &b, 0);
        break;

        case 4:
          printf("\nTitle to search: ");
          gets(b.title);
          modify_book(fp, &b, 1);
        break;

        case 5:
          fclose(fp);
        return 0;

        default:
          printf("\nWrong choice\n");
        break;
      }
   }
   return 0;
}

void read_data(book *ptr)
{
  printf("\nTitle: ");
  gets(ptr->title);

  printf("Authors: ");
  gets(ptr->auth);
```

```
  do
  {
    printf("Code [> 0]: ");
    scanf("%d", &ptr->code);
  }
  while(ptr->code <= 0);
}

void find_book(FILE *fp, char title[])
{
  book b;

  fseek(fp, 0, SEEK_SET);
  while(1)
  {
    if(fread(&b, sizeof(book), 1, fp) != 1)
      break;
    else
    {
      if(strcmp(b.title, title) == 0)
      {
        printf("\nT:%s A:%s C:%d", b.title, b.auth, b.code);
        return;
      }
    }
  }
  printf("\n%s doesn't exist", title);
}

void modify_book(FILE *fp, book *ptr, int flag)
{
  book b;

  fseek(fp, 0, SEEK_SET);
  while(1)
  {
    if(fread(&b, sizeof(book), 1, fp) != 1)
      break;
    else
    {
      if(strcmp(b.title, ptr->title) == 0)
      {
        /* Now, fp points to the next entry. We call fseek() to make
           it point at the current entry. */
        fseek(fp, -sizeof(book), SEEK_CUR);
        if(flag == 0)
          fwrite(ptr, sizeof(book), 1, fp);
        else
        {
          b.code = -1; /* Set the code to -1. */
          fwrite(&b, sizeof(book), 1, fp);
        }
        return;
      }
    }
  }
}
```

```
    printf("\n%s doesn't exist", ptr->title);
}

void show_books(FILE *fp)
{
  book b;

  fseek(fp, 0, SEEK_SET);
  while(1)
  {
    if(fread(&b, sizeof(book), 1, fp) != 1)
      return;
    else
    {
      if(b.code != -1)
        printf("T:%s A:%s C:%d\n", b.title, b.auth, b.code);
    }
  }
}
```

17.17 The permutation method is an example of symmetric cryptography, in which the sender and the receiver share a common key. The encryption key is an integer of n digits ($n \leq 9$). Each digit must be between 1 and n and appear only once.

In the encryption process, the message is divided into segments of size n. For example, since this is the last exercise, suppose that we are using the key 25413 to encrypt the message "This is the end!!". Since the key size is 5, the message is divided into four segments of 5 characters each. If the size of the last segment is less than the key size, padding characters are added. Suppose that the padding character is the *, as shown in Figure 17.3.

The characters of each segment are rearranged according to the key digits. For example, the second character of the original message is the first one in the encrypted segment, the fifth character goes to the second position, the fourth character in the third position, etc., as shown in Figure 17.3. The same is repeated for the rest segments.

The receiver uses the same key to decrypt the message. For example, the first character of the first encrypted segment corresponds to the second character of the original message, the second character corresponds to the fifth one, the third character to the fourth one, etc.

Write a program that reads a string up to 100 characters, the encryption key, the padding character, and uses that method to encrypt and decrypt the string.

```
#include <stdio.h>
#include <string.h>
int main()
{
```

FIGURE 17.3
The original and the encrypted message.

```c
char pad_ch, key_str[10], in_str[110] = {0}, out_str[110] = {0};
  /* The size of arrays is more than 100 characters, to cover the
  case of padding characters in the last segment. */
int i, j, tmp, seg, key_len, max_key_dig, msg_len;

tmp = 1;
while(tmp)
{
  tmp = 0; /* The loop ends only if tmp remains 0. */
  printf("Enter 1 up to 9 different key digits: ");
  gets(key_str);
  key_len = strlen(key_str);

  if(key_len < 1 || key_len > 9)
  {
    printf("Error: Length should be 1 to 9 different digits\n");
    tmp = 1;
    continue;
  }
  max_key_dig = '0'; /* This variable holds the key digit with the
    highest value. */
  for(i = 0; (tmp != 1) && i < key_len; i++)
  {
    if(key_str[i] < '1' || key_str[i] > '9')
    {
      printf("Error: Only digits are allowed\n");
      tmp = 1;
      break;
    }
    if(key_str[i] > max_key_dig)
      max_key_dig = key_str[i];
    /* Check if each digit appears once. */
    for(j = i+1; j < key_len; j++)
    {
      if(key_str[i] == key_str[j])
      {
        printf("Error: Digits should be different\n");
        tmp = 1;
        break;
      }
    }
  }
  if(tmp == 0)
  {
    max_key_dig -= '0';
    if(key_len != max_key_dig)/* For example, the key 125 is not
      acceptable, because the value of the highest digit (i.e. 5)
      must be equal to the key length, that is 3 in this case. */
    {
      printf("Error: Digits should be from 1 to %d\n", key_len);
      tmp = 1;
    }
  }
}
```

```
    printf("Enter padding character: ");
    pad_ch = getchar();

    getchar();
    while(1)
    {
      printf("Enter text: ");
      gets(in_str);

      msg_len = strlen(in_str);
      if(msg_len >= key_len)
        break;
      else
        printf("Error: Text length must be more than the key size\n");
    }
    seg = msg_len/key_len;
    tmp = msg_len - (seg*key_len);

    if(tmp != 0)/* If it isn't zero, means that the message length is
      not divided exactly by the length of the key and padding
      characters must be added. Notice that the replacement starts from
      the position of the null character. */
    {
      seg++;
      for(i = 0; i < key_len-tmp; i++)
        in_str[msg_len+i] = pad_ch;
    }
    for(i = 0; i < seg; i++)
    {
      for(j = 0; j < key_len; j++)
      {
        tmp = key_str[j]-'1'; /* We subtract the ASCII value of
          character '1', in order to use the variable tmp as an index
          to the original message. */
        out_str[i*key_len+j] = in_str[i*key_len+tmp];
      }
    }
    printf("Encrypted text:%s\n", out_str);
    for(i = 0; i < seg; i++)
    {
      for(j = 0; j < key_len; j++)
      {
        tmp = key_str[j]-'1';
        in_str[i*key_len+tmp] = out_str[i*key_len+j];
      }
    }
    printf("Decrypted text: %s\n", in_str); /* Any padding characters
      appear at the end of the original text. */
    return 0;
}
```

Appendix A

Precedence Table

Table A.1 lists C operators from the highest to the lowest order of precedence. Operators listed in the same line have the same precedence. The last column indicates the order in which operators of the same precedence are evaluated.

TABLE A.1

C Operators' Precedence Table

Operator Category	Operators	Associativity
Primary	`()  []  ->  .  ++(postfix)  --(postfix)`	Left to right
Unary	`!  ~  ++(prefix)  --(prefix)  *(dereference)  &(address)  sizeof`	Right to left
Cast	`()`	Right to left
Multiplicative	`*(multiplication)  /  %`	Left to right
Additive	`+  -`	Left to right
Bitwise Shift	`<<  >>`	Left to right
Relational	`<  <=  >  >=`	Left to right
Equality	`==  !=`	Left to right
Bitwise AND	`&`	Left to right
Bitwise XOR	`^`	Left to right
Bitwise OR	`|`	Left to right
Logical AND	`&&`	Left to right
Logical OR	`||`	Left to right
Conditional	`? :`	Right to left
Assignment	`=  +=  -=  *=  /=  %=  &=  ^=  |=  <<=  >>=`	Right to left
Comma	`,`	Left to right

Appendix B

ASCII Tables

This appendix presents the standard (0-127) and extended (128-255) ASCII character sets.

Standard ASCII character set

Char	Dec	Hex	Char	Dec	Hex	Char	Dec	Hex	Char	Dec	Hex
(nul)	0	0x00	(sp)	32	0x20	@	64	0x40	`	96	0x60
(soh)	1	0x01	!	33	0x21	A	65	0x41	a	97	0x61
(stx)	2	0x02	"	34	0x22	B	66	0x42	b	98	0x62
(etx)	3	0x03	#	35	0x23	C	67	0x43	c	99	0x63
(eot)	4	0x04	$	36	0x24	D	68	0x44	d	100	0x64
(eng)	5	0x05	%	37	0x25	E	69	0x45	e	101	0x65
(ack)	6	0x06	&	38	0x26	F	70	0x46	f	102	0x66
(bel)	7	0x07	'	39	0x27	G	71	0x47	g	103	0x67
(bs)	8	0x08	(	40	0x28	H	72	0x48	h	104	0x68
(ht)	9	0x09	)	41	0x29	I	73	0x49	i	105	0x69
(nl)	10	0x0a	*	42	0x2a	J	74	0x4a	j	106	0x6a
(vt)	11	0x0b	+	43	0x2b	K	75	0x4b	k	107	0x6b
(np)	12	0x0c	,	44	0x2c	L	76	0x4c	l	108	0x6c
(cr)	13	0x0d	-	45	0x2d	M	77	0x4d	m	109	0x6d
(so)	14	0x0e	.	46	0x2e	N	78	0x4e	n	110	0x6e
(si)	15	0x0f	/	47	0x2f	O	79	0x4f	o	111	0x6f
(dle)	16	0x10	0	48	0x30	P	80	0x50	p	112	0x70
(dc1)	17	0x11	1	49	0x31	Q	81	0x51	q	113	0x71
(dc2)	18	0x12	2	50	0x32	R	82	0x52	r	114	0x72
(dc3)	19	0x13	3	51	0x33	S	83	0x53	s	115	0x73
(dc4)	20	0x14	4	52	0x34	T	84	0x54	t	116	0x74
(nak)	21	0x15	5	53	0x35	U	85	0x55	u	117	0x75
(syn)	22	0x16	6	54	0x36	V	86	0x56	v	118	0x76
(etb)	23	0x17	7	55	0x37	W	87	0x57	w	119	0x77
(can)	24	0x18	8	56	0x38	X	88	0x58	x	120	0x78
(em)	25	0x19	9	57	0x39	Y	89	0x59	y	121	0x79
(sub)	26	0x1a	:	58	0x3a	Z	90	0x5a	z	122	0x7a
(esc)	27	0x1b	;	59	0x3b	[	91	0x5b	{	123	0x7b
(fs)	28	0x1c	<	60	0x3c	\	92	0x5c			124	0x7c
(gs)	29	0x1d	=	61	0x3d	]	93	0x5d	}	125	0x7d
(rs)	30	0x1e	>	62	0x3e	^	94	0x5e	~	126	0x7e
(us)	31	0x1f	?	63	0x3f	_	95	0x5f	(del)	127	0x7f

FIGURE B.1
ASCII table with the standard character set (0-127).

Extended ASCII character set

Char	Dec	Hex	Char	Dec	Hex	Char	Dec	Hex	Char	Dec	Hex
Ç	128	0×80	á	160	0×a0	└	192	0×c0	α	224	0×e0
ü	129	0×81	í	161	0×a1	⊥	193	0×c1	β	225	0×e1
é	130	0×82	ó	162	0×a2	⊤	194	0×c2	Γ	226	0×e2
â	131	0×83	ú	163	0×a3	├	195	0×c3	π	227	0×e3
ä	132	0×84	ñ	164	0×a4	─	196	0×c4	Σ	228	0×e4
à	133	0×85	Ñ	165	0×a5	┼	197	0×c5	σ	229	0×e5
å	134	0×86	ª	166	0×a6	╞	198	0×c6	μ	230	0×e6
ç	135	0×87	º	167	0×a7	╟	199	0×c7	τ	231	0×e7
ê	136	0×88	¿	168	0×a8	╚	200	0×c8	Φ	232	0×e8
ë	137	0×89	⌐	169	0×a9	╔	201	0×c9	Θ	233	0×e9
è	138	0×8a	¬	170	0×aa	╩	202	0×ca	Ω	234	0×ea
ï	139	0×8b	½	171	0×ab	╦	203	0×cb	δ	235	0×eb
î	140	0×8c	¼	172	0×ac	╠	204	0×cc	∞	236	0×ec
ì	141	0×8d	¡	173	0×ad	=	205	0×cd	φ	237	0×ed
Ä	142	0×8e	«	174	0×ae	╬	206	0×ce	ε	238	0×ee
Å	143	0×8f	»	175	0×af	╧	207	0×cf	∩	239	0×ef
É	144	0×90	░	176	0×b0	╨	208	0×d0	≡	240	0×f0
æ	145	0×91	▒	177	0×b1	╤	209	0×d1	±	241	0×f1
Æ	146	0×92	▓	178	0×b2	╥	210	0×d2	≥	242	0×f2
ô	147	0×93	│	179	0×b3	╙	211	0×d3	≤	243	0×f3
ö	148	0×94	┤	180	0×b4	╘	212	0×d4	⌠	244	0×f4
ò	149	0×95	╡	181	0×b5	╒	213	0×d5	⌡	245	0×f5
û	150	0×96	╢	182	0×b6	╓	214	0×d6	÷	246	0×f6
ù	151	0×97	╖	183	0×b7	╫	215	0×d7	≈	247	0×f7
ÿ	152	0×98	╕	184	0×b8	╪	216	0×d8	°	248	0×f8
Ö	153	0×99	╣	185	0×b9	┘	217	0×d9	·	249	0×f9
Ü	154	0×9a	║	186	0×ba	┌	218	0×da	·	250	0×fa
¢	155	0×9b	╗	187	0×bb	█	219	0×db	√	251	0×fb
£	156	0×9c	╝	188	0×bc	▄	220	0×dc	ⁿ	252	0×fc
¥	157	0×9d	╜	189	0×bd	▌	221	0×dd	²	253	0×fd
Pts	158	0×9e	╛	190	0×be	▐	222	0×de	■	254	0×fe
ƒ	159	0×9f	┐	191	0×bf	▀	223	0×df		255	0×ff

FIGURE B.2
ASCII table with the extended character set (128–255).

Appendix C

Library Functions

This appendix provides a brief description of the C standard library functions.

<assert.h>
```
void assert(int exp);
```

If the value of `exp` is 0, a diagnostic message is displayed and the program terminates. If not, it does nothing.

<ctype.h>
```
int isalnum(int ch);
```

Checks if `ch` is alphanumeric (a-z, A-Z, 0-9). If it is, it returns a nonzero value, 0 otherwise.

```
int isalpha(int ch);
```

Checks if `ch` is alphabetic (a-z, A-Z). If it is, it returns a nonzero value, 0 otherwise.

```
int iscntrl(int ch);
```

Checks if `ch` is a control character. If it is, it returns a nonzero value, 0 otherwise.

```
int isdigit(int ch);
```

Checks if `ch` is a digit. If it is, it returns a nonzero value, 0 otherwise.

```
int isgraph(int ch);
```

Checks if `ch` is a printable character other than the space. If it is, it returns a nonzero value, 0 otherwise.

```
int islower(int ch);
```

Checks if `ch` is a lowercase character (a-z). If it is, it returns a nonzero value, 0 otherwise.

```
int isprint(int ch);
```

Checks if `ch` is a printable character (including the space character). If it is, it returns a nonzero value, 0 otherwise.

```
int ispunct(int ch);
```

Checks if ch is a punctuation character. If it is, it returns a nonzero value, 0 otherwise.

```
int isspace(int ch);
```

Checks if ch is a white-space character. If it is, it returns a nonzero value, 0 otherwise.

```
int isupper(int ch);
```

Checks if ch is an uppercase character (A–Z). If it is, it returns a nonzero value, 0 otherwise.

```
int isxdigit(int ch);
```

Checks if ch is a hexadecimal digit (a–f, A–F, 0–9). If it is, it returns a nonzero value, 0 otherwise.

```
int tolower(int ch);
```

If ch is an uppercase letter, it returns the corresponding lowercase letter. Otherwise, it returns ch unchanged.

```
int toupper(int ch);
```

If ch is a lower-case letter, it returns the corresponding upper-case letter. Otherwise, it returns ch unchanged.

<locale.h>
```
struct lconv *localeconv();
```

It returns a pointer to a structure of lconv type, which contains information about local settings.

```
char *setlocale(int category, char *locale);
```

Sets the portion of the program's locale settings specified by category and locale.

<math.h>
```
int abs(int a);
```

It returns the absolute value of a.

```
double acos(double a);
```

It returns the arc cosine of a in the range 0 to π radians.

```
double asin(double a);
```

It returns the arc sine of a in the range $-\pi/2$ to $\pi/2$ radians.

```
double atan(double a);
```

It returns the arc tangent of a in the range $-\pi/2$ to $\pi/2$ radians.

```
double atan2(double a, double b);
```

It returns the arc tangent of a/b in the range $-\pi$ to π radians.

```
double ceil(double a);
```

It returns the smallest integer that is greater than or equal to a.

```
double cos(double a);
```

It returns the cosine of a.

```
double cosh(double a);
```

It returns the hyperbolic cosine of a.

```
double exp(double a);
```

It returns the result of e^a, where e is the logarithmic base, that is, $2.7182...$

```
double fabs(double a);
```

It returns the absolute value of a.

```
double floor(double a);
```

It returns the largest integer that is less than or equal to a.

```
double fmod(double a, double b);
```

It returns the floating-point remainder of a/b.

```
double frexp(double a, int *ptr);
```

Finds the mantissa (m) and exponent (n) values of a so that $a = m \times 2^n$. It returns the mantissa.

```
long labs(long a);
```

It returns the absolute value of a.

```
double ldexp(double a, int n);
```

It returns the result of $a \times 2^n$.

```
double log(double a);
```

It returns the logarithm of a to base e.

```
double log10(double a);
```

It returns the logarithm of a to base 10.

```
double modf(double a, double* ptr);
```

Splits the value of a into integer and fractional parts. It returns the fractional part.

```
double pow(double a, double b);
```

It returns the result of a^b.

```
double sin(double a);
```

It returns the sine of a.

```
double sinh(double a);
```

It returns the hyperbolic sine of a.

```
double sqrt(double a);
```

It returns the square root of a.

```
double tan(double a);
```

It returns the tangent of a.

```
double tanh(double a);
```

It returns the hyperbolic tangent of a.

```
<setjump.h>
int setjmp(jmp_buf env);
```

Saves the current stack environment in env.

```
void longjmp(jmp_buf env, int val);
```

Restores the environment stored in env.

```
<signal.h>
int raise(int sig);
```

Sends the sig signal to the executing program. It returns 0 if successful, a nonzero value otherwise.

```
void (*signal(int sig, void (*func)(int)))(int);
```

Establishes the function that func points to as the handler of the sig signal. signal() takes as parameters an integer and a pointer to another function, which takes an integer parameter and returns nothing. It returns a pointer to a function, which takes an integer parameter and returns nothing.

```
<stdarg.h>
void va_start(va_list arg_ptr, type);
```

Makes arg _ ptr to point to the beginning of the variable argument list.

```
type va_arg(va_list arg_ptr, type);
```

Gets the value of the argument pointed to by arg _ ptr.

```
void va_end(va_list arg_ptr);
```

Terminates the processing of the variable argument list.

```
<stdio.h>
void clearerr(FILE *fp);
```

Resets the error indicator of the file pointed to by fp.

```
int fclose(FILE *fp);
```

Closes the file pointed to by fp. It returns 0 if successful, EOF otherwise.

```
int feof(FILE *fp);
```

Checks if the end of file is reached. If it is, it returns a nonzero value, 0 otherwise.

```
int ferror(FILE *fp);
```

Checks if an error has occurred on the file pointed to by fp. If it is, it returns a nonzero value, 0 otherwise.

```
int fflush(FILE *fp);
```

Flushes the content of the intermediate memory associated with the file pointed to by fp. It returns 0 if successful, EOF otherwise.

```
int fgetc(FILE *fp);
```

Reads a character from the file pointed to by fp. It returns the character read or EOF to indicate a read error or an end-of-file condition.

```
int fgetpos(FILE *fp, fpos_t* pos);
```

Stores the current position of the file pointed to by fp into fpos_t variable. It returns 0 if successful, a nonzero value otherwise.

```
char* fgets(char *str, int n, FILE *fp);
```

Reads characters from the file pointed to by fp until the new line character is met or n–1 characters are read or end of file is reached, whichever comes first. The characters are stored in the array pointed to by str. It returns str or NULL to indicate a read error or an end-of-file condition.

```
FILE* fopen(const char *filename, const char *mode);
```

Opens the file specified by filename according to the mode argument. It returns a pointer associated with the file or NULL if the file can't be opened.

```
int fprintf(FILE *fp, const char *mode,…);
```

Writes data to the file pointed to by `fp` according to the format specified by `mode`. It returns the number of the characters written in the file or a negative value if an error occurs.

```
int fputc(int ch, FILE *fp);
```

Writes the `ch` character to the file pointed to by `fp`. It returns the character written or `EOF` if an error occurs.

```
int fputs(const char *str, FILE *fp);
```

Writes the string pointed to by `str` in the file pointed to by `fp`. It returns a non-negative value if successful, `EOF` otherwise.

```
int fread(void *buf, int size, int count, FILE *fp);
```

Reads from the file pointed to by `fp`, `count` elements of `size` bytes each, and stores them in the array pointed to by `buf`. It returns the number of the elements successfully read.

```
FILE* freopen(const char *filename, const char *mode, FILE *fp);
```

Closes the file associated with `fp` and opens a new file specified by `filename` according to the `mode` argument. It returns a pointer associated with the new file or `NULL` if the file can't be opened.

```
int fscanf(FILE *fp, const char *mode, ...);
```

Reads data from the file pointed to by `fp` according to the format specified by `mode`. It returns the number of the data items successfully converted and assigned to respective arguments or `EOF` to indicate a read error or an end-of-file condition.

```
int fseek(FILE *fp, long offset, int origin);
```

Moves the file pointer indicated by `fp` to a new location that is `offset` bytes from the position specified by origin. It returns 0 if successful, a nonzero value otherwise.

```
int fsetpos(FILE *fp, const fpos_t *pos);
```

Moves the file pointer indicated by `fp` to the location specified by the `pos` parameter. It returns 0 if successful, a nonzero value otherwise.

```
long ftell(FILE *fp);
```

It returns the current position of the file pointer indicated by `fp` from the beginning of the file.

```
int fwrite(const void *buf, int size, int count, FILE *fp);
```

Writes `count` elements of `size` bytes each from the array pointed by `buf` into the file pointed to by `fp`. It returns the number of elements successfully written.

```
int getc(FILE *fp);
```

Similar to `fgetc()`.

```
int getchar();
```

Reads a character from `stdin`. It returns the character read or `EOF` to indicate a read error. It is equivalent to `fgetc(stdin)`.

```
char* gets(char *str);
```

Reads characters from `stdin` and stores them in the array pointed to by `str`. It replaces the new line character with the null character. It returns `str` or `NULL` to indicate a read error.

```
void perror(const char *str);
```

Displays a diagnostic error message.

```
int printf(const char *mode, ...);
```

Similar to `fprintf()`, except that the data are written to `stdout`.

```
int putc(int ch, FILE *fp);
```

Similar to `fputc()`.

```
int putchar(int ch);
```

Writes the `ch` character to `stdout`. It returns the character written or `EOF` if an error occurs. It is equivalent to `fputc(ch, stdout)`.

```
int puts(const char *str);
```

Writes the string pointed to by `str` to `stdout`. It returns a non-negative value if successful or `EOF` if an error occurs. It appends a new line character.

```
int remove(const char *filename);
```

Deletes the file specified by `filename`. It returns 0 if successful, –1 otherwise.

```
int rename(const char *oldname, const char *newname);
```

Renames the file specified by `oldname` to the name given by `newname`. It returns 0 if successful, a nonzero value otherwise.

```
void rewind(FILE *fp);
```

Sets the file pointer indicated by `fp` to the beginning of the file.

```
int scanf(const char *mode, ...);
```

Similar to `fscanf()`, except that the data are read from `stdin`.

```
void setbuf(FILE *fp, char *ptr);
```

The memory pointed to by `ptr` will be used as intermediate buffer to store the data before being written in the file pointed to by `fp`.

```c
int setvbuf(FILE *fp, char *ptr, int mode, int size);
```

Similar to `setbuf()`, except that the memory size is specified by the `size` argument.

```c
int sprintf(char *ptr, const char *mode, …);
```

Similar to `fprintf()`, except that the data are stored in the array pointed to by `ptr`. It appends the null character.

```c
int sscanf(const char *ptr, const char *mode,…);
```

Similar to `fscanf()`, except that the data are read from the array pointed to by `ptr`.

```c
FILE* tmpfile();
```

Creates a temporary file that will be automatically deleted when the file is closed or the program terminates. It returns a file pointer associated with the file or `NULL` if the file can't be created.

```c
char* tmpnam(char *str);
```

Creates a temporary file name that can be used to open a temporary file. The name is stored in `str`. It returns a pointer to the created name or `NULL` if the name can't be created.

```c
int ungetc(int ch, FILE *fp);
```

Pushes the `ch` character back to the file pointed to by `fp`. It returns `ch` if successful, `EOF` otherwise.

```c
int vfprintf(FILE *fp, const char *mode, va_list arg);
```

Similar to `fprintf()`, except that a variable argument list pointed to by `arg` will be written in the file.

```c
int vprintf(const char *mode, va_list arg);
```

Similar to `vfprintf()`, except that the data are displayed to `stdout`.

```c
int vsprintf(char *ptr, const char *mode, va_list arg);
```

Similar to `vfprintf()`, except that the data are stored in the array pointed to by `ptr`.

```c
<stdlib.h>
void abort();
```

Causes abnormal program termination.

```c
int atexit(void (*ptr)());
```

Specifies a function to be called if the program terminates normally. The `ptr` argument points to that function. It returns 0 if successful, a nonzero value otherwise.

```
double atof(const char *str);
```

Converts the string pointed to by `str` to a floating-point number. It returns the number if successful, 0 otherwise.

```
int atoi(const char *str);
```

Similar to `atof()`, except that the string is converted to an integer.

```
long atol(const char *str);
```

Similar to `atof()`, except that the string is converted to a **long** integer.

```
void* bsearch(const void *key, const void *base, int num, int width,
int(*cmp)(const void *elem1, const void *elem2));
```

Performs a search according to the binary algorithm for the value pointed to by `key` in a sorted array pointed to by `base`, which has `num` elements, each of `width` bytes. The `cmp` parameter is a pointer to a function that takes as parameters two pointers to two array elements and returns a value according to the following:

- <0 if `*elem1` is less than `*elem2`
- 0 if `*elem1` is equal to `*elem2`
- >0 if `*elem1` is greater than `*elem2`

If the `*key` value is found, `bsearch()` returns a pointer to the respective array element. Otherwise, it returns NULL.

```
void* calloc(int num, int size);
```

Allocates memory for an array of `num` elements, each of `size` bytes. Each element is initialized to 0. It returns a pointer to the beginning of the memory block or NULL if the memory can't be allocated.

```
div_t div(int a, int b);
```

It calculates the quotient and the remainder of `a/b` and stores them in the respective fields of the returned `div_t` structure.

```
void exit(int status);
```

Causes the program termination. The `status` code indicates the termination status. The value 0 indicates a normal exit.

```
void free(void *ptr);
```

Releases the allocated memory pointed to by `ptr`.

```
char* getenv(const char *name);
```

Checks if the string pointed to by name is contained in the system's environment list. If a match is found, it returns a pointer to the environment entry containing the name, NULL otherwise.

```
ldiv_t ldiv(long int a, long int b);
```

Similar to div(), except that the quotient and the remainder are stored in the fields of a ldiv _ t structure.

```
void* malloc(int size);
```

Allocates a memory block of size bytes. It returns a pointer to the beginning of that memory or NULL if the memory can't be allocated.

```
int mblen(const char *str, int count);
```

It determines the validity of a multibyte character pointed to by str and returns its length.

```
int mbtowc(wchar_t* wchar, const char *mbchar, int count);
```

Converts count bytes of the multibyte character pointed to by mbchar to a corresponding wide character and stores the result into the memory pointed to by wchar. It returns the length of the multibyte character.

```
int mbstowcs(wchar_t* wchar, const char *mbchar, int count);
```

Converts count bytes of the multibyte character string pointed to by mbchar to a string of corresponding wide characters and stores the result into the array pointed to by wchar. If the conversion is completed successfully, it returns the number of the converted multibyte characters, –1 otherwise.

```
void* qsort(void *base, int num, int width, int(*cmp)(const void *elem1,
const void *elem2));
```

Sorts the array pointed to by base, which has num elements each of width bytes, according to the quick sort algorithm. cmp points to a function similar to the one declared in bsearch().

```
int rand();
```

Generates and returns a random integer between 0 and RAND _ MAX.

```
void* realloc(void *ptr, int size);
```

Changes the size of the allocated memory block pointed to by ptr. That memory has been allocated from a previous call to malloc(), realloc(), or calloc(). The size parameter declares the size of the new memory block. It returns a pointer to the beginning of the new memory block or NULL if the new block can't be allocated.

```
void srand(unsigned int seed);
```

Uses seed to set a starting point for generating a sequence of random values produced by rand() calls.

`double strtod(const char *str, char **endp);`

Converts the string pointed to by str to a floating-point number. The endp points to the first character that can't be converted. If the conversion is completed successfully, it returns the converted number, 0 otherwise.

`long strtol(const char *str, char **endp, int base);`

Converts the string pointed to by str to a long integer. The endp points to the first character that can't be converted. The base parameter defines the radix of the number. If the conversion is completed successfully, it returns the converted number, 0 otherwise.

`unsigned long strtoul(const char *str, char **endp, int base);`

Similar to strtol(), except that it returns an **unsigned long** integer.

`int system(const char *str);`

Executes the operating system's command pointed to by str.

`int wctomb(char *mbchar, wchar_t wchar);`

Converts the wchar wide character into the corresponding multibyte character pointed to by mbchar. It the conversion is completed successfully, it returns the number of the bytes stored in wchar, –1 otherwise.

`int wcstombs(char *mbchar, const wchar_t *wcstr, int count);`

Converts count bytes of a sequence of wide characters pointed to by wcstr to a corresponding sequence of multibyte characters pointed to by mbchar. It the conversion is completed successfully, it returns the number of the converted multibyte characters, –1 otherwise.

`<string.h>`
`void* memchr(const void *str, int val, int count);`

Searches for the first appearance of the val character into the count characters of the string pointed to by str. If it is found, it returns a pointer to its first occurrence, NULL otherwise.

`int memcmp(const void *ptr1, const void *ptr2, int count);`

Compares count bytes of the memory blocks pointed to by ptr1 and ptr2. The returned value indicates their relationship.

`void* memcpy(void *dest, const void *src, int count);`

Copies `count` bytes from the memory pointed to by `src` to the memory pointed to by `dest`.

`void* memmove(void *dest, const void *src, int count);`

Similar to `memcpy()`, except that the copy operation will work properly even if the two memory blocks overlap.

`void* memset(void *dest, int val, int count);`

Sets the `count` bytes of the memory pointed to by `dest` equal to `val`.

`char* strcat(char *str1, const char *str2);`

Appends the string pointed to by `str1` to the string pointed to by `str2` and adds the null character. It returns a pointer to the concatenated string.

`char* strchr(const char *str1, int val);`

Searches the `val` character in the string pointed to by `str`. If it is found, it returns a pointer to its first occurrence, `NULL` otherwise.

`int strcmp(const char *str1, const char *str2);`

Compares the strings pointed to by `str1` and `str2` and returns a negative, zero, or positive value to indicate if the first string is lexicographically less than, equal, or greater than the second one.

`int strcoll(const char *str1, const char *str2);`

Compares the strings pointed to by `str1` and `str2` according to the current locale rules and returns the comparison result.

`char* strcpy(char *dest, const char *src);`

Copies the string pointed to by `src` into the array pointed to by `dest`. It returns a pointer to the destination string.

`int strcspn(const char *str, const char *set);`

Searches the string pointed to by `str` for a segment consisting of characters not in the string pointed to by `set`. It returns the length of the longest `str` segment that doesn't contain any character in `set`.

`char* strerror(int err);`

It returns a pointer to the string that corresponds to the value of `err`.

`int strlen(const char *str);`

It returns the length of the string pointed to by `str`, excluding the null character.

`char* strncat(char *str1, const char *str2, int count);`

Appends `count` characters of the string pointed to by `str2` to the string pointed to by `str1` and adds the null character. It returns a pointer to the concatenated string.

`int strncmp(const char *str1, const char *str2, int count);`

Compares `count` characters of the strings pointed to by `str1` and `str2` and returns the comparison result.

`char* strncpy(char *dest, const char *src, int count);`

Copies `count` characters of the string pointed to by `src` to the string pointed to by `dest`. It returns a pointer to the destination string.

`char* strpbrk(const char *str, const char *set);`

Checks if any of the characters in the string pointed to by `set` is contained in the string pointed to by `str`. If it is, it returns a pointer to the first occurrence of the character from `set` in `str`, NULL otherwise.

`char* strrchr(const char *str1, int val);`

Searches the `val` character in the string pointed to by `str`. If it is found, it returns a pointer to its last occurrence in `str`, NULL otherwise.

`int strspn(const char *str, const char *set);`

Similar to `strcspn()`, except that it returns the length of the `str` segment that consists entirely of characters in `set`.

`char* strstr(const char *str1, const char *str2);`

Checks if the string pointed to by `str2` is contained in the string pointed to by `str1`. If it is, it returns a pointer to its first occurrence in `str1`, NULL otherwise.

`char* strtok(char *str, const char *set);`

Searches the string pointed to by `str` for a segment consisting of characters not in the string pointed to by `set`. If it is found, it returns a pointer to the first character of the segment, NULL otherwise.

`int strxfrm(char* dest, const char *src, int count);`

Transforms `count` characters of the string pointed to by `src` based on the locale character set and stores the transformed string in the string pointed to by `dest`. It returns the length of the transformed length.

<time.h>
`char* asctime(const struct tm *ptr);`

Converts the date and time information stored in a structure of type `tm` pointed to by `ptr` to a string in the form `"Fri Aug 23 10:56:53 2013"`. It returns the constructed string.

`clock_t clock();`

It returns the elapsed time since the beginning of the program execution. To convert it to seconds, divide that value with the constant CLOCKS _ PER _ SEC.

```
char* ctime(const time_t *ptr);
```

Converts the calendar time pointed to by ptr to a string in the form "Fri Aug 23 10:56:53 2013", adjusted to the local time zone settings. It returns the constructed string.

```
double difftime(time_t t1, time_t t2);
```

It returns the time difference between t1 and t2, measured in seconds.

```
struct tm* gmtime(const time_t *ptr);
```

Breaks down the calendar time pointed to by ptr to the fields of a tm structure and returns a pointer to that structure.

```
struct tm* localtime(const time_t *ptr);
```

Similar to gmtime(), except that the time is converted to a local time.

```
time_t mktime(struct tm *ptr);
```

Converts a broken-down time stored in the structure tm pointed to by ptr to a structure of type time_t and returns that structure.

```
int strftime(char *str, int size, const char *fmt, const struct tm *ptr);
```

Formats the time information stored in the structure tm pointed to by ptr according to the format string pointed to by fmt and stores size characters in the array pointed to by str. It returns the number of the stored characters.

```
time_t time(time_t *ptr);
```

Returns the current calendar time, which is the number of seconds elapsed since midnight (00:00:00), January 1, 1970. The return value is also stored in the structure time_t pointed to by ptr.

Appendix D

Hexadecimal System

This appendix provides a brief description of the hexadecimal system. The base of the hexadecimal system is the number 16. The numbers 0–9 are the same of the decimal system. The numbers 10–15 are represented by the letters A to F. The correspondence of the hexadecimal system to the binary and decimal systems is shown in Table D.1.

As shown, each hexadecimal digit (0 to F) is represented with four binary digits. For example, the hexadecimal number F4A is written in binary as `1111 0100 1010`.

To find the decimal value of a hexadecimal number that consists of n digits (e.g., $d_{n-1} \ldots d_2 d_1 d_0$), we apply the formula

$$\texttt{Decimal} \; = \; \sum_{i=0}^{n-1} (d_i \times 16^i).$$

For example, the decimal value of the hexadecimal number F4A is $(\text{F} \times 16^2) + (4 \times 16^1) + (\text{A} \times 16^0) = 3840 + 64 + 10 = 3914$.

TABLE D.1

Correspondence between Hexadecimal, Binary, and Decimal Systems

Hexadecimal (hex)	Binary (bin)	Decimal (dec)
0	0000	0
1	0001	1
2	0010	2
3	0011	3
4	0100	4
5	0101	5
6	0110	6
7	0111	7
8	1000	8
9	1001	9
A	1010	10
B	1011	11
C	1100	12
D	1101	13
E	1110	14
F	1111	15

Bibliography

Harbison, S.P. and Steele, G.L., *C: A Reference Manual*, 4th edn., Prentice Hall, Englewood Cliffs, NJ, 1995.

Kernighan, B.W. and Ritchie, D.M., *The C Programming Language*, 2nd edn., Prentice Hall, Englewood Cliffs, NJ, 1988.

King, K.N., *C Programming: A Modern Approach*, W.W. Norton & Company, New York, 1996.

Koenig, A., *C Traps and Pitfalls*, Addison-Wesley, Reading, MA, 1989.

Plauger, P.J., *The Standard C Library*, Prentice Hall, Englewood Cliffs, NJ, 1992.

Index